APPLICATIONS OF BIOTECHNOLOGY IN FORESTRY AND HORTICULTURE

APPLICATIONS OF BIOTECHNOLOGY IN FORESTRY AND HORTICULTURE

Edited by
Vibha Dhawan

Tata Energy Research Institute
New Delhi, India

PLENUM PRESS • NEW YORK AND LONDON

Library of Congress Cataloging in Publication Data

International Workshop on Applications of Biotechnology in Forestry and Horticulture (1988:
New Delhi, India)
Applications of biotechnology in forestry and horticulture / edited by Vibha Dhawan.
        p.   cm.
"Proceedings of an International Workshop on Applications of Biotechnology in Forestry and
Horticulture, held January 14–16, 1988, in New Delhi, India"—T.p. verso.
Includes bibliographical references.
ISBN-13: 978-1-4684-1323-6   e-ISBN-13: 978-1-4684-1321-2
DOI: 10.1007/978-1-4684-1321-2
    1. Trees—Propagation—In vitro—Congresses. 2. Trees—Biotechnology—Congresses. 3. Ar-
boriculture—Congresses. I. Dhawan, Vibha. II. Title.
SD403.5.I57   1988   89-39108
634.9—dc20   CIP

Proceedings of an International Workshop on Applications of Biotechnology in Forestry and Horticulture,
organized by the Tata Energy Research Institute, New Delhi,
January 14–16, 1988, in New Delhi, India

© 1989 Plenum Press, New York
Softcover reprint of the hardcover 1st edition 1989
A Division of Plenum Publishing Corporation
233 Spring Street, New York, N.Y. 10013

PREFACE

Major and exciting changes have taken place recently in various aspects of biotechnology and its applications to forestry. Even more exciting is the prospect of major innovations that the entire field of biotechnology holds for plant growth in general. The importance of these developments for the forestry sector is considerable, particularly since forestry science has not received the kinds of technical and R&D inputs that, say, agriculture has received in the past few decades. Yet the problems of deforestation as well as stagnation in yields and productivity of existing forests throughout the world are becoming increasingly apparent, with consequences and ecological effects that cause growing worldwide concern. Policies for application of existing knowledge in biotechnology to the field of forestry and priorities for future research and development are, therefore, of considerable value, because it is only through the adoption of the right priorities and enlightened policies that scientific developments will move along the right direction, leading to improvements in forestry practices throughout the world.

It was against this backdrop that the Tata Energy Research Institute (TERI) organised a major international workshop on the "Applications of Biotechnology in Forestry and Horticulture" at New Delhi in January 1988. The present volume covers the proceedings of this international workshop. The underlying objective for publishing the proceedings is to ensure that large-scale dissemination of the knowledge generated and experiences exchanged in the workshop takes place beyond the distinguished group of participants who were present. It is intended that this volume would serve as a comprehensive survey and study of the state-of-the-art in this entire field.

The volume itself is divided into five sections. The first section covers the current status of forestry with a mix of papers presented by biotechnologists and researchers, ranging from those working in well-known scientific establishments to practitioners of forestry science and those responsible for programs of afforestation. This section serves to provide the background against which the potential and likely directions of specific techniques and developments in biotechnology may be seen as sources of increased productivity and output.

Section II presents a set of papers dealing with the applications of tissue culture in forestry and horticulture. This section provides by far the most extensive coverage of the subject published in recent literature, and contains contributions not only from experts in the developed countries but also from those working at the frontiers of this science in developing countries. Additionally, the coverage presents not only the work done in research and development units, but also describes the experiences of commercial organisations that are involved in large-scale tissue culture programs with a range of plant species. The section covers both techniques of tissue culture and their applications to tree improvement programs.

Section III goes into some of the issues involved in scaling up tissue culture practices for application on a commercial scale. This section, therefore, deals with some of the engineering and process aspects of tissue culture techniques. It also explores the economic and commercial issues that need to be considered and analysed in large-scale production facilities. The experience of tissue culture of ornamental plants is also covered with a view to seeing the applicability of experiences gained with these species to tree species and horticultural plants.

Section IV concentrates on nitrogen fixation studies in forestry. This section has been included with the realisation that in several tropical regions of the world, soil conditions require a vigorous and extensive program on leguminous species that are relevant not only because they would minimise inputs of fertilisers and nutrients to be provided generally at high cost, but also because in certain cases it is through the plantation of these species alone that soil improvements and restoration of degraded lands can be brought about.

The final section of the book deals with genetic engineering of forest species. While very little has happened in the development of genetic engineering techniques and their applications to forestry species, the prospects for the future appear bright and attractive. The papers included in this section present the limited but valuable experience that has been gained through work already done on the subject as well as provide some pointers to developments that would be of future relevance.

It is well known that developments in the subjects covered in this volume are taking place rapidly, and in such a publication, comprehensive as it may be, these could have value and currency only for a short period. Progress in the fields covered in this book is taking place rapidly, and it is important that an updating of knowledge be carried out from time to time. Yet the merit of a volume of this nature lies in the fact that it covers the entire field in a comprehensive manner and permits the formulation of directions and plans that take into account the inter-linking up of various scientific activities that must be seen in a wider perspective to ensure proper priorities for R&D and widespread applications in the field.

DR. R.K. PACHAURI
Director
Tata Energy Research Institute
New Delhi

ACKNOWLEDGEMENTS

I would like to thank several people for their help in the preparation of this book. I am thankful to Mr. Sanjay Saxena, Ms. Namita Bhatia, Ms. Omita Goel, Mr. Yateendra Joshi and Ms. Sharmila Sengupta for critically going through the manuscript. The mammoth task of typing and preparing camera-ready manuscripts on the laser printer was performed by Ms. N. Jayalakshmi, Ms. Deepa Mary Philip and Mr. B. Venkatakrishnan. I am also grateful to Mr. M. Karuppasamy of the Computer Center who offered expert advice from time to time.

The financial assistance, for organising the conference, from the National Science Foundation, USA and the Department of Biotechnology, Government of India, India is gratefully acknowledged.

CONTENTS

I.   **CURRENT STATUS OF FORESTRY** 1

     1.   **Applications of Biotechnology in Forestry** 3

        S. Ramachandran

     2.   **Applications of Biotechnology in Forestry and Rural Development** 9

        J. Burley

     3.   **Forestry in India: Problems and Prospects and the Role of Tissue Culture** 21

        T. N. Khoshoo

     4.   **Forest Tree Breeding and Mass Cloning for Tree Improvement in Indian Forestry** 45

        P. D. Dogra

     5.   **Afforestation in India** 51

        A. N. Chaturvedi

II.   **APPLICATIONS OF TISSUE CULTURE IN FORESTRY AND HORTICULTURE** 55

     6.   **Artificial Seed Production and Forestry** 57

        Keith Redenbaugh and Steven E. Ruzin

7. **Biotechnological Application of Plant Tissue Culture to Forestry in India** 73

A. F. Mascarenhas, S.S. Khuspe, R.S. Nadgauda, P.K. Gupta, E.M. Muralidharan and B.M. Khan

8. **Tissue Culture of Plantation Crops** 87

S. Bhaskaran and V. R. Prabhudesai

9. ***In vitro* Strategies for Tropical Fruit Tree Improvement** 109

R. E. Litz

10. **The Context and Strategies for Tissue Culture of Date, African Oil and Coconut Palms** 119

A. D. Krikorian

11. ***In vitro* Strategies for Sandalwood Propagation** 145

V. A. Bapat and P. S. Rao

12. **Micropropagation of *Ficus auriculata* Lour** 157

Nirmala Amatya and S. B. Rajbhandary

13. **Genetic Variation in Tissue Culture as a Consequence of the Morphogenic Process** 165

Robert D. Locy

14. **Performance Criteria in Response Surfaces for Metabolic Phenotypes of Clonally Propagated Woody Perennials** 181

Don J. Durzan

15. **Alteration of Growth and Morphogenesis by Endogenous Ethylene and Carbon Dioxide in Conifer Tissue Cultures**.. 205

Prakash P. Kumar and Trevor A. Thorpe

16. **Storage of Forest Tree Germplasm at Sub-zero Temperatures** 215

M. R. Ahuja

III. **COMMERCIAL EXPLOITATION OF TISSUE CULTURE IN HORICULTURE** 229

17. **Tissue Culture of Ornamental Plants** 231
    L. J. Maene

18. **Tissue Culture of Orchids in Thailand** 245
    Uthai Charanasri

19. **Acclimatization of Tissue Culture-Raised Plants for Transplantation to the Field** 249
    Sant S. Bhojwani and Vibha Dhawan

20. **Pilot Plant Tissue Culture Unit** 257
    R. D. Lai and P. Mohan Kumar

21. **Large-scale Production of Plants Through Micropropagation: Problems, Prospects and Opportunities for India** 265
    Jitendra Prakash

22. **Cost Analysis of Micropropagated Plants** 275
    R. K. Pachauri and Vibha Dhawan

IV. **NITROGEN FIXATION STUDIES IN FORESTRY** 283

23. **Micropropagation and Nodulation of Tree Legumes** 285
    Vibha Dhawan

24. **Role of Mycorrhizae in Forestry** 297
    H. S. Thapar

25. **Ectomycorrhizal Effects on Nodulation, Nitrogen Fixation and Growth of *Alnus glutinosa* as Affected by Glyphosate** 309
    L. Chatarpaul, P. Chakravarty and P. Subramaniam

V.   GENETIC ENGINEERING OF FOREST SPECIES 321

26. **Genetic Engineering of Tree Species**.. 323
    H. K. Srivastava

27. **Genetic Manipulation of Woody Species**... 331
    Malathi Lakshmikumaran

28. **Genetic Transformation System in Douglas-fir**
    Pseudotsuga menziesii 339
    Promod K. Gupta, Abhaya M. Dandekar and D. J. Durzan

29. **Biological and Economic Feasibility of Genetically**
    Engineered Trees for Lignin Properties and
    Carbon Allocation 349
    Roger Timmis and Patrick C. Trotter

VI.   CONCLUDING REMARKS 369

30. **Concluding Remarks** 371

Contributors 375

Taxonomic Index 379

I

CURRENT STATUS OF FORESTRY

1

APPLICATIONS OF BIOTECHNOLOGY IN FORESTRY

S. Ramachandran

Land is the most fundamental resource of a country and as a result of the high rate of population growth in India, the pressure to produce more food, fodder, timber and fuelwood from the available land is intense. Land use, therefore, needs judicious planning. India has about 130 Mha of wastelands. Wastelands are defined as lands that are degraded and are lying unutilised due to constraints such as salinity, alkalinity, waterlogging, wind and water erosion, etc. Since wastelands are degraded and unsuitable for agriculture, they can be utilised to meet the growing demand for fuelwood, fodder, small timber and industrial wood. Not only do we have to reclaim the wastelands but we should also try and protect our forest cover which is being depleted at the rate of 1.3 Mha per year. Efforts are also required to step up the existing per hectare yield of 0.2 m^3 to $2\text{-}3 \text{ m}^3$. The need of the hour is a quantum jump in existing afforestation targets. The plan is to afforest wastelands at a rate of 5 Mha per annum. Such large-scale plantations will demand a large amount of planting material (seeds or seedlings) and in order to achieve higher yields, it is important that

S. Ramachandran * Department of Biotechnology, C.G.O. Complex, Lodi Road, New Delhi - 110 003, India.

these planting materials are of good quality. This means a high yielding plant which is also resistant to different diseases and pests. Such plants can be obtained by conventional methods of tree breeding and non-conventional methods of genetic engineering and somatic hybridisation. Unlike annual crops, the breeding behavior of trees is complex and their life-cycle long. Therefore, it takes many years to produce a variety with desirable traits. However, the development of modern non-conventional methods of genetic engineering, tissue culture and somatic hybridisation has offered new ways to enhance forest productivity more rapidly. By these methods, it has become possible to develop such varieties of ornamental and crop plants, with many desirable traits such as fast growth, high yield, disease resistance, herbicide and stress tolerance, etc. in a much shorter time than that taken by the conventional methods. However, this quick, non-conventional methodology of producing plants is very expensive. Since both the methods have their own advantages and disadvantages, it will be prudent to combine the conventional methods and the new methods of biotechnology to increase productivity, both qualitatively and quantitatively, in order to achieve the desired goals.

CHOICE OF TREES

Forests constitute the base for agriculture, animal husbandry, industrial development and, thus, all-round development. A dense forest cover is also required for a healthy environment, as deforestation leads to environmental degradation, soil erosion, recurrent floods, droughts and desertification. India's forest area is about 64.20 Mha which constitutes about 19.52 percent of the total geographic area of the country (National Natural Resource Monitoring System, 1987, of the Department of Space, Government of India). With less than 2 percent of the world's forest area, we are called upon to meet the requirements of nearly 15 percent of the world's population. With vast areas of wastelands lying unutilised, there is tremendous potential for expanding plantations in India. The annual target of afforesting 5 million hectares of land set by the Government of India continues to be a distant dream. However, it would be possible to achieve this target if concerted efforts are made by the different agencies associated with the task. The plantation program now involves not only the forest department but also the voluntary agencies, individual societies (social forestry and farm forestry) and industrial houses. Though the task is difficult, it needs to be executed efficiently. The success of plantation efforts will depend upon the clearly laid out tasks, choice of species, requirements of the rural folk, protection to be provided from adverse climatic conditions, grazing and thefts and the availability of required quality seeds or saplings of each species. To meet the demand of industries, productive forests of identified species will have to be raised. Selection of suitable species for different problem soils and agro-climatic conditions is absolutely essential for effective reclamation of wastelands. The following are the tree and shrub species that could be planted, depending upon the existing agro-climatic conditions.

Temperate Regions

Fuelwood   : *Acacia mearnsii, Acer campbellii, Acer caesium, Alnus* sp., *Celtis australis, Quercus* sp., *Terminalia ciliata, Tsuga dumosa*

Fodder   : *Acacia mearnsii, Aesculus indica, Celtis australis, Grewia optiwa*

Timber   : *Abies pindrow, Alnus* sp., *Cedrus deodara, Cryptomeria japonica, Juglans regia, Picea smithiana, Pinus* sp., *Salix alba*

Industrial Wood : *Eucalyptus globulus, Pinus patula*

Moist Tropical Regions

Fuelwood   : *Acacia catechu, Albizia lebbeck, Cassia siamea, Casuarina equiseti-folia, Dalbergia sissoo, Hibiscus integrifolia, Melia azadirachta*

Fodder   : *Acacia catechu, Dalbergia sissoo, Gmelina* sp., *Terminalia* sp.

Timber   : *Adina cordifolia, Albizia* sp., *Artocarpus chaplasha, Chukrasia ve-lutina, Dalbergia latifolia, Dalbergia sissoo, Gmelina* sp., *Kydia ca-lycina, Michelia champaca, Phoebe attennata, Shorea robusta, Tec-tona grandis, Terminalia* sp.

Industrial Wood : *Eucalyptus* sp., *Populus* sp.

Dry Tropical Regions

Fuelwood   : *Acacia catechu, Acacia nilotica, Acacia tortilis, Albizia lebbeck, Ano-geissus latifolia, Anogeissus pendula, Cassia siamea, Casuarina equisetifolia, Dalbergia sissoo, Diospyros melanoxylon, Eucalyptus* sp., *Gmelina robusta, Madhuca longifolia, Prosopis* sp., *Terminalia indica*

Fodder   : *Acacia nilotica, Acacia tortilis, Ailanthus excelsa, Albizia lebbeck, Al-bizia procera, Areca catechu, Dalbergia sissoo, Prosopis* sp., *Termi-nalia* sp.

Timber   : *Albizia lebbeck, Albizia procera, Dalbergia sissoo, Gmelina arborea, Gmelina robusta, Holoptelea* sp., *Pongamia pinnata, Shorea robusta, Stercularia* sp., *Tectona grandis*

Industrial Wood : *Bombax ceiba, Diospyros melanoxylon, Eucalyptus* sp.

FOREST BIOTECHNOLOGY:
THE KEY TO EXPANDING FOREST RESOURCES

At present, most of the planting material for different afforestation programs comes from sources that are genetically diverse and that give poor yields. Therefore, in order to enhance productivity, it is essential to have a highly efficient production technology including high yielding certified planting material. In this context, tissue culture and other biotechnological tools can play an important role in boosting productivity. Tissue culture technology, together with mycorrhizae, nitrogen fixing microorganisms and institutional support, etc. have been used to enhance the productivity of forest trees in the United States, Brazil, Japan and some West European countries. Success has also been obtained in the case of orchids, ornamentals and a number of vegetable and fruit species. Species of strawberry and apple are now produced commercially by tissue culture. In India, the technique of raising plants through tissue culture has been perfected for a few tree species, e.g., *Bamboo, Dalbergia, Eucalyptus, Leucaena, Prosopis, Santalum, Sesbania* and elites of *Tectona*. In order to achieve the quantum jump in the production of biomass for fuel, fodder, timber and industrial woods using the new tools of biotechnology, the following species have been identified for mass propagation through tissue culture: *Acacia nilotica, Alnus nepalensis, Hardwickia binata, Madhuca latifolia, Prosopis cineraria, Tamarindus indica, Dendrocalamus strictus, Bambusa arundinacea, Bambusa vulgaris, Tectona grandis, Shorea robusta, Dalbergia latifolia, Santalum album, Populus deltoides.*

It is necessary that greater emphasis be given to the following three areas for overall applications of biotechnology towards the enhancement of biomass production:

1. *Improvement in overall growth of forests.* This involves improvement in regeneration techniques, ensuring better survival of young saplings, development of improved genetic stocks and reduction in the damage caused by fire, pests, etc. Planting with genetically improved stock alone can increase the timber yield per hectare by about 25 percent over current rates of growth.

2. *Advances in the products manufactured from forest resources.* Here, the potential is limited only by the researchers' vision. A substantial reduction in the susceptibility of wood to decay could improve the useful life of products ranging from those used in construction to utility poles.

3. *Understanding the impacts of non-forestry related activities on forest land used in construction.* More research is required to understand the effect of acid rain on the growth and productivity of the forests. This work will lead to solutions to mitigate the impact of this by-product of industrial growth.

Organisational integration at the level of the individual scientist, department and organisation must be taken into account if biotechnology is to have any realistic application in enhancing biomass production. Optimally efficient research integration is determined, in part, by the organisational setting involved and the technologies employed. If research and product development capabilities are not contained within the same organisation, they must be established through licensing arrangements, collaborative research or research network.

The integration of new technologies can be achieved through multi-disciplinary teams whose objectives are in agreement with those of conventional plant breeders. Collaborative research also necessitates the support of plant breeders for evaluation and testing. In the following areas, both applied as well as the basic aspects of biotechnology need to be integrated with conventional methods, in order to make an impact on forestry:

* Mass propagation of elite plants through tissue culture
* Scaling up the production of artificial seeds
* Cryopreservation of the gene pools of elite trees
* Exploiting somaclonal variants for tree improvement
* Microspore and anther culture for fixation of heterosis
* Development, in some important tree species of a genetic transformation system using *Agrobacterium tumefaciens* and regeneration of transformed plantlets by tissue culture
* Genetic manipulation of trees through protoplast fusion or somatic hybridisation
* Large-scale field evaluation of tissue culture raised plants.

2

APPLICATIONS OF BIOTECHNOLOGY IN FORESTRY AND RURAL DEVELOPMENT

J. Burley

ABSTRACT

The major effects of recent biotechnological developments will be seen in: (i) enhancing the planting and survival of trees on sites that are naturally difficult environmentally, degraded by poor husbandry or previously unused for forestry; (ii) increasing the yields of both traditional and new products; (iii) increasing the transfer and use of disease-free stock; (iv) improving the yield of wood-based products by efficient conversion processes including the use of micro organisms; and (v) facilitating the identification and conservation of genetic resources. Specific technologies that are outlined in this paper include:

J. Burley * Oxford Forestry Institute, Oxford University, South Parks Road, Oxford OX1 3RB, England. Dr. Burley is Director of the Oxford Forestry Institute, University of Oxford, South Parks Road, Oxford OX1 3RB, England. He did not attend the workshop but has visited India many times over the past 20 years to advise on tree breeding. This paper was initially prepared for the Oxford International Symposium 1987 *Implications of New Biology* sponsored by the M.O.A. Foundation of Japan. It was published in the Commonwealth Forestry Review 66(4), 1987, and is reprinted here with the permission of the Editor.

*rejuvenation through clonal propagation including cuttings, micrografting and tissue
  culture;
* haploid generation;
* somatic embryogenesis;
* protoplast fusion and gene transfer;
* cryopreservation of germplasm;
* microbial breakdown of wood components;
* forest soil microbiological improvement and risk detection.

The stage of development of all these techniques and their current economic
application to forest trees varies between species but the potential is high.

BACKGROUND

Place of Trees and Forests in Man's Welfare

In the past 10 years, there has developed widespread concern among politicians,
administrators and laymen about the fate of the world's trees and forests which are
faced with two major threats; (i) the alarming rate of the loss of tropical forests
(caused largely by the increase in human population and its associated demands for
subsistence and cultural products); and (ii) the insidious and overt damage to temperate forests by aerial pollution (itself caused mainly by industrial processes and transport systems). The recent quinquennial Congress of the International Union of Forestry Research Organizations (IUFRO) in Ljubljana, Yugoslavia, 1986, recognized
these as two of the current major threats to the environment and welfare of mankind.
This awareness is the result of campaigns by foresters, ecologists and others concerned
with the interactions of natural resources being presented effectively to the public
through the media.

Land Use Systems Incorporating Trees

There are many causes of the previous public apathy to forests; in developing
countries, particularly among rural populations, one reason is the belief that trees are
God-given, have always been in place and do not require planned management or
replacement. In developed countries, one of the causes is the low value per unit of the
volume of the major industrial wood products while another is the aesthetic antipathy
to the uniformity portrayed by plantations. There is also the belief that plantations are
associated with environmental deterioration; the slight evidence so far obtained in a
few extreme cases does not justify using the environmental argument to support
politically inspired campaigns against tree planting, such as those currently occurring

in Spain, India and many other tropical countries. In fact, most national administrations and many informed laymen now accept the important role of trees and forests in meeting national economic needs and in safeguarding the current environment, current and future productivity of agricultural and forest lands, and future genetic resources. There is a wide range of land use systems in which trees are incorporated, some traditional and some recent innovations. These include the management of natural vegetation from desert scrub through savannah woodland to high forest.

Forests cover approximately one-third of the earth's land surface and approximately 3 billion hectares are closed forests while 1.3 billion hectares are open woodland. Another 1 billion hectares are forest regrowth on fallowed crop lands or degraded forest land. Together, these forested lands account for 40 percent of the world's land area (see WRI and IIED, 1986). The rate of loss of tropical forests is variously estimated between 5 and 20 Mha per year, but the most widely accepted estimate is 11 Mha.

The loss of indigenous forest can be compensated partly by plantations, particularly since their productivity can be over 10 times as great as that of natural forests. Extensive plantations now exist in both temperate and tropical regions, aimed largely at the industrial products of saw-timber and pulp and paper products, but the bulk of these is in the temperate regions of North America, Europe and the USSR. On a worldwide scale plantation rates approach 14 Mha with only 6 M of these in the tropics (FAO, 1985). There is clearly a pressing need for more extensive and productive plantations and their establishment and management should incorporate appropriate new technology.

Between the two extremes of natural forest management and industrial plantation, there are many land use systems that incorporate various intensities of tree planting and management. Chief among these are: (i) village woodlots, planned as mini-plantations in developing countries to provide fuelwood and light construction material for rural communities; (ii) shelter belts designed to protect rural environments, enhance the quality of life for man and his domestic animals, and to sustain and increase agricultural productivity; and (iii) agroforestry systems that deliberately incorporate trees in an intimate mixture with agricultural crops on farm land; these may be in spatial or temporal mixtures and at varying densities of either crop.

This latter group of three land use systems has major applications in developing countries. The plants used will need to be hardy and capable of surviving the rigours of the local environment, producing quick results; they will need to be cheap and available in quantity, and new technologies will be required to provide fast growing, hardy stock capable of surviving the rigours of natural environments and of poor management and culture. However, in Europe there will be increasing demand for land use systems that can profitably occupy the agricultural land that is coming out of production as a result of the EEC policies and over-production.

Products and Services from Trees and Forests

Many people, particularly in developed countries, may see the forest solely as a source of obvious wood products such as saw-timber for construction and furniture, or comminuted products for pulp and paper. The economic value of these products is, in fact, massive. World trade in forest products exceeds 75 billion pounds annually with 50 billion pounds traded among the developed countries alone. The United States, Japan and the United Kingdom are the three major importers of wood products; in the latter, this represents over 5 billion pounds, an import figure only exceeded by the agricultural sector.

However, in developing countries, other products are perhaps more important even if not traded formally or recorded precisely. Thus, over half the wood used in the world each year is burned for domestic heating or cooking and in some countries over 90 percent of the national fuel consumption originates from woody biomass. Currently nearly 100 M people in 23 countries are facing acute scarcity of fuelwood and fuelwood deficits are affecting 1 billion people in 37 countries. By the year 2000, a further 0.5 billion people in 15 countries are predicted to suffer fuelwood deficits. These deficits have the effect of changing dietary, eating and cooking habits with consequent increases in malnutrition and intestinal disorders.

In addition to these solid wood products, forests provide a wealth of non-wood goods and services including liquid fuels, chemical feedstocks, plant extractives for medicinal and perfumery products, paints, varnishes, adhesives, drinks, human food and animal fodder. The genetic resources of other plants in forests have current and so far undiscovered applications. Forests are the habitats for wildlife that form a significant proportion of the diet and income of many rural peoples.

The services offered by woody vegetation include protection of soil against water and wind erosion, moderation of water flows, enhanced storage of ground water, provision of shade and shelter, encouragement of amenity and tourism, employment and income generation, import substitution, risk reduction for poor farmers, rehabilitation of abandoned or degraded farm land, and generally improved human and animal nutrition and health. Not all of these benefits can be quantified easily but their socio-economic importance is widely accepted by governments and development agencies and all of them can be subject to improvements achieved through the new technologies.

Historical Applications of Biotechnology in Forestry

Agriculture and forestry may be considered the original biotechnologies since they have traditionally involved man's manipulation of genetic resources with or without tools and machines, using techniques and technology systems that are intended to meet mankind's needs. However, three main phases of technological development merit special consideration since they indicate areas in which the newer biological developments may have application. These are classical selections within natural populations, creation of new populations through controlled breeding and propagation, and the wide range of techniques of processing wood to form the described end products.

Genetic Selection

The principles of classical plant and animal selection apply equally to forest trees. Outstanding phenotypes are identified in the best populations of uniformly aged and managed plantations (after the genetically superior populations have themselves been identified in field provenance trials); their genotypic values are estimated through some form of progeny testing and the best individuals are incorporated in freely pollinating breeding populations as the parents of the next generation.

Trees are different from annual plants in three respects : time required for maturity, space required for field tests and knowledge currently available about population genetic structure. Nevertheless, genetic improvements in yield of 20-30 percent have been obtained in many industrial species in the first generation of breeding (Zobel and Talbert, 1984).

When the commercial rotation is long (for instance with industrial conifers such as Sitka spruce in Britain, where the rotation may be 40 years or more), the gain per year is obviously not high; however, where early selection methods are available or where growth rates are naturally faster, as in the tropics, the investment return is acceptable. Classical first generation tree breeding programmes are now in progress in over 40 countries and with few exceptions they are producing significant gains in volume and value of product; the exceptions are those programmes that involve species with little natural variation and therefore, low heritabilities of the productive traits.

Research at the Oxford Forestry Institute (OFI) has identified populations and individual trees of several tropical species and is now providing breeding populations for developing countries to implement the research; the cost of the research over 20 years has been approximately 0.5 million pounds but, if all countries use the improved material at the current rate of planting unimproved samples, the benefit by the year 2000 will be approximately 1 billion pounds.

Controlled Breeding

Once individual superior genotypes have been identified, they can be mated together in various combinations to produce progeny for the next generation. This requires a knowledge of the natural breeding system, the phenology of individual trees, methods of collection, storing, testing and applying pollen to female flowers, and collecting seed from the controlled crosses. Although this information may be lacking for some species, there is nothing very technological about it. Rather, it is expensive to undertake controlled breeding on a large scale and most programs use open-pollinated seed orchard populations to produce the next generation.

Nevertheless, several gains arise from controlled breeding:information on individual and population genetic parameters is enhanced, specific combining abilities can be estimated in addition to the normal general combining abilities, and the parentage of each progeny is known so that crosses that produce superior offspring can be repeated. Both open-pollinated or controlled crossing populations are faced with meeting several objectives including the maintenance of genetic variability for future generations of selection and improvement of the technical qualities of wood material produced.

For rural development forestry, there may be many end products, each of which can be influenced by several anatomical, chemical or morphological features of the tree and these demand the use of multiple trait selection indices and multiple breeding populations (Namkoong, Barnes and Burley, 1980) based on species of so-called multipurpose trees (Burley, 1987).

Utilization of Forest Products

When wood is used in the solid form, its production from the standing tree requires mainly mechanical techniques such as sawing or veneer peeling; reconstituted wood products like paper or chip board require chemical or mechanical processes such as pulp digestion and chip adhesion. Likewise, when wood is converted to energy, it is usually through thermochemical methods such as pyrolysis, gasification and hydrogenolysis. However, none of these requires strictly biological processes. Nevertheless, newer biological processes are currently under investigation or already available and these are outlined below.

SPECIFIC RECENT TECHNOLOGIES RELEVANT TO FORESTRY

Clonal Propagation Including Tissue Culture

Many horticultural species have long been propagated vegetatively by cuttings, grafts or buddings and thousands of distinct varieties exist for hard and soft fruits and industrial crops such as cocoa, coffee and tea. Among forest trees, however, until recently, vegetative propagation has been restricted to a small number of genera (e.g. poplar and willow cuttings) or to very specific purposes (e.g. clonal seed orchards of industrial conifers and some hardwoods, and even here it was commonly by grafting rather than by cuttings).

There have been two main reasons for this. Firstly, as trees age, it becomes more difficult or impossible to root propagules; yet, to date, juvenile-mature correlations have been so poor that early selection of superior genotypes has not been precise and only mature trees have been reliably selected. Secondly, most forest tree populations are near to the wild type genetic structure with extensive genetic variation and tree breeders wish to avoid reducing the genetic base by cloning with all the implicit risk of pest and disease attack.

However, the first of these two objections can be overcome with modern developments in clonal propagation. For some species, rejuvenation of old trees can be achieved by repeated coppicing and coppice shoots can then be induced to root easily; in Brazil and the Congo, large areas of plantations are now established each year by this means. The trees and their products are more uniform to manage, harvest and use. For those species that cannot root from cuttings, the micropropagation and tissue culture techniques now available offer potential means of overcoming the juvenility problem, facilitating the bulk supply of valuable genotypes, and permitting the international exchange of improved material without the risk of transferring pests or diseases. The potential rate of increase in numbers of propagules of a given genotype can be phenomenal with one plant giving rise to several millions in one year although it is impractical to achieve such a rate because of limitations of labour and space.

Work is currently in progress in Australia, Brazil, Canada, Europe, Japan and the USA *inter alia* with a wide range of species including some 44 angiosperms and 19 gymnosperms (Bajaj, 1986; Dodds, 1983a); rejuvenation of tropical pines and temperate oaks and valuable ornamental woody plants is in progress at OFI through micropropagation. In Britain, clonal propagules of Sitka spruce may exceed a million in 1987. Propagation of 100-year-old trees of redwood and teak has been achieved (Boulay et al, 1979; Gupta et al, 1980) but generally, juvenile material has been used. Micrografting is a technique that inserts that apical dome of a select tree on to the tip of a rootstock and it encourages rejuvenation; it has been successful with apple, cherry and citrus species, rubber and eucalyptus (Jonard, 1986).

The second objection above is, of course, a valid implication of the risk of clonal propagation. In practice, no programme should rely on one or a few clones and most use at least 100 clones in the main planting pool. (However, the planting of poplars in France has been dominated by one clone, I-214.) Clones themselves can be known genotypes of high specific combining ability and thus maximize the rate of genetic gain and advance. Further, tissue culture systems based on a few clones will still provide genetic variation through somaclonal variation and somatic hybridization.

Clonal propagation through tissue culture offers the possibility of juvenile screening for growth rate and disease susceptibility of resistance; at OFI, current work includes the development of model *in vitro* systems to examine the disease defense and resistance mechanisms of tree species. These will then be extended to evaluate all selected genotypes before bulking and distribution into national breeding populations. Tissue cultures and micropropagules will also permit the evaluation of trees for tolerance of difficult soils (acid, alkaline, saline or affected by heavy metals), adverse climatic factors (extremes of temperature and rainfall) or aerial pollutants. The latter is of prime concern to temperate, industrialized countries and, at the IUFRO Congress in 1986, one member spoke out against investing funds in research on tree resistance to pollutants since, if successful, it would reduce the need for the major known polluters to address the causes. However, the evaluation of trees for difficult environmental conditions is a major imperative for tropical countries where billions of hectares are below their optimum potential productivity because of soil reaction or nutrient status; in India alone, there are some 150 Mha of wastelands, much of it with high salinity.

Whether or not tissue culture screening identifies disease resistance, it is a means of propagating virus-free planting stock. Meristem culture is well established for agricultural and horticultural species and could have application in tree species such as poplar that are commonly affected by virus diseases (Boxus and Druart, 1986).

One should distinguish between direct stimulation of axillary buds through micropropagation and indirect shoot regeneration through callus culture. The former is genetically more stable while the latter may undergo genetic erosion, mutation and changes in ploidy level at each periodic subculturing. However, the advantage lies in the large number of propagules that can be obtained since each cell of a callus is poten-tially capable of regenerating a plant. To date, some 200 woody species have been established in callus culture but few have so far produced plants while some cell lines have shown great instability (Dixon, 1986).

While the bulk of the tree products are solid or reconstituted wood, a significant proportion of the economic value derived from trees is made up of chemical deriva-tives; many of these are tapped directly from the standing tree, such as rubber, resin and gum, or extracted from wood, bark or leaves, such as essential oils (Burley and Lockhart, 1985; Caldecott, 1987). A possible application of tissue and callus culture is the mass production of such chemical feedstocks *in vitro*. The Unilever Company's major programme for palm oil production was based largely on the tissue culture of selected Nigerian clones, bulked in the UK, and transferred to Malaysia (although problems were encountered in sex determination within the cultures).

The following three techniques have relatively high risks of mutation:

Somatic Embryogenesis

One further use of *in vitro* techniques is the culture of embryos. These may be derived as normal sexually produced embryos in plants that would otherwise abort (e.g. sterile interspecific hybrids such as the Leyland cypress widely used in Britain) or produce recalcitrant seeds (that cannot be stored or germinated after storage). Alter-natively, they may be somatic embryos derived from callus cultures. Bajaj (1986) be-lieved that it would be feasible to encapsulate the embryos in a coating that resembles a seed coat and thus store them or sow them. In the USA, a patent has already been taken out for this technique (Abo-El Nil, 1980).

Generation of Haploids

In vitro techniques can also be used to propagate haploid plants from excised anthers, single pollen grains and haploid protoplasts. These can be used to shorten breeding programmes with forest trees that are normally highly heterozygous, to ob-tain diploid pure lines and to create specific heterotic combinations. Bajaj (1986) reported 24 woody species in which haploid callus, embryos, shoots or whole plants had been obtained through the culture of excised anthers or pollen.

Protoplast Fusion

The basic principles and methods are outlined by Dodds (1983b) and NAS (1984). Somatic hybridization through protoplast fusion permits the synthesis of crosses that would not normally be possible, e.g., between different genera or species that produce infertile offspring. As an alternative to sexual reproduction, it combines the entire nuclear and non-nuclear genomes of the parents. Protoplast culture has been successful in only 12 tree species to date, resulting in cell division, colony formation and whole plantlet formation in citrus (Vardi et al., 1975).

This genetic engineering technique offers the possibilities of : (i) production of amphidiploids; (ii) transfer of partial nuclear information between species; and (iii) transfer of cytoplasmic information. It will yield a greater range of genetic variation although it is clear that for most forest tree species the existing natural variation has often not been adequately explored or evaluated so that new variations are not at such a great premium as in advanced cultivars of agricultural crops.

The rate of development and application of these techniques to forest trees is rising rapidly. A recent IUFRO workshop received 17 papers concerned with somatic cell genetics of woody plants including two describing gene transfer through *Agrobacterium* and one with transfer of a herbicide tolerance gene to hybrid poplar.

Genetic Resources Conservation

In common with agricultural crops, the breeding of trees has to face the conflict between maximizing genetic gain and minimizing genetic erosion. Genetic variation must be broad to allow gain in future generations and selection for changing objectives, environments, management systems or diseases. Conversely, gain in any objective function is maximum with the highest selection intensity and the least variation among the selected parents and consequently their progeny. The multiple population breeding strategy referred to above allows for the maintenance of variability within systematic breeding programs.

However, advanced generation breeding risks losing genotypes that have ceased to be of direct interest in the current generation; also, the entire process of logging and deforestation in natural forests risks losing individual, population and species diversity. Some tree species are included in the Red Data Book of the International Union for the Conservation of Nature and Natural Resources (IUCN) but the major threat is the loss of populations within species that are not at risk as a whole taxon. For many of these seed, storage is impracticable (because they have recalcitrant seeds, because techniques are not yet known, or because they suffer enhanced mutation rates in storage) and tissue culture offers a possible means for long-term conservation.

Desirable germplasm may be conserved by two culture-based techniques. Tissue culture *per se* requires repeated subculturing and continuous monitoring of genetic behaviour and physiological status. An alternative is cryopreservation in which these tissue cultures are reduced to -196°C; to date, eight woody species have been preserved in this way and regenerated the desired germplasm, in some cases with apparently increased cold hardiness.

The cryopreservation of pollen is also possible but for this to be useful in genetic resources conservation, it also required techniques of diploidization or the conservation of female capability elsewhere.

Microbial Breakdown of Wood Components

One of the major problems facing paper makers and farmers alike is the disposal of lignin in wood and straw, respectively. Lignin inhibits straw breakdown in soil and must be removed in pulp manufacture to facilitate fibre separation and subsequent bonding in the paper. Various white-rot fungi have been shown to produce lignase enzymes, although at present the amounts of lignin destroyed are only 3-4 percent. The US Forest Products Laboratory is currently examining mutants and genetically engineered variants of one species, *Phanerochaete chrysosporium* (Suleski, 1987).

The same laboratory is currently working on the fermentation of xylose from hemicellulose in wood by *Candida tropicalis;* this leaves the cellulose intact and the fermentation products include ethanol, acetone, glycerol and butanediol, a precursor of synthetic rubber. These biotechnologies are cleaner environmentally and less demanding energetically than the traditional wood processing methods.

Micro-organisms have been involved in one of the oldest biotechnologies, fermentation and brewing, but today alcohol production is largely synthetic. However, with the progressive depletion of non-renewable energy sources, interest is being revived, and in some countries, significant amounts of alcohol are produced by fermentation of sugar or sugar contents and its consequent need of acid hydrolysis before fermentation. However, lignocellulosic sources including straw and wood wastes may become energy sources because of their abundance and lack of alternative markets. Research is needed on genetic variation among and within micro organism species in relation to quantity and type of enzymes produced and their temperature reactions. Pulp mill liquors may be used as bacterial feedstocks and in Finland, commercial plants have been established to produce protein.

Anaerobic digestion of plant biomass and sewage sludge in simple biogas generators is cheap and provides low purity methane for energy while leaving mineral nutrients and organic structure for soil improvement. Social and cultural constraints often inhibit the widespread adoption of small scale digesters owned by individuals or small communities. On the larger scale, higher purities of methane (90 percent) may be obtained with thermophilic digesters packed with immobilized organisms, thus eliminating the need for purification of the gas (NAS, 1982).

Applications of Biotechnologies to Forest Soil Microbiology

The genetic manipulations of nitrogen-fixing relationships between plants and rhizobia and actinomycetes were discussed by Postgate (1987). The principles clearly apply to the many leguminous trees now in use, particularly in the tropics and to the non-leguminous, nitrogen-fixing species such as alder, casuarina and some elms. Enhancement of the amount of nitrogen fixed has obvious benefits to rural populations in the Third World who cannot afford or are unable to import fertilizers. This is one of the major aims of agro forestry systems for adverse conditions such as high temperature, low rainfall, and extreme soil pH; for any condition there is potential for selection of fungal strain and for synthesis of tree-fungal association genotypes. This may be achieved *in vitro* and mass production of fungal spores for nursery inoculation may be feasible in culture. One of the common problems in establishing tree plantations is the existence in the soil of spores of mycelia of fungi that had infected the preceding crop. It is hard to determine the risk of infection by normal sampling and microscopic examination because of the difficulty of species identification. The OFI will shortly initiate a study of the feasibility of using monoclonal antibody techniques to overcome this problem (Haugen and Tainter, 1987).

CONCLUSIONS

Traditionally, research and development of new techniques has been conducted on annual plants, commonly those of immediate agricultural or horticultural importance and particularly for genetic studies where short generation times are desirable. Work on woody plants has tended to lag behind this unless a given species is seen to be obviously economically attractive, such as the industrial food and beverage crops. Forest trees have received little attention until recently when the socio-economic and environmental values of trees and forests began to be recognized. Now there is a worldwide burgeoning of interest and support for research on the maintenance, genetic improvement and efficient use of forest resources and products. The economic value of forest products alone justifies large investments in research.

It is unlikely that major breakthroughs in new biological technologies will arise from research using tree species (with the exception of those related to wood chemistry and to solving the problems of ageing), but there are clear applications of the new technologies to all tree species used in commercial forestry and to many used in rural development. As human populations and their demands increase and land available for production declines, concerted efforts are needed to maximize production on remaining sites, to rehabilitate degraded and abandoned or difficult sites, and to ensure sustainability of production and efficient use of all forest products and services.

REFERENCES

Abo-El Nil, M., 1980, Embryogenesis of gymnosperm forest trees, US Pat. No. 4:217,730, Washington D.C, USA.

Bajaj, Y.P.S. (Ed.), 1986, *Biotechnology in Agriculture and Forestry, 1. Trees 1.*, Springer-Verlag, New York, USA.

Boulay, M., Chaperon, H., David, A. and David, H. 1979, Micropropagation of forest trees, AFOCEL, Nangis, France.

Boxus, P. and Druart, P., 1986, Virus-free trees through tissue culture. In: *Biotechnology in Agriculture and Forestry, 1. Trees, 1.* (Y.P.S. Bajaj, ed.), pp. 24-30, Springer-Verlag, New York, USA.

Burley, J., 1987, Exploitation of the potential of multipurpose trees and shrubs in forestry. In: *Agroforestry: A Decade of Development* (H.A. Steppler and P.K.R. Nair, eds.), pp. 273-287, ICRAF, Nairobi, Kenya.

Burley, J. and Lockhart, L.A., 1985, Chemical extractives and exudates from trees, *Ann. Proc. Phytochem.* Soc. Eur., 26: 91-102.

Caldecott, J., 1987, Medicine and the fate of tropical forests, *British Med. J.,* 295: 229-230.

Dixon, R.K., 1986, Opportunities for MPTS biotechnology research networking. In: *Forestry Networks* (N. Adams and R.K. Dixon, eds.), pp. 5-14, Winrock International, Washington D.C, USA.

Dodds, J.H. (Ed.), 1983a, *Tissue Culture of Trees,* Croom Helm, London and Canberra.

Dodds, J.H., 1983b, The use of protoplast technology in tissue culture of trees. In: *Tissue Culture of Trees* (J. Dodds, ed.), pp. 103-112, Croom Helm, London and Canberra.

FAO, 1985, *Forest Resources 1980,* Food and Agricultural Organization, Rome, Italy.

Gupta, P.K., Nagdir, A.L., Mascarenhas, A.F. and Jagannathan, V., 1980, Tissue culture of forest trees: Clonal multiplication of *Tectona grandis* L. (teak) by tissue culture, *Plant Sci. Lett.,* 17: 259-268.

Haugen, L.M. and Tainter, F.H., 1987, Immunoassay, *J. Forest,* 89: 15-20.

Jonard, R., 1986, Micrografting and its applications to tree improvement. In: *Biotechnology in Agriculture and Forestry 1. Trees 1.* (Y.P.S. Bajaj, ed.), pp. 31-48, Springer-Verlag, New York, USA.

Namkoong, G., Barnes, R.D. and Burley, J., 1980, A philosophy of breeding strategy for tropical forest trees, Tropical Forestry Papers 16, 67, Commonwealth Forestry Institute, Oxford, UK.

NAS, 1982, *Priorities in Biotechnology Research for International Development,* National Academy Press, Washington, D.C, USA.

NAS, 1984, *Genetic Engineering of Plants: Agricultural Research Opportunities and Policy Concerns,* National Academy Press, Washington, D.C, USA.

Postgate, J., 1987, Prospects for new or improved biological nitrogen-fixing systems, Pap. Oxford International Symposium, (mimeo), MOA Foundation, Oxford, UK.

Suleski, J.C., 1987, Biotechnology promises new wood products, *J. Forest.,* 85: 38-43.

Vardi, A., Spiegel-Roy, P. and Galun, E., 1975, Citrus cell culture; isolation of protoplasts, plating densities, effect of mutagens and regeneration of embryos, *Plant Sci. Lett.,* 4: 231-236.

WRI and IIED, 1986, *World Resources 1986 : An Assessment of the Resource Base that Supports the Global Economy,* World Resources Institute and International Institute for Environment and Development, Basic Books, New York, USA.

Zobel, B.J. and Talbert, J.T., 1984, *Applied Forest Tree Improvement,* John Wiley and Sons, New York, USA.

3

FORESTRY IN INDIA : PROBLEMS, PROSPECTS AND THE ROLE OF TISSUE CULTURE

T.N. Khoshoo

ABSTRACT

Forestry in India is, indeed, at the crossroads. The effective forest cover is not more than 10.88 percent, which is too low to afford long-range ecological security and supply goods and services to people and industry. There is also the widening gap between supply of and demand for fuelwood, timber and pulpwood. A major program needs to be mounted to restore the forest cover by protecting forests in conservation areas and raise large-scale plantations and man-made forests. Under the circumstances, the role of tissue culture lies in enhancing the capability for the production of planting material of selected high-yielding types so as to boost production and productivity. However, tissue culture technology has to be part of an overall forest genetics and tree breeding program. The underlying philosophy in agro-forestry and industrial forestry has to be tree-crop farming. Among other things, this would involve using special ideotypes for producing timber and pulpwood, fuelwood and fodder with enhanced productivity.

T.N. Khoshoo * Tata Energy Research Institute, 7 Jor Bagh, New Delhi - 110 003, India.

Such an approach has to be strongly backed by change in laws on land-use and tenure and grazing and livestock, which are supportive of raising large-scale industrial plantations and generating silvi-pastoral agro-ecosystems and man-made forests.

The seed-raised plantations of superior types are essentially heterozygous and can be grown on poorer and less accessible land giving relatively lower yields and involving lower maintenance costs. On the other hand, tissue culture and/or clonal plantations of superior types are essentially genetically uniform with high yields and can be used on good land but have high maintenance costs. Both would enable the production of forestry-based goods and services on a sustainable basis, which in turn, would help conserve natural forests. In this manner, long-term environmental security and short- and long-term production imperatives can be properly intermeshed into integrated strategies.

INTRODUCTION

Due to a high rate of deforestation, forestry has become critical to the long-range ecological security of many developing countries. The loss of forest cover has many ramifications. One of these can be gauged from a very old saying of a Kashmiri saint -philosopher, Nand Rishi: *Ann poshi teli, yeli poshi van,* i.e., food will last only as long as the forests last. In fact, the African crisis has taught us that we need some minimal forest cover to secure sustained food and livelihood. The purpose of the present discussion is to look into the possibility of using a powerful technique like plant tissue culture in forestry to generate appropriate planting material in order to halt the galloping rate of deforestation and meet the needs of people and industry.

Plant tissue culture is an essential component of plant biotechnology. Apart from mass multiplication of elite plants, it also provides the means to multiply and regenerate novel plants from genetically engineered cells. The promising plants thus produced may be readily cloned in cultures under aseptic conditions. Additional advantages of tissue culture propagation of plants, popularly known as micropropagation, are economy of time and space and freedom from seasonal constraints.

For almost 25 years, micropropagation has been used for clonal multiplication of orchids and other ornamental species on a commercial scale. However, the success with forest trees has been rather slow and limited and to date, the only practical method of propagation is through seeds, which, for obvious reasons of genetic heterogeneity, may not be entirely satisfactory where high productivity is required. Even if the seeds are collected from superior plants, the progeny represents half sibs. The vegetative propagation of trees is beset with many problems. Often, by the time the tree attains the stage at which it can be evaluated for the desirable traits, it has already passed the stage at which it can be vegetatively propagated. It loses the ability to root and in such cases, tissue culture is the most valuable aid for cloning the selected trees.

Many laboratories all over the world are extensively working to apply *in vitro* methods of clonal propagation to both gymnospermous and angiospermous forest trees (DBT, 1988). So far, success in the micropropagation of elite trees has been largely restricted to those species that root readily *in vivo,* such as poplars, willows and rosaceous fruit trees. Reports concerning the *in vitro* propagation of difficult-to-root species deal mainly with embryonal or young seedling materials. The mature tissues and organs of these plants have remained recalcitrant. However, successful micro-propagation of mature trees of *Eucalyptus camaldulensis, E. citriodora, Leucaena leu-cocephala, Mangifera indica, Morus alba, Phoenix dactylifera, Prosopis cineraria, Santalum album* and *Tectona grandis* has been achieved. In a way, micropropagation, in its essential form, is available today for demonstration in forestry but its perfection requires more research. It has the potential to serve as a tool to exploit the gains achieved in tree improvement programs in the shortest possible time.

The role of tissue culture in clonal propagation of tree species varies with the situation: for hard-to-root species, it would be useful for routine propagation. In others, it may help to accelerate the rate of multiplication achievable through the conventional methods. The latter method applies to rare, highly productive genotypes from new selections or introductions. Since tissue culture is a relatively expensive technique and not always easy, the selection of the species for micropropagation should be done very carefully. It must involve the collective wisdom of forest scientists and managers in addition to tissue culturists and must be based on some objective considerations.

In India, few systems (bamboo, eucalyptus, sandal, teak) have been reasonably well worked out for *in vitro* clonal propagation. It would be worthwhile to treat these plants as systems for case studies to evaluate the economics and the problems encountered in developing the protocols for large-scale *in vitro* propagation (DBT, 1988). It is only after the gap between the laboratory experiments and the commercial production of tissue culture-raised plants has been bridged for the relevant species that one can think of large-scale use of tissue culture in forestry.

The foregoing represents a very brief state-of-the-art account of tissue culture technology with reference to forest trees. This needs to be related to the short- and the long-term needs of forestry in a developing country like India. It is, therefore, necessary to look into the scope and limitations of application of tissue culture to forestry in India. For this, it is essential to review the prevailing situation in the area of forestry with particular reference to:

* The extent and nature of forest cover
* Role and goals of forestry against a country's particular socio-economic milieu
* Strategies for enhancing the production/productivity
* The relative short- and long-term advantages and disadvantages of the seed-raised plantations, as compared to clonal/tissue-raised plantations.

FOREST COVER

Nearly 80 percent of the Indian Sub-continent was under forest cover in 1000 AD. Over the years, the cover has been reduced and many estimates, often irreconcilable, have been put forward by the agricultural and forestry revenue records, Directorate of Economics and Statistics and non-governmental groups (Khoshoo, 1986; Bisht, 1988). Part of the controversy stems from the fact that a distinction is not made between land under the control of State and Central Forest Departments, and *actual* forest cover. The former is nearly 24 percent of the total land mass of India, and is composed of all types of land, ranging from well-wooded to totally barren. However, when one talks of forest cover, one means closed forest; ideally, the ground should not be visible from the air. The density in a closed forest ranges from 0.5 to 0.8, while the theoretical maximum is 1.0 which is practically unattainable.

In order to arrive at a reliable estimate of the forest cover in India, the National Remote Sensing Agency (NRSA) of the Department of Space, Government of India, was charged with this responsibility. NRSA found that the forest cover in India in 1980-82 was 13.94 percent (Khoshoo, 1986; Table 1). However, the revised figure now released by the National Natural Resource Monitoring System (NNRMS) of the Department of Space, Government of India, reveals that the forest cover in 1980-82 was 19.52 percent (64.22 Mha), indicating a substantial increase of 5.58 percent (18.36 Mha). This is not due to any accelerated rate of afforestation, but is the result of the reconciliation of figures between the Departments of Space and Forestry and Wildlife of the Government of India (G. Behera, personal communication).

A close look at Table 1 shows that after reconciliation between the two Departments,there has only been a marginal increase of 0.34 Mha in the closed forest areas, which is the result of the inclusion, after random ground check, of roadside plantings, farm forestry and even young plantations. The increase of 0.14 Mha in mangroves has also been marginal. In contrast, the increase in open forest/degraded

Table 1: Change in forest cover (in Mha) from 1972-75 to 1980-82

Forest type	Forest cover in 1972-75	Forest cover in 1980-82	
		Figures released by	
		NRSA, 1984	NNRMS, 1987(in %)
Closed Forest	46.10	35.43	35.77 (10.88)
Open/Degraded Forest	8.80	10.00	27.65 (8.41)
Mangroves	0.30	0.27	0.41 (0.12)
Coffee Plantations	-	-	0.37 (0.11)
Total	55.20 (16.83%)	45.70 (13.94%)	64.20 (19.52%)

Source: NRSA, 1984; NNRMS, 1987

areas has registered an increase of over 270 percent and the area is now 27.65 Mha. In addition, 0.37 Mha of coffee plantations have also been shown as forests (Table 1). The marginal increase in the area under closed forests and mangroves, the sharp increase in the area under open degraded forests, and the inclusion of coffee plantations are not of any consequence whatsoever. *Obviously, in real terms, the forested area today is only 35.77 Mha which is 10.88 percent of the land mass of India. This represents the effective forest cover,* and includes plantings made on roadsides and farms as well as young plantations. There is a difference between plantations and forests. However, the latter in course of time may become man-made forests. It is, therefore, open to question if the actual forest cover is even 10.88 percent.

There is often a tendency to treat information on forest cover as classified. Such a view is indeed untenable, because with the present-day advances made in remote sensing technology, such information is also available elsewhere. Furthermore, since our future is intimately linked with forest cover, such information should not be withheld from the public. It may also be pointed out that under the Forest Policy of 1952, the area under forest cover, for the country as a whole, has to be about 33 percent (108.57 Mha) with 60 percent of it in the hilly regions. *In reality, the country is deficient by about 72.80 Mha as far as effective forest cover is concerned.* Such a target cannot be reached even if all the open/degraded forest areas (27.65 Mha) are developed into closed forest. There would still be a shortage of 45.15 Mha, or 45.56 Mha if mangroves are included. However, if we separate hilly regions, where the area under forest cover is required to be 60 percent, the area required under forest cover will be far more than 45.56 Mha.

The foregoing analysis is based on the premise that there has been no degradation after 1980-82. However, a comparison of closed forest cover in 1972-75 and 1980-82 (46.10 Mha and 35.43 Mha, respectively) indicates that the reduction in forest cover has been about 1.5 Mha per year (Table 1). If such degradation has been continuing, the present day effective forest cover may be of the order of 26.77 Mha, i.e., 8.13 percent. In short, the situation is far from being ideal and a gigantic task lies before the country, which would involve raising billions and trillions of saplings for planting over 81 Mha of land for purposes of greening the forest areas which have no worthwhile forest cover at present.

The causes of such an *ecological deficit* can be traced to the disastrous Forest Policy of 1894 during the British Raj (Khoshoo, 1986). Forestry was subservient to agriculture and was also declared as a revenue-generating sector. A lot of forest destruction took place during the initial period of the Raj to create a railway network for the speedy movement of troops to quell the uprisings in different parts of the country. During the post-Independence era (1947 onwards), the same philosophy has more or less continued in practice. Furthermore, there has been affluence or greed-related as well as poverty or need-related forest destruction. Unlike agriculture, forest destruction has long-range implications and its effects may be invisible in the short range. Forestry has also not been in the country's mainstream of R & D. Today, the rate of forest regeneration is far less than the rate of deforestation, with the result that there is a major ecological backlog or deficit, or an environmental drag which has very serious and all-encompassing social, economic and long-term environmental implications. Essentially, our present plight in forestry is the result of lack of knowledge and political will.

GOALS OF FORESTRY

The principal goals of forestry in the Indian context are:

* Affording long-range ecological security for the conservation of climate, water, soil and bio-diversity
* Meeting the needs for goods and services, including firewood, charcoal and fodder, for rural/tribal communities and the urban poor
* Meeting the wood requirements of the people and industry for timber, pulp, fiber and silvi-chemicals
* Amelioration of degraded areas and wastelands so as to enhance the pro-ductive capacity of such derelict lands and to improve general aesthetics.

Emanating from these principal goals are four *mutually supportive* types of forestry:

* Conservation Forestry
* Agro-forestry
* Industrial Forestry
* Environmental/Revegetation Forestry

Conservation Forestry

This is most relevant to all water regimes/watersheds/ catchments; representative ecosystem and biosphere reserves (located in different biomes); centres of diversity; national parks and sanctuaries and fragile ecosystems. In these regions, exploitation of wood and non-wood resources should be prohibited, unless warranted on scientific and technical grounds, for maintaining the health of the concerned forest, and that too, not more than the Mean Annual Increment (MAI). The restoration and repair of such areas has to be done with local and indigenous species, and on no account should exotics ever be introduced in conservation areas.

Conservation forestry benefits all people because it is linked to the stabilisation of micro-climate, conservation of soil, water and bio-diversity, source of non-wood products and other amenities.

Agro-forestry

Here, the objective is the integration of agriculture, forestry and animal husbandry to meet food, fuel and fodder needs through a well chalked out Agri-Silvi-

Pastural, Agri-Silvicultural or Silvi-Pastural model of development. The basic idea is to aim at the intensification and diversification of biomass production in rural areas.

There should be no objection to the use of exotic trees in agro-forestry, if their use is warranted on the grounds of land use and end-use to meet the needs of location-specific edaphic conditions and the demands of the local population, respectively. Here, the beneficiaries are the rural and urban poor whose needs of fuelwood and small timber are also met and it may serve as a source of some income as well. Agro-forestry practised on a sustainable basis would ultimately relieve pressure on natural forests and thereby help in forest conservation.

Industrial Forestry

Here, the objective is to meet the needs of timber, pulpwood and fiber. It is a commercial venture based on wood quality and input-output considerations. The objective has to be tree-crop agriculture. The immediate clients are wood-based industries. Industrial forestry has to be related to land use and end-use considerations. Since these are commercial ventures, production and productivity are chief considerations, and if warranted on other grounds, fast growing exotics are most welcome.

India can afford to be oblivious to the needs of industrial/commercial forestry only at the cost of ecological security. Industrial forestry should not be ignored on account of emotional considerations. Realism demands that there be a crash program on industrial forestry in order to save our forest wealth.

Suggestions regarding the import of timber, firewood, pulpwood, etc., can only help to avert the most critical immediate situation but do not offer a permanent solution, for more reasons than one. Firstly, they only help to shift forest degradation to other countries (most probably developing ones) which is not ethical. Secondly, the kind of money required for import may not always be available. Thirdly, wood can become a political weapon like food and oil, and its prices will keep soaring. The best strategy would be to give very high priority to industrial forestry and take steps to extend all help by suitable modification of land laws, etc.

Environmental/Revegetation Forestry

The objective here is to green derelict and wastelands in order to ameliorate, and finally, restore them. The process can be started by creating natural wilderness areas by using the principles of plant colonisation. Owing to litter fall, a decomposer chain will start, followed by soil conservation and increased water retention. This would go a long way in improving the quality of these lands. Starting with plantations of tolerant species which would lead to some improvement of soil, there is a distinct possibility of growing less tolerant species, resulting in further improvement in soil characteristics. In the succeeding cycles, it would be possible to grow increasingly less tolerant species.

The four major uses of forestry outlined above are neither *mutually exclusive,* nor is one at the expense of the other. *They are mutually supportive.* Furthermore, it may be pointed out that wasteland development, though laudable, cannot be expected to be the panacea for all our food, fodder, timber and fuelwood problems. Wastelands are essentially derelict lands, and for several years to come, these will be less productive. While forestry is expected to perform miracles on wastelands, the prime agricultural land continues to be used for non-food purposes such as human settlements, industries, road and rail systems, airports, etc. The land laws are either too weak or non-existent, as a result of which such ventures are permitted on prime agricultural land.

Table 2: Supply and demand (Mm^3)

Category	Supply (recorded) 1984	Demand Mean of high and low levels 2000
Fuelwood	14.00	225.00
Timber (including sawn, match and round wood and panel products)	7.96	42.12
Pulp and Paper	6.74	13.68
Total	28.70	280.80

Source: India's Forests. 1984 Central Forestry Commission. Government of India. New Delhi.

DEMAND AND SUPPLY

There are many different estimates of the demand for and supply of fuelwood, timber and pulpwood. Table 2 is based on a publication of the Government of India, entitled *India's Forests* (CFC, 1984). It is evident that there is a significant gap between demand and supply in all the three major uses of wood, viz. fuelwood, timber, and pulpwood and paper, in addition to the large unrecorded demand for wood for the fruit industry and for the armed forces. Furthermore, demand figures for other wood-based industries is given in Table 3. This covers timber used in small rural industries as well as in sports goods, pencils, railway sleepers, truck bodies, etc.

Additional fuelwood requirement by A.D. 2000 (taking 1984 supply figures as the baseline) is the maximum (Table 2), and includes both firewood (domestic use) and fuelwood (wood burned in industrial installations). The demand, as calculated by the Advisory Board on Energy (1985) is of the order of 375.5 Mt (565.25 Mm^3). There is yet another estimate by the Planning Commission (PC, 1982).

The data are not accurate, as they do not include the firewood used by those who have the legal right to collect firewood from the forests in the form of head-loads. Therefore, the recorded demand for fire/fuelwood is not realistic. Furthermore, the approach to the problem has become highly emotional and, therefore, subjective in character. The demand is highly exaggerated and is usually calculated on the basis of population, at the rate of 20 kg per household per month. In actual practice, much of the demand of the rural poor is met from agricultural and forestry residues and wastes. There is, however, no doubt that firewood is a genuine need of the landless and the rural poor, at least during some part of the year, e.g., the winter months.

Table 3 : Demand projections for other wood-based industries (Mm^3)

Item	1985	2001
Agricultural implements	6.020	7.713
Bullock carts	11.300	12.000
Katha *(Acacia catechu)*	0.222	0.264
Shuttles and bobbins	3.090	3.500
Sports goods	0.260	1.010
Shoe lasts	0.118	0.168
Pencils	0.021	0.024
Railway sleepers	0.350	0.500
Truck bodies	0.124	0.157
Total	21.505	25.336

Source : K.G. Venkatraman. personal communication.

Table 4 shows that there is a shortage of 211 Mm^3 of fuelwood and if this has to be met on a rotational cycle of 10 years (as in the case of acacias and *Prosopis*) and with a productivity of 0.7 m^3 per year per hectare from the wastelands, the country would need about 300 Mha to grow fuelwood alone. Timber, pulp and paper needs are far more realistic. Timber would require 7 Mha at a rotational cycle of 50 years and a productivity of 5 m^3 per hectare per year, while pulp and paper would need 1 Mha in a rotational cycle of eight years (as in major pulpwoods such as eucalyptus) and a productivity of 7 m^3 per hectare per year. Other wood-based industries (Table 3) would also need 1 Mha at 5 m^3 per year productivity with a 50-year rotation (Table 4). While fu-

elwood may be grown on wastelands, timber, pulp, paper and other wood-based industries (Table 4) would need relatively fertile land. The reason for the unusually long
rotation for timber species (such as teak and sissoo) and other wood-based industries
is that it takes 15 years or more for the hardwood to form, and it is only with time that
it acquires the requisite properties of strength. In general, trees are harvested for timber when their diameter is about 30 cm or more, which coincides with the age of 50
years and more. The requirement of land under forest plantations to meet the demand
is about 309 Mha (Table 4), while the total geographic area of the country is only 329
Mha. Such an area-demand, when taken in conjunction with the administrative ban
imposed by the Government of India on felling trees, indicates that there is no prospect of meeting even a part of the demand from existing sources. It may be pointed out
that such ban orders are, in effect, administrative orders. The ban orders may not
always be legally binding on the States, because forestry is a State subject. Furthermore, such orders are not applicable to *panchayats* (village communes), where felling
goes on unabated because it is the Revenue Department which exercises a degree of
control over these areas. The Divisional Forest Officer only acts as an adviser to the
District Magistrate. In practice, most panchayats, at least in the state of Uttar Pradesh,
sell trees without any advice or permission from the forest department.

The conservation areas will not yield much wood, except in very exceptional
cases where wood can be collected if warranted on ecological grounds of the
concerned forest. Even so, this cannot be more than 1 to 1.5 percent, which is the Mean
Annual Increment (MAI). Besides meeting the fuelwood and small timber needs of
rural areas, agro-forestry can, to some extent, meet the requirements of nearby urban

Table 4 : Land area needed to meet the demand

Category	Additional requirement	Rotation of harvesting	Production in m^3/ha/year	Area needed to meet the demand (Mha)
Fuelwood (Table 2)	211.00	10	0.7 (wastelands)	300
Timber (Table 2)	34.16	50	5 *	7
Pulp and Paper (Table 2)	6.94	8	7 *	1
Other Wood-Based Industries (Table 3)	3.83	50	5 *	1
Total	255.93	-	-	309

* Reasonably fertile land.

areas as well. Industrial forestry, however, which is still in the plantation stage at present, can neither meet the timber needs of the people nor the pulpwood needs of the paper industry. The reason is low productivity: these plantations are based on unselected stocks and are, therefore, genetically heterogeneous with no intensive management. In the foreseeable future, revegetation of the wastelands cannot wipe out the colossal deficit in wood demand. This would only help to green the barren areas and, in course of time, stop soil erosion, help in water retention and ultimately result in soil amelioration.

Therefore, it is not possible to meet the demand from the existing forest cover and with the present pattern of production technology. Furthermore, land is going to be an increasingly scarce resource as there are many competing demands due to population explosion. Production has to depend less on present standing forest stock and plantations and more on future plantations with very high productivity. However, the prevailing view that *only* wastelands need to be used for purposes of forestry will mean that the yields will be significantly low.

One of the options available for averting the serious wood crisis is to import wood for various purposes. The question that arises is : from where and at what price? There is no doubt that wood import has been kept on Open General Licence for the past several years. Can we depend on imports to meet our wood needs for all times to come? Is it possible to sustain such a policy and, above all, is it ethical to shift forest degradation to some other developing country which is in need of foreign exchange?

It is evident that what is required is a mix of several options: one of these can be the import of wood, but this should be supplemented with home production on a sustainable basis. To do this, a basic change is required in our approach towards forestry from an emotional and subjective approach which, in practical terms, is negative, to a positive approach stressing wood production on a sustainable basis. This, in due course, will turn out to be the best form of conservation of our natural forests. In other words, to save forests, we need to generate man-made plantations and forests, because we can no longer depend only on natural forests for our wood needs on account of the escalating demands of our increasing population. Today, the rate of regeneration is far less than the rate of deforestation.

Therefore, what is needed is a quantum jump in wood production through a well thought out strategy. This is possible if there is a strong infiltration of science and technology into forestry research and development, followed by training and education and demonstration and extension. These efforts have to be backed by an appropriate policy formulation on land use, forestry, grazing and livestock, water, irrigation, energy, etc. All these are intimately connected with forestry and cannot be taken in isolation. Such a holistic approach in policy formulation will help in better protection of forests and plantations from grazing animals and unauthorised cutting and lopping. If such a strategy based on science and technology is chalked out and implemented, it will not only help boost production, but also assist in the amelioration and conservation of soil, water, and the microclimate in general.

Furthermore, on account of shortage of land, there is an urgent need for the integration of agriculture, forestry and animal husbandry. This is of paramount importance, as all these are land-based ventures and land is a finite resource. Furthermore, these three uses of land are critical to the well-being of people at the rural subsistence level. As pointed out earlier, what is required is detailed land use planning at the district level (Khoshoo, 1986). Also, it is not correct to apportion only wastelands for forestry purposes and use prime agricultural land for non-agricultural and non-forestry uses. We have to admit that the forest productivity of wastelands will remain low in the foreseeable future.

Forestry today is not in the mainstream of science and technology of the country. It needs to be cross-fertilised with newer advances in biology. There are little or no elementary data on the collection and evaluation of germplasm and basic information on breeding biology, let alone the breeding methodology to boost productivity. It is not possible to meet the wood needs through the so-called people's forestry with no scientific and technical inputs. Today, it is not a question of planting a tree at a location for an end-use, but planting *the* tree at *the* location for *the* end-use. The only way out is to *treat trees as crops*. The earlier we realise and practise this, the better it is for our long-range ecological security.

As indicated above, if the country wishes to plan for self-sufficiency in matters of timber needs for diverse uses (Tables 2 and 3), we need to plant about 309 Mha of land. However, two things are clear: first, a large amount of planting material, i.e., billions and trillions of saplings would be needed; second, in view of the competing land resources, the country does not have so much land available for forestry. Therefore, there is a need to go vertical and raise plantations based on high-yielding quality materials which, once identified, can be multiplied on a large scale through the application of tissue culture technology and/or clonal propagation, among other options. In simpler terms, we need to treat forest trees as tree-crops.

FOREST TREE-CROPS

A crop has been defined as a plant which, on cultivation, yields a useful product or products (Simmonds, 1985). Implicit in this statement are three things :

First, a crop has a period of domestication involving conscious or unconscious selection under human influence. Examples of this among forest trees are *Populus spp., Tectona grandis, Shorea robusta, Eucalyptus spp., Cryptomeria japonica,* some pines and many fruit trees. All these were harvested at one time from wild stock, but today these are harvested from early domesticates.

Second, associated with domestication are genetic changes. While small genetic changes are seen in forest trees wherever they have undergone a period of domestication, the large changes are seen in fruit trees such as mango, apple, pear, etc.

Third, a crop gradually ousts the wild progenitor and the latter may be ultimately lost. The best examples of this among arboreal species are: cultivated palms (arecanut, coconut, date, oil, sago), nuts (almond, walnut), spice trees (clove, nutmeg), and fruit trees (apple, citrus, peach, pear, mango). The demand for these is met entirely from the domesticates and not from the wild types, on account of the high production/productivity from limited land resources. In such cases, these species are now available only from domesticated stocks, which have spread beyond the frontiers of their centers of diversity, so much so that in some cases, such centers have even become questionable (e.g., coconut).

There is no doubt that in India, we have a very high degree of awareness about trees in general, but hardly any, if at all, about trees as crops. There is an urgent need to build such knowledge, particularly for species of trees that have now become relevant. Some of the areas are :

* Botany, phenology, anatomy, ecology, physiology, genetics, breeding, pathology, wood science, silviculture, timber engineering, chemistry, bioconversion, logging, etc.
* Conservation of tree genetic resources in *ex situ* form in seed orchards, seed and tissue banks, and arboreta; and in *in situ* conditions in ecosystems.
* A policy for tree introductions based on scientific and technological considerations so as to avoid future controversies, like the eucalyptus.

While genetic upgrading of tree species is most desirable even though it takes considerable time, the most immediate need is to direct proper attention to the identification and development of elites based on proper scientific and technological criteria, particularly for agro-forestry and industrial forestry. Depending on the quality of the initial planting material, for the same resources, effort and infrastructure, yields could be enhanced considerably. A procedure for screening, testing and selection of proper location-specific high-yielding planting material is also required. The three phases could run simultaneously or could sequentially follow a particular design and management system, followed by a specified system of analysis so as to make results comparable to one another. A large number of species or clones could be mass screened in small plots for a short rotation (about 1/15 to 1/5th of the normal). In the next phase, a smaller number of species/clones could be tested in relatively larger plots and for longer rotation (1/4th-1/3rd), followed by a selection of the best type(s) for a particular end-use (timber, pulp, fiber) and for a particular location, for use under normal plantation conditions or as single trees on borders, small wood lots, strip plantations on roads, railway tracts, canal banks, etc.

Multipurpose Trees

A lot of populism and mistaken ideas have distorted the concept of multipurpose trees. Although one would like to put a tree to many diverse uses (wood for timber, fuel, charcoal, pulp, fiber, etc. edible fruits and seeds; leaves for fodder, seed oil, latex, gum exudates, etc.), the same tree cannot be put to all these uses at the same time. For instance, if leaves are constantly used as fodder, there would be reduced photosynthesis leading to reduced wood output.

Each use (fuel, charcoal, timber, pulp, fiber, fodder, latex, etc.) involves specific parameters including chemicals. Therefore, for enhanced productivity, *a single purpose must dominate* even when there are by-products. A tree is *multipurpose* at the overall specific level. However, specific plant types are needed for specific end-uses. Thus, the aim has to be to develop single-purpose clones of an overall *multipurpose* species. There is, therefore, an urgent need to breed trees for a single-purpose with specific plant type or ideotype.

IDEOTYPES

The concept of ideotypes in forestry has come from agricultural crops. For instance, the Mexican dwarf wheats can withstand close planting; the plant is fertiliser responsive; its stems are erect and non-lodging; its leaves capture maximum sunlight and apportion more photosynthates to the grain rather than to the vegetative parts; etc. These characteristics have come from different wheat varieties, e.g. the dwarf character came from Norin 10, a Japanese accession. The ideotype concept is, no doubt, followed intuitively by foresters and not by design. The ideotypes of trees in natural population have evolved through natural selection for an altogether different purpose and may not be advantageous in intensively managed plantations.

The conceptualisation, identification and assembling of a particular set of characteristics and the development of tree ideotypes require a high degree of coordination between tree physiologists, geneticists, breeders and silviculturists (Dickmann, 1985). The ideotypes for various end-uses will differ: for each end-use, they may differ depending upon whether the trees are to be grown in a plantation or under agro-forestry conditions. This difference is due to the interplay of a host of characteristics ranging from morphological, anatomical, physiological, genetical, canopy and root architecture, chemical, to silvicultural. Furthermore, yield parameters have to be related to phenology, photosynthesis, sink dynamics, competition, ageing, etc. Many of these characteristics are strongly correlated. The aim is to develop ideotypes for specific end-uses with *high harvest* index.

An instance of mismatch between end-use and the ideotype is found in the eucalyptus cultivated in India. The ideotype characteristics of this species are suitable to meet the requirements for timber or pulpwood rather than for firewood, although for the latter, even a timber species can be chopped and used for energy generation. The latter would be a waste of valuable biomass. To grow eucalyptus for firewood under social forestry programs in India is basically a wrong end-use, and it is no wonder that eucalyptus biomass has, therefore, found an outlet for use as timber and pulpwood.

Ideotype for Cultivars for Timber Plantations

In the ideotypes for timber, the partitioning of nutrients has to be in favour of wood in the main stem, rather than for branch-wood, bark and even reproductive organs. Inflorescence in such a type should be small, with few flowers. In a plantation, such types should suppress the neighbouring trees to a lesser extent in competition. At the same time, it should have an efficient light interception on account of a good vertical crown with high foliage vs. branch-wood ratio (Dickmann, 1985). The general characteristics of such a tree would be :

* Straight stem
* Narrow green crown
* Thin bark
* Good quality timber : tough and stiff with high tensile strength; straight grain and good thermal insulation properties; easy working with simple tools and high strength/weight ratio
* Rapid growth
* Branches thin, slow growing, small and arising at 90° to the trunk
* More leaf area per unit of branch-wood weight
* Longer lasting foliage
* Few flowers
* Tolerant to a particular kind of environmental stress (heat, frost, drought, wetness etc.)
* High survival under specific agro-ecological conditions.

The advantages of such an ideotype are the elimination of expensive thinning operations; easier pruning because the branches would be thin and arising at 90° to the stem, good quality bole with fairly uniform timber; and easy harvesting of timber, leaving very small logging residue which could be easily handled.

The possible disadvantages are that, in comparison to normal spreading trees, the thin crown of the ideotype would capture less light; the low competitive ability may encourage the growth of unwanted weeds, particularly in the early stages of development and the long stems may lead to thinner trees.

The characteristics enumerated above for the ideotypes for timber would also apply to pulpwood with suitable anatomical and chemical characteristics of the fiber. It may be pointed out that pulpwood for writing paper would be in demand in the foreseeable future, notwithstanding the developments in microchip and TV technology.

Fuelwood

The need for fuelwood is most urgent in the rural subsistence sector, and for this purpose, marginal wastelands would be available for a considerable length of time in the future. These are chronically low-nutrient soils. For fuelwood species, therefore, there is a need for :

* High genetic variability leading to wide agro-ecological adaptability
* Rapid growth
* Wood with medium to high density
* High calorific value

* Straight grain
* Burns steadily without any toxic smoke and sparks
* Stem thin and of medium length
* Thornless
* Not excessively branched
* Ability to coppice and pollard
* Easy vegetative propagation
* Minimum bark
* Ability to tolerate competition
* Amenable to high density/short rotation for high biomass production.

Fodder Trees

Tree fodder (leafy shoots, twigs and fruit) was popular in ancient times but down the ages, attention shifted to fodder annuals (grasses and legumes). On account of use in agro-forestry, interest in fodder trees has been revived as an important element in sustainable rural development. However, it may be pointed out that tree fodder is rarely the sole feed because its nutritive quality is lower than that of herbaceous forages. These are used as a mix in the dry season in order to obtain optimal performance. It is necessary to work out a calendar of lopping, stripping and cropping down to almost ground level, vis-a-vis the age and size of the tree. For instance, a good *Prosopis cineraria* tree gives nearly 25-45 kg of dry leaf forage per year when properly manipulated. Some of the general characteristics of fodder trees are :

* Small trees/shrubs
* Highly branched
* Capacity of regrowth after periodic lopping and leaf-stripping without affecting phenology and metabolism
* Palatability with livestock-specificity (e.g. cattle, goat, sheep, camel, etc.)
* Digestibility
* Proper chemical characteristics, particularly protein content.

ENHANCING THE PRODUCTIVITY OF TREE-CROPS

Such a strategy goes well with the concept of *forest tree-crop farming*. It is also in line with the contemporary developments that are taking place in the developed countries. Historically, human beings met all their wood needs from natural forests because forests were far in excess of the human needs. This stage was akin to the foraging, hunting and gathering food, fuel and fiber needs from the wild, which has been abandoned a long time ago except amongst the tribals. We have today well organised agriculture and animal husbandry systems. Such a change in forestry in India is long overdue to meet the wood needs through production/industrial forestry as also agro-forestry, which help to create alternative sources and thus help in saving the natural ones.

The foregoing broad descriptions of the ideotypes of timber and pulpwood, fuelwood and fodder trees give an idea about the characteristics to be looked for while making selections from the natural variants for a specific purpose, so as to maximise the productivity of the end-product. Having identified elites, there would follow a strategy to enhance their productivity. A five-point strategy to attain such a goal is given below which is indeed interconnected and can be taken up simultaneously :

* Optimising silvicultural and nutritional requirements, including use of fertilisers, irrigation and bacterial and mycorrhizal inoculations
* Disease and pest control
* Weed control
* Application of advanced forest tree breeding methodology for evolving superior genetic strains
* Judicious use of tissue culture and biotechnology

The knowledge about tree genetics and breeding is particularly deficient. The objective is to produce tree populations that are *more profitable to grow than their progenitors*. Obviously, yield is the most important criterion. Basic knowledge of breeding systems and population genetics of the trees is, indeed, critical to their success. Such knowledge would help in chalking out the exact breeding methodology. This would involve close collaboration between tree geneticists, breeders and silviculturists. The former would evolve improved trees, which the latter would pack into plantations in a suitable design. The relationship is similar to that of crop geneticists who evolve cultivars and agronomists who cultivate these in the field.

The present practice involves the use of seeds produced as a result of open pollination from superior mother trees and these are technically half sibs. Two basic approaches can be adopted for tree genetics and breeding.

First, higher genetic gains can be obtained from full sibs with selective cross pollination. This is followed by mass selection involving progeny testing, or the development of synthetics through inter-population crossing. The basic idea is to increase desirable genotypes in a population and at the same time, maintain a broad genetic base for future increase in productivity. Such a method of tree improvement is based on the principles of organic evolution, and genetically such plantations approach the situation found in nature because there is a high degree of genetic diversity.

The advantages of such an approach are that it affords insurance against loss of recognised and unrecognised genes that may benefit forestry in the event of future climatic changes (global warming, ozone holes, pollution, etc.), pests and diseases and known and unknown needs for marketable products. Such an approach is appropriate because it involves low maintenance costs on account of high genetic diversity.

The other approach is to stress highly specialised economic yield by using clonal and/or tissue culture techniques. However, in such cases, there would not be much genetic variability. In fact, there would be an inverse correlation between economic yield and genetic diversity. It would, therefore, be costly to maintain such tree plantations on account of low genetic variability. It would also involve the protection of the

Table 5 : Comparison of two types of genetically superior plantations

	Seed-raised	Clonal/Tissue culture-raised
1.	Genetically variable	Uniform
	- Gene complexes both for production (short-term economic yield) and future adaptability	- Stress on short-term economic yield at the expense of future adaptability
	- Immediate fitness and long-term genetic flexibility	- Immediate fitness only
2.	Low maintenance costs	High maintenance costs
	- Low inputs and moderate yield	- High inputs and high yield
3.	Advisable under poly-cultural forestry	Advisable under monocultural conditions
	- Species-rich; various species serve different purpose/need	- Low genetic diversity
4.	Productivity	
	- Low (per species basis)	- High
	- High on cumulative basis	
5.	Advisable for conservation/revegetation/environmental forestry and wasteland development	Advisable for industrial forestry, but selectively for other types of forestry

plantation against future climatic extremes and possible diseases. The two approaches have been compared in Table 5.

Two models are possible for clonal/tissue-raised plantations (Khoshoo, 1986). These are :

Brazilian Model : Blocks of 50 hectares are planted with a single super-clone of eucalyptus. The stands are uniform, even isogenic and have to be protected from diseases and pests, thereby increasing maintenance costs.

German Model : Clonal mixtures of 100 super-clones of the Norway spruce are raised. Thus, the stands are not genetically uniform but are mosaics of unrelated super-clones, thereby lowering the maintenance costs. By this method, clonal/tissue-raised plantations can be made genetically heterogeneous, thereby reducing the genetic vulnerability by having an appropriate mix of genetically diverse cultivars. The objective is to impart ecological resilience to the plantations.

A comparison between seed-raised and clonally-propagated forest tree plantations was also made by Matheson and Lindgren (1985) for transferring genetic gains made through long-term breeding programs. The important conclusions they arrived at are as follows:

* Clonal propagation provides greater advantages than the seed option on account of the shorter time-lag between selection made in the breeding population and raising sizeable field populations, the reason being that multiplication through clonal propagation is expeditious.
* Seed option is not superior to clonal option in terms of gains obtained. This, however, does not stand to reason because the authors are familiar with only industrial forestry and have not taken into account the high maintenance costs involved under the conditions prevailing in a developing country such as India where forestry has yet to become a full-fledged industry.
* Clonal propagation cannot make any headway without a strong breeding program to develop superior genotypes through genetic recombination.
* The gain from clonally propagated plantations is directly related to the effectiveness of a breeding program. The better the breeding program, the greater is the gain through the clonal option.

CONCLUDING REMARKS

Population explosion of both livestock and humankind has resulted in overgrazing and the destruction of forests. This destruction, which is both poverty- and affluence-related, calls for strategies to ensure long-range ecological security and supply of goods and services on a sustainable basis to rural and urban people and industry. This is possible only if a quantum jump is made in production of wood and productivity of forests. To realise this, two things are needed. First, a strong R&D base involving intensive collaborative effort (biotechnology offers one such possibility to boost productivity), and second, appropriate land laws and regulations governing forest tree farming and harvesting of the produce.

From Tables 1 and 4, it is clear that India has to launch a major effort to restore the forest cover, not only to ensure long-range ecological security and supply of goods and services to rural communities, but also to meet the timber needs of the people and those of the forest-based industry. It is also clear that if the needs have to be met and the gaps between demand and supply bridged, what is required is a land area under forest cover almost equal to the size of India. Obviously, the widening gap can be bridged only by enhancing the overall production and productivity. That there has to be a very strong infusion of science and technology in forestry and that forestry has to come into the scientific and technological mainstream of the country, is implicit.

The difference between forests and plantations has to be recognised for conservation purposes. While we need *forests* of the indigenous species, to meet the commercial/industrial wood needs, there is no other way except to raise scientifically managed *plantations,* wherein trees are treated as crops. However, there is a difference between crop agriculture and tree-crop forestry on account of the duration of the generation. It is short (6-11 months) in agriculture, while it is long (7-50 years) in forestry. Obviously, one of the expensive items in the production costs is the unusually high cost on account of the *waiting* period in forestry. In situations where tissue culture cannot be applied, mass selection involving parental selection with progeny testing or the development of synthetics from inter-plantation crossing is relevant, and is safe as a general strategy.

On account of the long generation-time and high environmental costs involved, one cannot afford to make mistakes in forestry. For development and subsequent propagation of elite trees through clonal multiplication and/or tissue culture, what is required is a strong backing of forest tree genetics and breeding. The identification and production of good clones is the first step which has to be followed by intensive selection among a large number of clones involving prolonged trials to identify the really superior clone(s).

Based on appropriate protocols for screening, testing and identification of species/provenances, individual elite trees/shrubs have to be identified for applications of tissue culture approaches in order to circumvent specific problems which, in individual cases, are barriers to higher yield.

There is a general lack of mutants in forest trees. Therefore, genetic engineering, as relevant to forestry, may be initiated on selected species/clones for the benefit of the forest industry, for developing insect and pathogen resistance, more efficient photosynthesis, superior quality and quantity of lignin, desirable pulp characteristics, improved wood texture and silvi-chemicals,and resistance/tolerance to environmental stresses such as drought and acidified atmosphere. Such transformed plants may be effectively segregated, both during their development and field-testing, so as not to contaminate the natural forest areas and man-made plantations through cross-pollination.

Tissue culture/clonal propagation can render exemplary service in the multiplication of at least the initial material. This is particularly true in the following cases :

* Hybrids, including F_1 with heterotic vigour
* Individuals with additive/non-additive genetic variation
* Mutants
* Polyploids/haploids
* Sterile individuals
* Short seed viability
* Rare individuals
* Expensive wood quality

* Making clones pathogen-free
* Individual trees that are slow-growing or difficult and expensive to multiply through conventional means
* Limited stock material
* Conservation of elite germplasm at relatively low cost

A prolonged callus phase has to be avoided in order to produce uniform plantations, because callus induces genetic changes (somaclonal variation), the causes of which are not yet properly understood. Somaclonal variation can be a source of new variation, but such clones will need long-term rigorous testing. Therefore, embryogenesis without a prolonged callus phase offers the best option for multiplication because it can be essentially large-scale, and tap root development is easy.

Research also needs to be intensified on nutritional aspects, together with silvibiological nitrogen fixation and mycorrhizal association for the enhancement of productivity.

Conservation of forest tree/shrub genetic resources should be undertaken through both *in situ* conservation of different forest types in selected ecosystems/biomes, as well as through *ex situ* conservation in the form of seed and clonal orchards, seed banks and tissue and organ storage.

Since the tissue culture strategy is costly and plants need considerable aftercare, such an approach cannot be applied on a mass scale for raising forests or for use in general afforestation programs, or even for wasteland development. At present, its use is limited to industrial plantations because here inputs are high and profits can also be high.

The potential of tissue culture approach in forestry, though apparent, has not yet been applied on a large scale in a *practical sense* even in the developed countries. However, intensive work must continue in this direction.

The seed-raised plantations of superior types are essentially heterozygous and low-yielding, can be grown on poorer, less accessible land and involve low maintenance costs. On the other hand, tissue culture/clonal plantations of superior types are essentially genetically uniform with high yield: these can be used on good land but involve high maintenance costs. This approach offers an attractive option because it may be better than a good seedling population of an outbreeder, though on a cumulative basis (management costs included), the margin may not be very high.

Taking into account the prevailing financial and other constraints in India, we need to opt for a wider genetic base for some time to come. This should include gene complexes for both immediate production and long-term adaptation on less fertile and derelict land. However, wherever possible, the tissue culture approach should be used to raise high yielding and high value plantations in which inputs as well as profits are high. However, the tissue culture approach cannot be used as general therapy. Like a good perfume, it has to be used selectively and on special occasions

The wood-needs for different purposes are enormous and there is an ever-widening gap between demand and supply. That the involvement of people is backed by scientific and technological expertise is most imperative. This would help meet the local needs on one hand, and of forest-based industry on the other. In order to ensure such an involvement on a sustainable basis, the government needs to update policies on forestry, food production, fodder and livestock, fuelwood, land use and tenure, water and irrigation.

Since people are becoming increasingly involved in raising decentralised nurseries and planting trees under the agro-forestry programs, it is necessary to ensure good productivity and assured returns from such plantings on farmers' fields or on community and common lands. It is essential for the quality of seed to be ensured through the introduction of seed certification so that fair returns are assured to the farmers/community, and their stake in the program is also enhanced. However, it may be pointed out that wastelands would be economically unfit for sustained timber production.

Forestry, unlike agriculture, has a long gestation period and affects inter-generational equity. Therefore, it is necessary to have a long range view on forestry R&D and policies. Funding has to be ensured on a long-term basis. This would also help in bridging the gap between the techniques developed in the research establishments and the technology needed for large-scale application.

There is also an urgent need to modify or change the relevant land laws/Acts/ old administrative procedures which are at present acting as deterrents in making forestry a movement of both the government (state and central) and the people. Individuals and institutions need to be motivated to take tree-crop farming. The land use practices, land distribution patterns, land tenure, etc. regarding private, community, and revenue and forestry land, need to be looked into very thoroughly. The real situation of these lands on ground is far different from the one in the files. If the government is serious in involving people in the forestry revolution, the land reforms have to be undertaken so that land laws are supportive for people to be involved meaningfully and develop a stake. A government policy on forestry needs to be developed, taking into account the policies on grazing and livestock, water conservation, soil conservation and land use, planning and tenure, together with a package of incentives/disincentives.

Location-specific land-use patterns need to be evolved, taking into account the social, economic and environmental needs of a community which has evolved a particular agro-ecosystem. While a drastic revision of laws is going to take time, as an interim measure, forest tree-crops farming may be exempted from the Land Ceiling Act like other crops (tea, coffee, rubber, arecanut, etc.). Furthermore, such exemption may be granted to any agency that has the capability to deliver the goods. At present, bank finance is not forthcoming because banks insist on land ownership and lease deeds. Present laws do not allow this on account of land ceiling. The whole process has become a vicious cycle which has to be broken so that forest tree-crop farming is included in the list of exemptions. If this is accepted, only executive orders are needed to bring it in force.

Another aspect relates to the Rural Tree Protection Act under which a farmer is not permitted to fell a tree except for bonafide domestic use. While there may be a basic rationale behind not granting such blanket permissions, the system, as it exists today, has become a source of harassment to a farmer as he cannot sell the trees when he needs money. Attention has to be paid to this aspect. Thus, appropriate regulations are needed to fell trees on private land, otherwise only revenue and community land would be available for public involvement in forestry. There is no dearth of programs which are supportive of forestry. Some of these are: Small Farmers Development Agency (SFDA), Marginal Farmers Development Agency (MFDA), Tribal Development Agency (TDA), District Rural Development Agency (DRDA), Drought Prone Area Programme (DPAP), National Rural Employment Programme (NREP), National Rural Labour Employment Guarantee Programme (NRLEGP) and even Food for Work Program.

All these need to be critically reviewed and may be a comprehensive program/ agency could be floated, which would also act an appropriate delivery system in the extension chain. In fact, the extension wing of forestry has been the weakest aspect of forestry system in India.

ACKNOWLEDGEMENTS

My grateful thanks are due to Mr. A.N. Chaturvedi and Dr. Deepak Pental of TERI for several fruitful discussions on the subject.

REFERENCES

Advisory Board on Energy, 1985, *Towards a perspective on energy demand and supply in India in 2004/ 05,* Advisory Board on Energy, Government of India, New Delhi.

Bisht, S.M., 1988, The Government decides which areas are forests, *The Times of India,* July 27, 1988.

CFC, 1984, *India's forests,* Central Forestry Commission, Government of India, New Delhi, pp. 1-114

DBT, 1988, Report of the task force on *Production of biomass (fuel, fodder, timber and commercial/ industrial woods) using tissue culture technology,* Department of Biotechnology, Government of India, New Delhi.

Dickmann, D.I., 1985, The ideotype concept applied to forest trees. In: *Attributes of Trees as Crop Plants,* (M.G.R. Cannell and J.E. Jackson, eds.), pp. 89-101, Institute of Terrestrial Ecology, Huntingdon (U.K.).

Khoshoo, T.N. 1986, *Environmental priorities in India and sustainable development,* pp. 1-224, Presidential Address, Indian Science Congress Association, Calcutta.

Matheson, A.C. and Lindgren, D., 1985, Gains from clonal and clonal seed-orchard options compared for breeding programs, *Theor. Appl. Genet.,* 71: 242-249.

PC, 1982, Report of the Fuelwood Study Committee, Planning Commisision, New Delhi.

Simmonds, N.W. 1985, Perspectives of evolutionary history of tree crops. In: *Attributes of Trees as Crop Plants,* (M.G.R. Cannell and J.E. Jackson, eds.), pp. 3-12, Institute of Terrestrial Ecology, Huntingdon (U.K.).

4

FOREST TREE BREEDING AND MASS CLONING FOR TREE IMPROVEMENT IN INDIAN FORESTRY

P.D. Dogra

ABSTRACT

World-wide activity in forestry shows forest genetics to be the most effective component in tree plantation programs because it produces genetically improved trees that show better growth, tree form, site adaptability, wood quality, crop security and product uniformity. In tree improvement, genetic and silvicultural techniques are combined at both the nursery-and planting-site-levels (e.g., genotype/site-reactions). The intra-specific genetic variation found within the distributional range of wild tree species forms the basis for selection. This includes provenance, race and plus-tree selection and their site performance evaluations, followed by progeny tests and advanced-generation selective breeding and hybridisation. Genetic control is exercised

P.D. Dogra * Biomass Research Centre. National Botanical Research Institute (CSIR). Lucknow 226 001. India.

through seed (sexual) and vegetative propagules (asexual reproduction). Improved germplasm thus produced is multiplied by the tree-breeder through clonal banks, seed zones, seed production areas and seed orchards. Outstanding genotypes are cloned true-to-parent only on a small-scale and mostly to aid breeding through methods of vegetative propagation. Genetically improved plants for plantations are, however, required in large numbers. They can be cloned on a large-scale by developing a tissue culture technique that can produce a large number of genetically stable and uniform trees identical to the elite parent. The situation regarding tree-improvement and mass-cloning research in Indian forestry is discussed with particular reference to the tissue culture technique.

INTRODUCTION

The genetic and environmental patterns of variation encountered within a species, distributional range are correlated. Thus, genetic variation within species is associated with geographically different regions and climates to which its varieties, races, provenances or genotypes are adapted. The genetically based, habitat-correlated variation evolved through the centuries is found in the natural distributional range. This enables the species to grow in its various niches. It is the genetic variation that forms the basis for selection and tree-improvement work. This includes provenance race and plus tree selection, their site performance evaluations, followed by progeny tests, selective breeding, advanced-generation breeding and hybridisation work. Genetic control is exercised through seed (sexual) and vegetative propagules (asexual reproduction). Improved germplasm, thus produced, is multiplied by the tree-breeder through clonal banks, seed zones, seed production areas and seed orchards.

Outstanding genotypes are cloned true-to-parent only on a small-scale and mostly as an aid to breeding through vegetative propagation methods. However, once the selected germplasm is fully field-tested to give high genetic gain, it needs to be multiplied on a large-scale for plantation work. The best selections are often those of individual trees (plus trees) which are progeny tested and evaluated for good growth, tree form, wood quality and disease resistance. These need to be mass-cloned for use in clonal trials, germplasm conservation, establishment of seed orchards, tree breeding and sometimes for direct plantation work. There is ample evidence that quick and permanent genetic gain can be achieved in a short period by selecting the best germplasm from nature (plus trees, provenances) and multiplying or mass-cloning it for utilisation in forest plantation work (Dogra, 1981a). Cloning by vegetative propagation methods (rooted cuttings, grafting, root suckers, layering) cannot be done on a large-scale. Genetically improved trees for plantations are, however, required in large numbers. This can be achieved only by developing aseptic techniques to produce a large number of genetically stable and uniform trees identical to the elite parent tree.

In India, both indigenous and fast growing exotic tree species are being planted extensively. Improved seed or germplasm is, however, not being used in these plantations because genetically improved reproductive materials are neither properly identified nor produced in systematic tree improvement programs for use by tree growers. For example, planting of *Eucalyptus* may mean planting one out of any five or six species without regard to the provenance of the seed used. The most widely planted species in India today is the naturally evolved and highly successful *E. tereticornis,* an unimproved land race of Mysore provenance, often known by the name of *Eucalyptus hybrid.* Most *Eucalyptus* plantation evaluations suffer because intra-specific genetic variation is not taken into account while judging site/genotype performance. In indigenous species, the seeds used for plantations are often unselected without provenance control and/or procured from an inferior source tree or seed stand. So far, detailed field studies on genetic diversity present within natural distributional ranges of forest tree species in India have not been done except in a general way for *Pinus wallichiana* and *Tectona grandis* (Dogra, 1972; 1981a,b,c; 1985).

There are more than 500 Indian tree species of forestry value, most of which belong to the broad-leaved genera such as *Acacia, Adina, Albizzia, Amoora, Dalbergia, Dipterocarpus, Hopea, Juglans, Morus, Pterocarpus, Salix, Shorea, Tectona, Terminalia* and few conifers such as *Abies, Cedrus, Picea* and *Pinus.* Many of these are widely distributed and have large reserves of genetic variation and wild gene resources (Dogra, 1986). *Acacia nilotica,* for example, is widely distributed in northern Africa, Arabia, Pakistan and India. In India, it is commonly found in Rajasthan, Gujarat and northern Deccan, but is distributed throughout northern and central India, being absent only in the humid and cold regions. The species is now classified into nine sub-species: of these, five occur in Pakistan and India, viz. *indica, cupressiformis, adstringens, hemispherica* and *subalata,* which have arisen from variation and inter-specific hybridisation (Ali and Quaiser, 1980).

Tectona grandis L.f. (teak) is widely distributed in India, Burma, Thailand, Laos and Indonesia. Distribution in the Indo-Burma area is not uniform. In India, teak is widely distributed with a northern limit in the western Aravallis that extends to Jhansi and continues down in a south-eastern direction towards the Mahanadi river. There are large gaps in its distribution (Brandis, 1906; Troup, 1921) and in unconnected areas of its range it shows discontinuous variation. The eastern Burma-teak is very different and genetically more uniform than the western India-teak. The Indian-teak differentiates into many varieties, races and local forms. It is ecologically divided into five types on the basis of rainfall, tree associations, soil and other characteristics. These are classified as very dry, dry, semi-moist, moist and very moist types (Seth and Khan, 1958; Champion and Seth, 1986). The dry and moist types differ with respect to period of leaf shed and the deciduous or evergreen nature of their populations (Brandis, 1906). Differences are seen to occur in seed germination between seed origins of dry (Nagpur) and moist (Masale Valley, Mysore), types (Keiding and Knudsen, 1974). Morphological variation is seen in shoot, seed form, seed characters and wood quality in different races. On the basis of these, 'Teli' race and eight different provenances can be identified with key characters (Bor, 1939). The size and the type of seed and seed coat also shows distinct variation in some widely separated provenances belonging to geographically different and extremely dry or wet regions of the Indian Peninsula (Dogra and Tandon, unpublished). Distinct differences in growth and tree form linked to seed origin have been experimentally demonstrated in seed origin trials conducted in Uttar Pradesh (Lohani, 1980). Distributed in India over wide and geographically diverse areas-ranging from moist to arid hills and plains teak has genetically diversified into many varieties, races and ecotypes. These are adapted to different rainfall regions in which the distinction between dry and wet types is marked. These differences are shown in an evaluation of Indian teak provenances made through international provenance trials by Keiding et al. (1986). The Burma-Thailand region produces the best teak but it does not show much variation. It is the teak of the Indian region that shows the greatest genetic diversity and for the teak breeder, India is an important region for the collection of germplasm for work in genetic improvement.

Investigations carried out on chromosome numbers of over 650 Himalayan hardwood genera, show the presence of chromosomal races within some tree species (Mehra, 1976). Khoshoo and Singh (1963) reported the chromosome number of 37 varieties of *Zizyphus jujuba* Lam., which are widely distributed and naturalised, and of some selections of *Z. nummularia* (Burn. f.), Wright and Arn. (*syn. Z. rotundifolia* Lam.). In *Z. jujuba*, 34 varieties were tetraploids, one was pentaploid and the remaining two were octaploid. In *Z. nummularia*, both tetraploid and hexaploid trees were found. Mehra (1976) observed intra-specific polyploidy in *Terminalia chebula* (2x, 4x), *T. belerica* (2x, 4x), *Syzygium cumini* (2x, 3x, 5x, 6x), *Callicarpa tomentosa* (4x, 5x), *Putranjiva roxburghii* (2x, 6x), *Betula utilis* (2x, 4x) and *Salix tetrasperma* (2x, 4x). Intra-specific aneuploid races are reported in *Toona ciliata* (n = 26, 28) and *Cassia fistula* (n = 12, 14; see Mehra, 1976).

Pinus wallichiana is widely distributed along the entire length of the Himalayas both in the outer wet and inner dry zones and in an altitudinal range of 1,200-3,800 m, which is greater than that of any other Himalayan conifer. It shows both a continuous and a discontinuous distribution, especially in the eastern regions. It occurs in small disjunct and scattered patches in Assam and in the north and east of the Brahmaputra, extending upto northern Burma and Yunnan in China. The species is shown to have seven conspicuously different broad provenances (Dogra, 1972). Those of the dry extreme north-west are separated by over 3,200 km and are very different from those of the wet eastern regions. Variation also exists between different altitudinal provenances of both the outer moist zone and the inner dry or arid zone.

The Indian forest tree species is yet to be subjected to vigorous selection and tree improvement programs which have, unfortunately, been neglected so far. Their tree improvement potential, though vast, remains untapped. The seed used for most forest tree plantations in India is genetically not improved.

Considerable tree plantation is being carried out with exotic species of *Acacia, Eucalyptus, Populus* and *Prosopis* but the germplasm used to raise these trees has a narrow and undefined genetic base: this is inadequate for tree improvement. A wide range of germplasm (superior provenances) of exotic species successful in India need to be imported judiciously from donor countries. A broad genetic base must be built for successful exotics if tree improvement programs are to produce superior populations and genotypes. This has been done for *Populus deltoides* in the Terai region of Uttar Pradesh, and selections of highly productive clones were tested, multiplied and released with genetic control by Chaturvedi (1982). The most promising ones can be multiplied by the tissue culture technique and field-tested once the methods for aseptic cloning are evolved. Several species of *Acacia, Dalbergia, Prosopis* and *Terminalia* are being grown on wastelands. These can be genetically improved for site adaptability and high productivity and multiplied in large numbers. The most commonly planted species of bamboo belong to *Arundunaria, Bambusa* and *Dendrocalamus.* Very little is known about the methodology of genetic improvement of bamboos, or of the cloning technology, either by conventional or by tissue culture techniques. These need to be developed.

For mass-cloning for afforestation, a tissue culture laboratory is not all that is required. A good tree breeding program is essential to produce elite trees for mass propagation. This means that superior provenances and plus trees of most planted species must be identified and selected from all parts of the country. They must be genetically controlled and their germplasm tested, multiplied and released. It is in these operations that tissue culture propagation programs can be successful. First generation forest tree seed orchards must be established both at the state and the national level. The seed orchard must be laid out properly (in terms of the number of units and area). Sites have to be selected and rules established for progeny testing of orchard clones. All these are practical steps which must be undertaken for tree improvement programs in which mass cloning through tissue culture would be most successful. Special emphasis must be laid on research on the "hardening process" of laboratory produced plants and on their field-testing methodology. Field-testing of tissue culture-propagated clones is essential to prove their worthiness. Production costs of tissue culture raised plants must be brought down, the genetic stability of clones guaranteed and techniques for transferring plantlets to the field perfected. Thus, micropropagation of trees has great potential if viewed realistically. The technique developed must produce plantlets that are genetically uniform within a clone for tissue culture to be of value for operational plantings. Genetic plant uniformity is the most needed and the most sought after quality in mass cloning of trees for plantation work. This has yet to be achieved.

REFERENCES

Ali, S.I. and Quaiser, M., 1980, Hybridisation in *Acacia nilotica* (Mimosoideae), *Bot. J. Linn. Soc.,* 80: 69-77.

Bor, N.L., 1939, Summary of results to date of All-India Cooperative Seed Origin Investigation. In: *Proc. 5-Silvicult. Conf. Paper II,* Item IV, Dehra Dun.

Brandis, D., 1906, *Indian trees,* pp. 7-24, Bishen Singh Mahendra Pal Singh, Dehra Dun.

Champion, H.G. and Seth, S.K., 1986, *A revised survey of forest types of India,* pp. 6-18, Govt. of India, Delhi.

Chaturvedi, A.N., 1982, Poplar farming in U.P., *U.P. Forest Bull.,* (Lucknow), 45. 1-42.

Dogra, P.D., 1972, Intrinsic qualities growth and adaptation potential of *Pinus wallichiana.* In: *Biology of Rust Resistance in Forest Trees* (R.T. Bingham, T.S. Hoff and G.I. McDonald, eds.), pp. 163-178, USDA Forest Service, Misc. Publ., No. 1121: Washington D.C.

Dogra, P.D., 1981a, Forest genetics research and application in Indian forestry-I, II, *Indian For.,* 107: 191-219, 263-288.

Dogra, P.D., 1981b, Variability in biology of flowering in blue pine provenances of north-western Himalayas in relation to reproductive barriers and gene flow. In: *Proc. Symp. Flowering Physiology IUFRO,* W.P. Reprod. Processes 17th IUFRO World Cong. (L. Krugman and M. Katsuta, eds.), pp. 8-16, Japan Forest Tree Breeding Assoc., Tokyo.

Dogra, P.D., 1981c, Natural variability and improvement potential of Indian trees. In: *Woodpower: New Perspectives on Forest Usage* (J.J. Talbot and W. Swanson, eds.), pp. 59-80, Pergamon Press, New York.

Dogra, P.D., 1985, Conifers of India and their wild gene resources in relation to tree breeding, *Indian For.,* 111: 935-955.

Dogra, P.D., 1986, Species diversity and gene conservation in Indian tree species in relation to forestry, *Indian For.* 112: 596-607.

Keiding, H. and Knudsen, F., 1974, Germination of teak in relation to international provenance testing, *Forest Tree Improvement, Arboretet Horsholm, Copenhagen,* 7: 19-29.

Keiding, H., Wellendorf, H. and Lauridsen, E.B., 1986, *Evaluation of an International Series of Teak Provenance Trials,* pp. 1-81, Publ. DANIDA Forest Seed Centre, Humlebaek, Arboretum Horsholm.

Khoshoo, T.N. and Singh, N., 1963, Cytology of north-western Indian trees *Zizyphus jujuba and Z. rotundifolia, Silvae Genet.,* 12: 158-174.

Lohani, D.N., 1980, Teak seed origin trials in U.P., U.P. Forest Dept. R & D Circle, Lucknow, *For. Leaflet.*

Mehra, P.N., 1976, *Cytology of Himalayan Hardwoods,* Saraswaty Press, Calcutta.

Seth, S.K. and Khan, M.A.W., 1958, Regeneration of teak, *Indian For.,* 84: 455-465.

Troup, R.S., 1921, *The Silviculture of Indian Trees,* Vol. 1-3, pp. 1170, Clarendon Press, Oxford.

5

AFFORESTATION IN INDIA

A. N. Chaturvedi

The forests in India, as they survive today, have a very uneven and unbalanced spatial distribution; large areas in different parts of the country are bereft of any vegetation. The growing pressure of population - both human and cattle - has accelerated the process of deforestation. The ameliorative power of forests, though known to people who practise shifting cultivation, has failed to be effective due to short cycles of forest regrowth.

The climatic conditions in India are extremely favourable to tree growth. Several tree species, both indigenous and exotic, have been successfully planted in India in afforestation programs. A matching of the species to specific site conditions is one of the key factors for successful afforestation. There is no miracle tree that can grow everywhere. No tree species harms the environment. Species of a low genetic base such as *Leucaena leucocephala* should be avoided in large-scale afforestation. Proper spacing and cultural practices are necessary. The ameliorative effects of different species and their rates of growth need to be studied before arriving at rotations of harvest. Protection of afforested areas from the local people and their domestic cattle is the most difficult aspect of afforestation. Afforestation can succeed

A.N. Chaturvedi * Tata Energy Research Institute, 90 Jor Bagh, New Delhi - 110 003, India

only with the social awakening of the local people, coupled with a high degree of technical skill. Large-scale afforestation programs based on the success of laboratory-level studies of small trials in protected areas over short periods would be extremely misleading.

Afforestation has achieved a lot of importance in India today. The main reasons for this awareness are : the decline in fertility of several sites, the reduced availability of water and increase in the minimum and maximum temperatures in the environment. In addition, the reduced availability of wood products including timber and firewood has pushed up the prices of these products. Fortunately, many farmers have now realised that growing trees on farmlands is competitive with several other land uses. This has led to an increase in the tree planting activity on farmlands.

The need for afforestation in degraded lands was first established in the report of Dr. Brandis, the first Inspector General of Forests, in 1879. He had stated that unless the ravine lands are afforested, the agricultural production will go down. Some afforestation activities were started a few decades after the publication of Dr. Brandis' report. The Fisher Forests in Etawah and the Ellen Forests in Kanpur, Uttar Pradesh, clearly showed what could be achieved on these degraded lands through afforestation. The environment and the moisture regime have greatly improved in both these lands. At present, the objectives of afforestation are, primarily:

* restoring the productivity of the presently degraded lands; and
* improving the deforested lands so that agricultural lands at lower altitudes continue to be productive.

The production of firewood and fodder, though no less important, is secondary. Several tracts in India have no forests but good agriculture. These will also suffer badly if degraded lands are not afforested. The remaining natural forests can be protected only by increasing the production of wood to meet the local and industrial demand.

When trees grow as a forest crop, they grow slowly in the beginning. The rate of growth then picks up to a certain level after which it starts declining. The growth curve of trees is sigmoid. During the period of establishment, most tree species can be made to grow even on apparently inhospitable sites. The mortality or stagnation may set in after the initial phase of establishment. Thus, many tree species appear to grow well in the first few years of introduction but die or stagnate later on. The latter growth depends on a good root system that establishes itself in the first few years. The high survival rates of trees at earlier stage do not necessarily indicate the suitability of the tree to that site. During the early phase of establishment, the demand for nutrients is very low. During the period of fast growth, the demand for moisture and nutrients is high. Consequently, during this phase, whatever nutrients are recycled, are consumed by the plants and the sites generally do not improve appreciably in their nutrient and moisture status. After the decline of the mean rate of growth, called the Mean Annual Increment (MAI), the trees recycle more nutrients than they take up. All sites improve during this phase. Trees spend the largest part of their life cycle in this phase of slow growth after crossing the MAI peak. Thus, forests improve sites only when they are maintained for longer periods. An early harvest of trees will invariably deplete the sites.

A large number of tree species, both indigenous and exotic, have been tested in India in different edapho-climatic conditions. No one tree species has succeeded every where. With the varied conditions of soil and climate, there cannot be a miracle tree for all sites. Trees with a narrow genetic base are extremely risky, as they are likely to be attacked by pests or diseases. The example of *Leucaena leucocephala* is too recent to be forgotten. This species has been a miserable failure in India in large-scale afforestation projects after showing reasonable success in protected small-scale trials. It has also been a complete failure in Philippines where large-scale plantations were set up but were later totally wiped out.

Unfortunately, not enough work on the program of seed orchards has been carried out in India. Consequently, seeds of unknown genetic make-up continue to be used in our afforestation programs. The yields are, therefore, poor even on fairly good sites. A good genetic input is thus crucial for our afforestation programs. Even a genetically superior clone, if planted on a poor site, is likely to grow poorly. Hence, a fast-growing clone needs to be planted in a reasonably fertile soil. The sites wherein most of our afforestation activity takes place, lack fertility and adequate moisture content. These are sites that have been over-used in the past; they are often mistakenly referred to as fallow land. The fact is that these sites are under very heavy pressure of grazing and are, consequently, used beyond their productive capacity. The sites need to be relieved from this pressure if they are to be successfully afforested.

A major source of recycled nutrients is the decomposing leaves. If such species that can be used for fodder are planted on the sites, then the process of their recovery slackens because the leaves are removed from the site. An afforestation program for fodder on degraded land has the least chance of success because such trees are the most difficult and the most expensive to protect. Many trials carried out over small plots in a protected environment, and with heavy inputs of irrigation and fertilisers, give high figures of productivity. Such figures are, however, unrealistic and not related to actual field conditions.

In forestry, field research programs cannot be judged on the basis of two or three years' performance. The comparatively long gestation period is a very severe constraint on realistic forestry research. Forestry field research is necessarily long term, though in recent years, the concept of 'long term' has changed from centuries to decades. In the trials of *Eucalyptus* provenances that were carried out all over India between 1970 and 1980, many Australian provenances showed high rates of growth up to three or four years; however, in the analysis carried out after eight years, many of these provenances turned out to be poor.

In recent years, there has been a lot of emphasis on clonal propagation of several tree species to increase productivity. Poplars are mostly planted through clonal cuttings. In clonal forestry, it is essential to change clones frequently. A clone which appears to be resistant to some pests and diseases becomes vulnerable to the same pests as the progenies increase. This is because the pests and diseases also modify their systems to break the clone's resistance to their attack.

Tissue culture of trees has been a subject of research in India for several decades. Teak was one of the earliest species on which tissue culture work was carried out. Tissue culture-raised plantlets of teak have been planted at several places in India. Nowhere have these plantlets shown any improvement over the seedlings which were planted at similar sites.

Tissue culture-raised plants also show wide variability in form and growth. Though several claims have been made about the production of millions of forest plants from a single explant, nowhere have large-scale tissue culture plants of forest species gone into field planting. Tissue culture plants are raised in protected environments, under controlled conditions of temperature, light and humidity: they need hardening to make them fit for field planting. This takes eight to ten months and needs a lot of space. When discussing the production of tissue culture plants for afforestation purposes, these constraints are often ignored. At present, tissue culture of forest trees is restricted to the laboratory. The use of this technique for afforestation requires that first, solutions be found to several problems, including bringing down the cost of production and maintenance of true characters of the parent. Above all, tissue culture plants need to be tested in field conditions for relatively long periods, as compared to seedlings and clones of the same parents. Tissue culture will be feasible when it is developed from plants in which a fair amount of forest genetical work has been done and the superiority of some progenies has been established. At the present stage of research, tissue culture-raised plantlets cannot be considered reliable planting material in afforestation programs.

SELECTED READINGS

Ahuja, M.R., 1986, Perspectives in plant biotechnology. *Curr. Sci.*, 55:217-224.

Chaturvedi, A.N., 1985, Firewood farming on degraded lands, *U.P. Forest Bull.*, 50:4-8.

Chaturvedi, A.N., 1987, Tissue culture of forest trees. *J. Trop. Forestry,* 3: 20-24.

Chaturvedi, A.N. and Khanna, L.S., 1982, *Forest Mensuration,* pp. 183-202, International Book Distributors, Dehra Dun.

Debergh, P.C., 1987, Improving micropropagation, Newsletter, *International Association for Plant Tissue Culture,* March 1987, pp. 2-4.

George, E.F. and Sherrington, P.D., 1984, *Plant Propagation by Tissue Culture: Hand Book and Directory of Commercial Laboratories,* pp. 20-24, Exegetics Ltd, Eversley, Basingstoke.

Ghosh, R.C., 1977, *Handbook on Afforestation Techniques,* pp. 35-108, Controller of Publications, Delhi.

Sommer, H.E. and Caldas, L.S., 1981, *In vitro* methods applied to forest trees. In: *Plant Tissue Culture : Methods and Applications in Agriculture* (T.A. Thorpe, ed.), pp. 24-30, Academic Press, New York.

II

APPLICATIONS OF TISSUE CULTURE IN FORESTRY AND HORTICULTURE

6

ARTIFICIAL SEED PRODUCTION AND FORESTRY

Keith Redenbaugh and
Steven E. Ruzin

INTRODUCTION

The encapsulation of somatic embryos as hydrated or desiccated artificial seed provides a potential method to combine the advantages of clonal propagation with the low-cost, high-volume capabilities of seed propagation. Calcium alginate has been used to encapsulate somatic embryos of several species, including alfalfa and celery. Currently, alfalfa artificial seeds can be sown directly in the greenhouse, although germination and conversion frequencies are still low. The potential advantages of artificial seed technology for tree genetic engineering are also great, but to date, no conversion of somatic embryos of tree species outside a tissue culture environment has occurred. In this article, the state-of-the-art of artificial seed technology, the advances in tree somatic embryogenesis and the application of artificial seeds to forestry are discussed.

Keith Redenbaugh and *Steven E. Ruzin* * Plant Genetics Inc., 1930 Fifth Street, Davis, CA 95616, USA.

ARTIFICIAL SEED TECHNOLOGY

An artificial seed is defined as an analogous to a true seed in which a somatic embryo is encapsulated with a coating material, which can be a thin or a thick covering. The resultant artificial seed can be hydrated or desiccated. In some instances, a hydrated coating may be essential if the species is of a recalcitrant (desiccation-intolerant) nature (Redenbaugh, 1988). The coating must not damage the embryo and should provide the necessary protection during storage, handling and planting so that germination and conversion occur freely. It may also be desirable for the coating to contain and deliver nutrients, hormones, beneficial micro-organisms, and/or efficacious chemicals. For many crops, single-embryo artificial seeds are required, although for others, this is not a requirement. Finally, artificial seeds will be accepted more readily if they can be handled and planted using the existing farm machinery.

Two types of artificial seeds have been reported in the literature: desiccated and hydrated. Janick's laboratory has been successful in coating carrot and celery somatic embryos in polyoxyethylene (Polyox), drying the embryo/Polyox mixture and obtaining the survival of embryos (Kitto and Janick, 1985a,b; Kim and Janick, 1987). Other laboratories have focused on the desiccation of somatic embryos without encapsulation, in an attempt to study the desiccation process and understand embryogeny better. Obendorf's laboratory (Obendorf and Slawinska, 1986; Slawinska and Obendorf, 1987) was able to improve soybean embryo conversion from 30 percent to 60 percent after embryo desiccation at 70 percent relative humidity. Gray (1987) desiccated orchard grass embryos, achieving eight percent conversion after seven days of storage and four percent after 21 days. However, without desiccation, the conversion was 32 percent. With grapes, Gray obtained 28 percent conversion after seven days of storage, 20 percent after 21 days, but only five percent without desiccation. Gray suggested that desiccation appeared to break the dormancy in grape somatic embryos. It also appeared that an adequate storage environment remained to be identified. Finally, Carman and his co-workers (1987) focused on wheat somatic embryos, finding that they could achieve 59 percent conversion after desiccation, although this was less than that achieved with non-desiccated embryos.

Redenbaugh and co-workers (1986) reported the encapsulation of somatic embryos in hydrated gels, the most suitable being sodium alginate complexed with calcium chloride or calcium nitrate, sodium alginate mixed with gelatin and complexed in calcium chloride, carrageenan mixed with locust bean gum and complexed with potassium or ammonium chloride, and Gelrite gelled in a microtiter plate after the temperature was lowered. The most useful encapsulation system was to drip two percent LF60 sodium alginate (Protan Company, Drammen, Norway) from a separatory funnel into a 100mM calcium nitrate solution. As the sodium alginate drops form at the tip of the funnel, a somatic embryo is inserted. The encapsulated embryos complex in calcium salt for 20 min, after which they are rinsed in water and then stored in a closed container. Without airtight storage, the capsule will dry out within 24 h. In our laboratory, we have encapsulated alfalfa, celery, carrot, lettuce, *Brassica,* cotton, and corn somatic embryos and obtained plants *in vitro*. However, most of our research was focused on alfalfa and celery.

Other laboratories have also used alginate or carrageenan to form artificial seeds. In two popular articles published in Japan, Yamakawa (1985) discussed the encapsulation of carrot and asparagus somatic embryos in alginate but did not indicate conversion frequencies. Lutz (1985) and his co-workers reported carrot encapsulation, without discussing the gel or whether plants were obtained or not. Kamada (1985) reported a five to ten percent conversion from alginate-encapsulated carrot somatic embryos. In a Japanese patent application, Hama (1986) implied that carrot embryos were encapsulated in either alginate or carrageenan and that germination in vermiculite was achieved. However, the frequency was not given. Gupta and Durzan (1986a and 1987) reported the storing of alginate-encapsulated loblolly pine and Norway spruce somatic embryos, but were unable to obtain conversion. Recently, at the Conference on the Genetic Manipulation of Woody Plants, held in Michigan, U.S.A., during June 22-25, 1987, Mascarenhas reported the encapsulation of *Eucalyptus* somatic embryos and indicated at a later conference (1988) that 50 percent germination was obtained. Demarly (personal communication) indicated artificial seed research on alfalfa, oil palm, pine and soybean, but without any specifications.

Alginate artificial seeds are spherical and transparent (Fig. 1). The somatic embryos readily germinate and emerge from the capsules, provided the embryos are mature and vigorous. Germination of poorly developed embryos may be further inhibited, even though without the capsule, germination and conversion may occur at a low

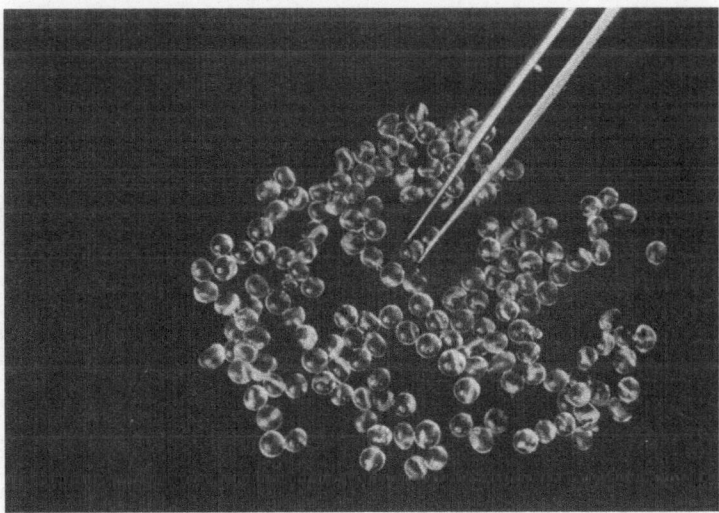

Fig.1. Alfalfa artificial seeds. Somatic embryos are in a 2% alginate gel. Capsules are 4 mm in diameter.

frequency. The alginate capsule is generally non-inhibitory for high quality somatic embryos such as alfalfa.

Because of the labor required and the low rate of capsule formation using the separatory funnel, research was conducted to develop and automate encapsulation (Gautz et al., 1987). An encapsulation machine has been used successfully to encapsulate alfalfa somatic embryos. In a separate series of experiments, blank alginate capsules were planted in Speeding ™ trays (Techniculture, Inc.) using a vacuum seeder. The blank capsules could also be planted in the field using a Stanhay planter. However, because of the rapid drying and the tackiness of the alginate capsules, a hydrophobic coating was required for mechanical handling. The coating, an Elvax 4260 copolymer (Dupont), was sufficient for producing a slow-drying, non-tacky coating that still allowed embryo conversion (Redenbaugh et al., 1987). Based on this system for producing artificial seeds, an estimated cost of 0.03 cents per unit was determined. Although this cost is considerably higher for many agronomic crops like alfalfa (0.0007 cents per seed), it is within the range of crops like hybrid broccoli (0.09 cents) or hybrid lettuce (0.012 cents). The cost may not actually be a deterrent for applications such as hybrid alfalfa, where existing technology is insufficient.

The term conversion was coined to aid the development of superior quality or seed-quality somatic embryos. For any practical application of artificial seeds, whole, vigorous plants must be produced. Thus, it is imperative that experimentation be designed for the selection of superior quality somatic embryos. Conversion is defined as the germination (radicle elongation) of a somatic embryo, followed by the development of a vigorous root system. The root is connected directly to a shoot that has

Fig. 2. Greenhouse germination of celery artificial seeds in sand. Note capsule at the base of the shoot.

produced two or more leaves. The hypocotyl region must lack hypertrophy and callus growth. Furthermore, a converted plant has a normal phenotype. Using this definition of conversion as an assay, we have increased *in vitro* alfalfa somatic embryo conversion from less than one percent to an average of 80 percent. The quality of the embryos has been established, and so we have directed our research away from an *in vitro* conversion assay to one that consists of potting mix in aluminium trays in either an incubator or a greenhouse (Fig. 2). The mix is non-sterile and contains no additives except fertiliser.

An additional concept that has greatly aided the improvement of artificial seed performance is "mass balance". Mass balance considers the amount of tissue at the beginning of the experiment (or production run) and the number of high quality plants produced at the end of it. Simply emphasising the number of embryos per gram of fresh callus or the number of embryo-producing calli is not adequate. In fact, treatments that lead to higher numbers of embryos may actually produce fewer superior quality embryos than another protocol. At this time, the artificial seed package, consisting of a calcium alginate bead coated with a hydrophobic Elvax polymer, appears to be sufficient. The limitation is the quality of somatic embryos. The steps needed for commercialising artificial seed technology are listed below:

1. Production of embryogenic tissue from transformed cells or tissue.
2. Large-scale production of synchronous somatic embryos (SE).
3. Maturation of SE.
4. Non-toxic encapsulation/coating process.
5. Artificial endosperm/megagametophyte, depending on species.
6. Storage capability of artificial seeds.
7. High frequency, direct greenhouse/nursery/field conversion, depending on production requirements.
8. Low genetic and epigenetic variation.
9. Appropriate expression of engineered trait.

Somatic Embryogenesis in Trees

Progress in tree somatic embryogenesis has been extensive in the 1980s and was recently reviewed (Tulecke, 1987). Generally, the induction and pattern of the development of somatic embryogenesis in tree species differ, depending on whether the species is angiospermous or coniferous.

Conifer somatic embryogenesis is analogous to zygotic cleavage polyembryony and for most examples to date, depends on an *in vitro* proliferation of non callus, suspensor-like cells derived from mature or immature zygotic embryos. In Norway spruce (*Picea abies*; Gupta and Durzan, 1986a), the cells, known as the embryonal suspensor mass (ESM), proliferated on BM-1 medium containing high concentrations of kinetin and BAP. Differentiation of ESM to cotyledonary-stage somatic embryos occurred on

Table 1: Somatic embryogenesis in trees of economic importance:

Pulp, paper and timber

Name	Common Name	Germi-nation	Conver-sion	Reference
Gymnosperms				
Abies balsamea	Balsam fir	no	no	Bonga, 1977
Larix decidua	European larch	yes	1% of calli	Nagmani and Bonga, 1985
Picea abies	Norway spruce	5-10%	no	Gupta and Durzan, 1986a
		24% of calli	no	Hakman and von Arnold, 1985
		15 % of calli	no	von Arnold, 1987
P. glauca	White spruce	yes	no	Lu and Thorpe, 1987
Pinus lambertiana	Sugar pine	1-2%	no	Gupta and Durzan, 1986b
P. taeda	Loblolly pine	yes	1-5/g callus	Gupta and Durzan, 1987
Pseudotsuga menziesii	Douglas-fir	no	no	Abo El-Nil,1980 Durzan, 1982
Angiosperms				
Albizia lebbeck	Indian walnut	yes	yes	Gharyal and Maheshwari, 1981
Eucalyptus citriodora	Gum	50%	yes	Mascarenhas, (in press)
Fraxinus americana	White ash	yes	yes	Preece et al., 1987
Juglans regia	English walnut	yes	yes	Tulecke and McGranahan, 1985
J. hindsii	Walnut	yes	yes	Tulecke and McGranahan, 1985
Leucaena diversifolia	Popinac	yes	no	Nagmani and Venketeswaran, 1983
Liquidambar styraciflua	Sweet gum	yes	yes	Sommer and Brown, 1980
Populus ciliata	Poplar	yes	yes	Cheema (Personal communication)
Pterocarya sp.	-	yes	yes	Tulecke and McGranahan, 1985
Quercus rubra	Red oak	no	no	Gingas and Lineberger, 1987

BM-1 medium containing 10 percent of the original kinetin and BAP. To date, however, few somatic embryos have been converted to whole plants.

Somatic embryogenesis in angiosperm tree species is analogous to adventive sporophytic polyembryony which is common in crassinucellate angiosperms (33 of the 39 angiosperms species listed in Table 1 and 2 are crassinucellate, species Webber,1940; Davis, 1966; Haccius and Lakshmanan, 1969) and exemplified by nucellar polyembryony in *Citrus* (Gurgel, 1952; Rangan, 1984), or adventive zygotic polyembryony exemplified by the development of somatic embryos from immature zygotic embryos of *Juglans* (Tulecke and McGranahan, 1985). The key tissue culture manipulation in a number of species was the use of immature zygotic embryos (*Liriodendron, Juglans,* etc.) or nucellar tissue (*Citrus,* etc.) as the original explant source. The cell proliferation stage in tissue culture is in the form of an undifferentiated callus.

SEG has, however, been obtained from non-ZE explant tissue (coffee leaf explants; Sondahl and Sharp, 1977), suggesting that somatic embryogenesis may be obtained from sexually immature trees where zygotic embryo or ovule tissue is not available, thus allowing an acceleration of the process of tree genetic engineering.

Difficulties in developing somatic embryogenesis systems for tree species are similar to those for herbaceous species. However, tree species may also exhibit a unique set of tissue culture-dependent variabilities. Genetic variation may be higher for tree species than cultivated herbaceous species. The size of the genome is quite large for many trees, which is particularly important for genetic engineering. A further limitation for tree improvement using genetic engineering is that mature material must be collected from the field rather than from a controlled environment. In addition, the problem of ecotypic or physiological variation from individual to individual is much greater due to the overall heterozygosity of most tree species. These additional variables further increase the potential difficulty of developing repeatable somatic embryogenesis in tree species.

For the production of artificial seeds in tree species, the recovery of plants (conversion) is crucial. However, many papers are unclear as to what results were actually obtained. To understand the progress of conversion of tree somatic embryos better, reports in the literature were carefully analysed and Tables 1, 2 and 3 were constructed. When a report clearly stated that complete plants were produced or showed convincing photographs of plants, then those species were grouped under the conversion category. At times, terms such as "plantlets" or "germination" were used or photographs showed structures that were less than a complete plant (such as having a root, shoot and cotyledons, but no leaves or needles). Such structures were grouped under the germination category. When available, conversion or germination frequencies were listed. In addition, some species that have only been reported at conferences but not yet in the literature, were included. The major value of this list is to indicate which species are the most developed in terms of somatic embryogeny. Possibly the first tree species for which viable artificial seeds will be produced are the ones such as mango, *Citrus* and coffee, which already have reasonable conversion frequencies.

Table 2: Somatic embryogenesis in trees of economic importance

Fruit trees

Name	Common Name	Germi-nation	Conver-sion	Reference
Citrus aurantifolia	Lime	yes	no	Mitra and Chaturvedi, 1972
C. aurantium	Sour orange	no	no	Kochba et al., 1982
C. limon	Lemon	yes	yes	Moore, 1985
C. medica	Citron	no	no	Tisserat and Murashige, 1977
C. paradisi	Grapefruit	yes	32%	Kochba et al., 1982
C. reticulata	Mandarin	yes	yes	Moore, 1985
		yes	yes	Vardi and Spiegel-Roy, 1982
C. sinensis	Sweet orange	yes	20%	Kochba et al., 1982
Cocos nucifera	Coconut	no	no	Pannetier and Buffard-Morel, 1982
Eriobotrya japonica	Loquat	no	no	Litz, 1985
Eugenia malaccensis	Malay apple	yes	no	Litz, 1984a
Malus sp.	Apple	yes	yes	Liu et al., 1983
Mangifera indica	Mango	yes	yes	Litz 1982, and in press
Musa 'ABB'	Plantain	no	no	Cronauer and Krikorian, 1983
Myrciaria cauliflora	Jaboticaba	no	no	Litz, 1984b
Phoenix dactylifera	Date palm	yes	no	Reynolds and Murashige, 1979
Prunus cerasus	Sour cherry	no	no	Durzan, 1985
Pyrus sp.	Pear	no	no	Janick, 1982
Syzygium jambos	Rose apple	yes	no	Litz, 1984a

Table 3: Somatic embryogenesis in trees of economic importance

Oils, pharmaceuticals, and ornamentals

Name	Common Name	Germi-nation	Conver-sion	Reference
Chamaedorea costaricana	Palm	no	no	Reynolds and Murashige, 1979
Coffea arabica	Arabian coffee	yes	yes	Sondahl and Sharp, 1977
C. arabica var. Catimor	Coffee	yes	20-27%	Garcia and Menendez, 1987
C. canephora	Robusta coffee	yes	no	Staritsky, 1970
Elaeis guineensis	Oil palm	yes	no	Rabeschault et al., 1972
Hevea brasiliensis	Para rubber	yes	no	Paranjothy and Othman, 1978
		yes	no	Carron and Enjalric, 1985
Howea forsteriana	Paradise palm	no	no	Reynolds and Murashige, 1979
Liriodendron tulipifera	Yellow poplar	yes	yes	Merkle and Sommer, 1986
Paulownia tomentosa	Paulownia	50%	no	Radojevic, 1979
Santalum album	Sandalwood	yes	yes	Bapat and Rao, 1979
Theobroma cacao	Cacao	yes	no	Pence et al., 1979
Thuja orientalis	Oriental arbor vitae	no	no	Konar and Oberoi, 1965
Veitchia merrilli	Christmas palm	yes	no	Srinivasan et al., 1985

Artificial Seed Application in Forestry

The use of biotechnological approaches in forestry may be greatly enhanced and considerable time could be saved by using artificial seed technology. Genetic engineering in forestry will be similar to that for field and horticultural crops. Desirable genes will be identified, cloned and inserted into the tissue (protoplasts, cells, pollen, zygotic embryos, needle tissue, etc.) of a few candidate tree genotypes. The putatively transformed tissue will be regenerated to plants and tested for expression of the genes. With annuals, the transformed individual plants can then be backcrossed with the original population for large-scale production of transformed seed within one to few years. However, for most tree species, after adequate gene expression is confirmed, scale-up production of transformed seeds deviates significantly at this point from that of annual and biennial crops because of the very long generation time for trees, particularly conifers.

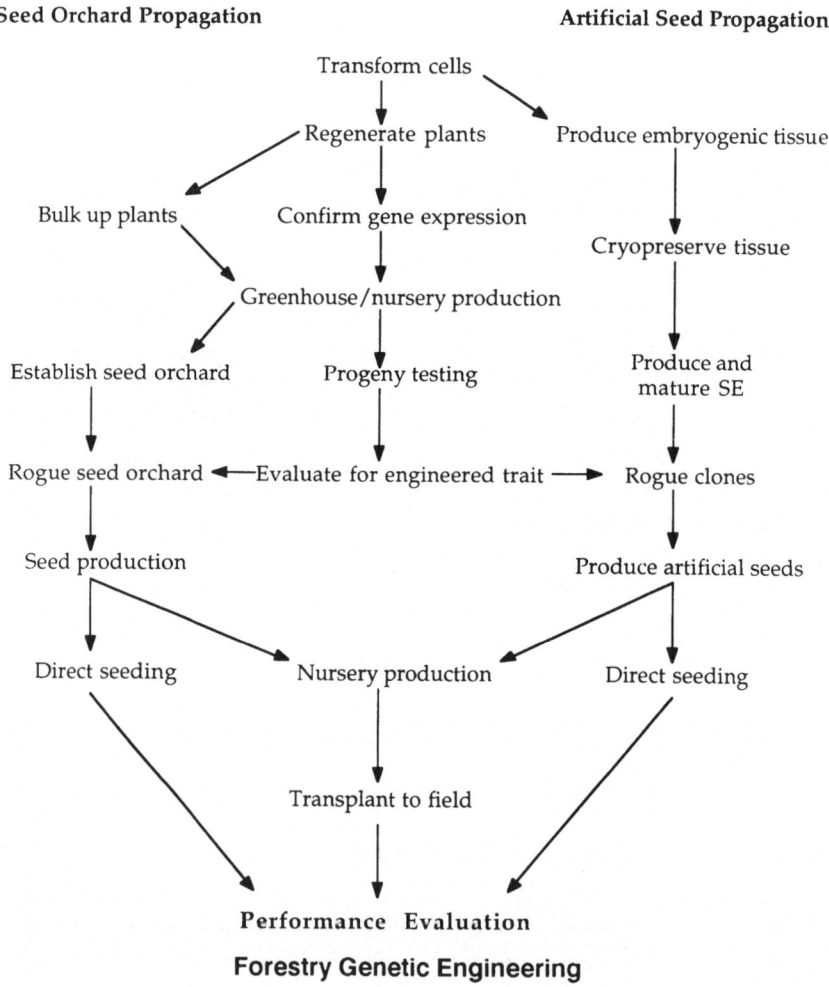

Fig. 3. Proposed scheme for improving forest trees using genetic engineering and artificial seed technology.

Most approaches for genetic engineering of tree species focus on transformation, followed by traditional forest seed production and tree evaluation. The general process is as follows:

1. regeneration of genetically engineered tissue;
2. confirming presence and expression of engineered gene;
3. bulking up propagules through vegetative propagation or seed production;
4. greenhouse/nursery establishment and growth;
5. seed orchard establishment with concurrent progeny testing, including evaluation of the engineered trait;
6. seed orchard roguing;
7. seed production in seed orchards and
8. either nursery production with transplanting into the field or direct seeding.

The final stage will be the evaluation of plantations for increased performance due to the engineered trait (Fig. 3, left column).

By combining genetic engineering techniques developed for annual species with traditional forestry propagation practices, the process of genetic engineering for tree improvement will still be extremely long for many species. Consequently, additional processes that can decrease the time required to produce improved genotypes need to be developed and applied. The processes unique to tree genetic engineering are dependent on the proposed method or methods of mass propagation (somatic embryogenesis vs seed).

Seed Propagation

1. Seed orchard establishment: For traditional forestry propagation practices, the steps after the initial transformation and regeneration events (Steps 1-4 above) will be to establish seed or ramet orchards containing those genetically engineered individuals that passed the first level of *in vitro* testing. As with herbaceous species, tree growth and maturation would be accompanied by R_1-generation and progeny testing for the presence and expression of the engineered trait. However, testing would necessarily be accomplished over a number of years in the progeny tests as against a single year (or perhaps two) for herbaceous crops. If the progeny tests do not show an improvement based on genetic engineering, then the seed orchard would lose a significant portion of its genetic gain and probably be cut down. If the genetically engineered trait is found to be unstable during meiosis, then the seed orchards would not be viable and alternative methods of propagation would need to be developed.

2. Seed orchard roguing: This process is analogous to culling out *in vitro* plantlets for herbaceous species but would occur over a longer period of time. Seed orchard roguing would occur as a consequence of progeny testing and/or observation of morphological or horticultural aberrations in the R_1-generation trees.

3. Seed orchard seed production: Seed production is another lengthy process using traditional forestry propagation, since a seed orchard takes 5-20 years to begin adequate cone production. The overall time from the transformation of the tree tissue to seed production would be 15 to 30 years (Fig. 3 left column).

Artificial Seed Propagation

Artificial seed propagation could potentially reduce the time needed to insert a desirable gene into a production forest, as compared to using seed as the propagation method. A considerable advantage would be to eliminate or minimise the requirements for seed production using the following process:

1. production of large-scale embryogenic tissue from genetically engineered cells;
2. concurrent plant regeneration, confirmation of transformation, and "progeny" testing;
3. cryogenic storage of potential superior lines;
4. scale-up production and maturation of somatic embryos;
5. encapsulation of somatic embryos as artificial seeds;
6. either greenhouse/nursery establishment, growth, and transplanting into the field or direct seeding.

Again, the final stage will be the evaluation of production plantations for increased yield/performance due to the engineered trait (Fig. 3, right column).

Unlike herbaceous species, the growth and maturity of trees far exceed the time required for tissue culture manipulations. Thus, *in vitro* steps and the R_1 and progeny testing would be asynchronous by a number of years. If the engineered trait is expressed only in the mature tree and is not stable during meiosis, then mass clonal propagation would be accomplished only through tissue culture methods (or, by traditional relatively low-volume ramet production). Again, the development of the tissue culture line and the required mass propagation would be years apart.

Cryopreservation of desirable genotypes is one of the key components for artificial seed propagation of tree species. It retains the genetic gains from genetically engineered tree species without having to establish clonal orchards. Once the progeny tests are completed, then embryogenic tissue corresponding to the superior clones can be thawed for rapid scale-up production via somatic embryogenesis. Because the time between the initial development of somatic embryos and acquiring the results of the progeny testing would be years or decades, the improved genotypes would need to be stored or continuously propagated over that time, to maintain the genetic line. Continuous *in vitro* propagation is labor-intensive and may induce undesirable genetic changes, so a method of germplasm storage such as cryopreservation is needed. Cryogenic storage of superior lines would probably be the preferred method of retaining the genetic gain represented in tissue cultures developed through genetic engineering.

Although the use of artificial seed technology should be extremely valuable for the rapid introduction of genetically engineered material into production forests, other uses of artificial seeds can be identified. The production and scale-up of hybrid (artificial) seed production should be possible. High costs of clonal propagation could potentially be reduced to that of true seeds (Redenbaugh et al., 1988), thereby providing new production methods for foresters. Clonal conifer plantations may then be economically feasible. If early selection techniques can be improved as artificial seed production becomes possible, then progress in tree breeding will be significantly accelerated.

ACKNOWLEDGEMENTS

We thank Keiichiro Ishizaki at Kirin Brewery Company Ltd. for suggesting the use of mass balance for somatic embryogeny.

REFERENCES

Abo El-Nil, M., 1980, Embryogenesis of gymnosperm forest trees, *U.S. Patent* 4: 217,730. Washington, D.C.

Bapat, V. and Rao, P.S. 1979, Somatic embryogenesis and plantlet formation in tissue cultures of sandalwood *(Santalum album* L.), *Ann. Bot. (Lond.),* 44: 629-630.

Bonga, J., 1977, Organogenesis in *in vitro* cultures of embryonic shoots of *Abies balsamea* (balsam fir), *In Vitro,* 13: 41-48.

Carman, J., Jefferson, N. and Campbell, W., 1987, ABA, O_2, temperature and relative humidity effects on embryoid maturation in *Triticum aestivum* L. calli, Annual Agronomy Meeting, Atlanta, GA, pp.148 (Abst).

Carron, M.P. and Enjalric, F., 1985, Somatic embryogenesis from inner integument of the seed of *Hevea brasiliensis* (Mull. Arg), *C.R. Acad. Sci., Ser. D,* 300: 653-658.

Corley, R., 1982, Clonal planting material for the oil palm industry, *Planter,* 58: 515-528.

Cronauer, S. and Krikorian, A., 1983, Somatic embryos from cultured tissues of triploid plantains *(Musa* 'ABB'), *Plant Cell Rep.,* 2: 289-291.

Davis, G.L., 1966, *Systematic embryology of the angiosperms,* John Wiley & Sons Inc., New York.

Durzan, D., 1982, Somatic embryogenesis and sphaeroblasts in conifer cell suspensions. In: *Proc. 5th Intl. Cong. Plant Tissue Cell Cult.,* (A. Fujiwara, ed.), pp. 113-114, Jpn. Assoc. Plant Tissue Cult., Tokyo.

Durzan, D., 1985, Tissue culture and improvement of woody perennials: An overview. In: *Tissue culture in Forestry and Agriculture,* (R. Henke, K. Hughes, M. Constantin and A. Hollaender, eds.), pp. 233-256, Plenum Press, New York.

Garcia, E. and Menendez, A., 1987, Somatic embryogenesis from leaf explants of coffee plants "catimor", *Cafe Cacao The,* 31: 15-22.

Gautz, L., Upadhyaya, S. and Garrett, R., 1987, Vibratory separation of gel encapsulated plant propagules, *Trans. ASAE,* 30: 652-656.

Gharyal, P.K. and Maheshwari, S.C. 1981, *In vitro* differentiation of somatic embryoids in a leguminous tree, *Albizzia lebbeck,* L., *Naturwissenschaften,* 68: 379-380.

Gingas, V. and Lineberger, R., 1987, Asexual embryogenesis in red oak *(Quercus rubra* L.), *Hortic. Sci.,* 22: 1131 (Abst. 674).

Gray, D., 1987, Quiescence in monocotyledonous and dicotyledonous somatic embryos induced by dehydration, *Hortic. Sci.,* 22: 810-814.

70   *Redenbaugh and Ruzin*

Gupta, P.K. and Durzan, D., 1986a, Plantlet regeneration via somatic embryogenesis from sub-cultured callus of mature embryos of *Picea abies* (Norway spruce), *In Vitro Cell. Dev. Biol.*, 22: 685-688.

Gupta, P.K. and Durzan, D., 1986b, Somatic polyembryogenesis from callus of mature sugar pine embryos, *Biotechnology*, 4: 643-645.

Gupta, P.K. and Durzan, D., 1987, Biotechnology of somatic polyembryogenesis and plantlet re-generation in loblolly pine, *Biotechnology*, 5: 147-151.

Gurgel, J.T.A., 1952, Poliembrionia e embriogenia adventicia em *Citrus, Mangifera* e *Eugenia*, I-Revisao da literatura, Dusenia, 3: 443-449.

Haccius, B. and Lakshmanan, K.K., 1969, Adventive-Embryonen-Embryoide-Adventiv-Kno-spen, Ein Beitrag zur Klarung der Begriffe, *Osterr. Bot. Z.*, 116: 145-158.

Hakman, I. and von Arnold, S., 1985, Plantlet regeneration through somatic embryogenesis in *Picea abies* (Norway spruce), *J. Plant Physiol.*, 121: 149-158.

Hama, I., 1986, Artificial seeds, Japanese Patent Application No. 40708/1986.

Janick, J., 1982, Adventive embryony in pear, *Acta Hortic.*, 124: 37-41.

Kim, Y.H. and Janick, J., 1987, Production of synthetic seeds of celery, *Hortscience*, 22: 89 (Abst).

Kitto, S. and Janick, J., 1985a, Production of synthetic seeds by encapsulating asexual embryos of carrot, *J. Amer. Soc. Hortic. Sci.*, 110: 277-282.

Kitto, S. and Janick, J., 1985b, Hardening treatments increase survival of synthetically-coated asexual embryos of carrot, *J. Amer. Soc. Hortic. Sci.*, 110: 283-286.

Kochba, J., Spiegel-Roy, P., Neumann, H. and Saad, S., 1982, Effect of carbohydrates on somatic embryogenesis in subcultured nucellar callus of *Citrus* cultivars, *Z. Pflanzenphysiol.*, 105: 359-368.

Konar, R. and Oberoi, Y., 1965, *In vitro* development of embryoids on the cotyledons of *Biota orientalis, Phytomorphology*, 15: 137-140.

Litz, R.E. 1982, Somatic embryos from cultured ovules of polyembryonic *Mangifera indica* L., *Plant Cell Rep.*, 1: 264-266.

Litz, R.E. 1984a, *In vitro* responses of adventitious embryos of two polyembryonic *Eugenia* spe-cies, *Hortscience*, 19: 720-722.

Litz, R.E. 1984b, *In vitro* somatic embryogenesis from callus of jaboticaba, *Myrciaria cauliflora, Hortscience*, 19: 62-64.

Litz, R.E. 1985, Somatic embryogenesis in tropical fruit trees. In: *Tissue culture in Forestry and Agriculture* (R. Henke, K. Hughes, M. Constantin and A. Hollaender, eds.), pp. 179-183, Ple-num Press, New York.

Litz, R.E. *In vitro* strategies for tropical fruit tree improvement. In: *Applications of Biotechnology in Forestry and Horticulture*, (V. Dhawan, ed.) Plenum press, New York (In Press).

Liu, J., Sink, K. and Dennis, F., 1983, Adventive embryogenesis from leaf explants of apple seedlings, *Hortscience*, 18: 871-873.

Lu, C.Y. and Thorpe, T., 1987, Somatic embryogenesis and plantlet regeneration in cultured immature embryos of *Picea glauca, J. Plant Physiol.*, 128: 287-302.

Lutz, J., Wong, J., Rowe, J., Tricoli, D. and Lawrence, R., Jr., 1985, Somatic embryogenesis for mass cloning of crop plants. In: *Tissue culture in Forestry and Agriculture*, (R. Henke, K. Hughes, M. Constantin and A. Hollaender, eds.), pp. 105-116, Plenum Press, New York.

Mascarenhas, A.F. Biotechnological application of plant tissue culture to forest species in India. In: *Applications of Biotechnology in Forestry and Horticulture*, (V. Dhawan, ed.) Plenum Press, New York (in press).

Merkle, S.A. and Sommer, H.E., 1986, Somatic embryogenesis in tissue cultures of *Liriodendron tulipifera, Can. J. For. Res.*, 16: 420-422.

Mitra, G. and Chaturvedi, H., 1972, Embryoids and complete plants from unpollinated ovaries and from ovules of *in vitro* grown emasculated flower buds of *Citrus* spp., *Bull. Torrey Bot. Club*, 99: 184-189.

Moore, G., 1985, Factors affecting *in vitro* embryogenesis from undeveloped ovules of mature *Citrus* fruit, *J. Amer. Soc. Hortic. Sci.*, 110: 66-70.

Nagamani, R. and Bonga, J., 1985, Embryogenesis in subcultured callus of *Latrix decidua, Can. J. For. Res.,* 15: 1088-1091.

Nagamani, R. and Venketeswaran, S., 1983, Morphogenetic responses of cultured hypocotyl and cotyledonary segments of *Leucaena, In Vitro,* 19: 265.

Obendorf, R. and Slawinska, J., 1986, Somatic embryogenesis from cotyledonary tissues of soybean and maturation to desiccation tolerant state, *In Vitro Cell. Dev. Biol.,* 22: 53(Abst).

Pannetier, C. and Buffard-Morel, J., 1982, First results of somatic embryo production from leaf tissue of coconut (*Cocos nucifera* L.), *Oleagineux,* 37: 349-354.

Paranjothy, K. and Othman, R., 1978, Embryoid and plantlet development from cell cultures of *Hevea.* In: *Abstract Book, 4th Int. Cong. Plant Tissue and Cell Cult.,* pp.42, Calgary.

Pence, V., Hasegawa, P. and Janick, J., 1979, Asexual embryogenesis in *Theobroma cacao* L., *J. Am. Soc. Hortic. Sci.,* 104: 145-148.

Preece, J., Zhao, J.I. and Kung, F., 1987, *In vitro* callus production and somatic embryogenesis of ash *(Fraxinus), Hortscience,* 22: 1131 (Abst. 675).

Rabeschault, H., Martin, J. and Cas, S., 1972, Recherches sur la culture des tissus de palmier a hiule (*Elaeis guineensis* Jacq.), *Oleagineux,* 27: 531-534.

Radojevic, L., 1979, Somatic embryogenesis and plantlets from callus cultures of *Paulownia tomentosa* Steued, *Z. Pflanzenphysiol.,* 91: 57-62.

Rangan, T.S., 1984, Clonal propagation: somatic embryos of *Citrus.* In: *Cell Culture and Somatic Cell Genetics of Plants, Vol. 1,* (I. K. Vasi, ed.) pp. 68-81, Academic Press, Orlando.

Redenbaugh, K., 1988, Artificial seed: Application to tropical crops, *Hortscience,* (In Press).

Redenbaugh, K., Slade, D., Viss, P. and Fujii, J., 1987, Encapsulation of somatic embryos in synthetic seed coats, *Hortscience,* 22: 803-809.

Redenbaugh, K., Fujii, J., Slade, D., Viss, P. and Kossler, M., 1988, Synthetic seeds - encapsulated somatic embryos, *1988 Annual Agronomy Meeting,* New Orleans, (In Press).

Redenbaugh, K., Paasch, B., Nichol, J., Kossler, M., Viss, P. and Walker, K., 1986, Somatic seeds: Encapsulation of asexual embryos, *Biotechnology,* 4: 797-801.

Reynolds, J. and Murashige, T., 1979, Asexual embryogenesis in callus cultures of palms, *In Vitro,* 15: 383-387.

Slawinska, J. and Obendorf, R., 1987, Soybean somatic embryo composition, respiration and water relations during maturation to a desiccation tolerant state, *1987 Annual Agronomy Meeting,* Atlanta, GA, pp. 153 (Abst.).

Sommer, H. and Brown, C., 1980, Embryogenesis in tissue cultures of sweetgum, *For. Sci.,* 26: 257-260.

Sondahl, M. and Sharp, W., 1977, High frequency induction of somatic embryos in cultured leaf explants of *Coffea arabica,* L., *Z. Pflanzenphysiol.,* 81: 395-408.

Srinivasan, C., Litz, R., Barker, J. and Norstog, K., 1985, Somatic embryogenesis and plantlet formation from Christmas palm callus, *Hortscience,* 20: 278-280.

Staritsky, G., 1970, Embryoid formation in callus tissues of coffee, *Acta Bot. Neerl.,* 19: 509-514.

Tisserat, B. and Murashige, T., 1977, Repression of asexual embryogenesis *in vitro* by some plant growth regulators, *In Vitro,* 13: 799-805.

Tulecke, W., 1987, Somatic embryogenesis in woody perennials. In: *Cell and Tissue Culture in Forestry, Vol. 2,* (J. Bonga, and D. Durzan, eds.), pp. 61-91, Martinus Nijhoff, Dordrecht.

Tulecke, W. and McGranahan, G., 1985, Somatic embryogenesis and plant regeneration from cotyledon tissue of walnut, *Juglans regia* L., *Plant Sci.,* 40: 57-63.

Vardi, A. and Spiegel-Roy, P., 1982, Plant regeneration from *Citrus* protoplasts: Variability in methodological requirements among cultivars and species, *Theor. Appl. Genet.,* 62: 171-176.

von Arnold, S, 1987, Improved efficiency of somatic embryogenesis in mature embryos of *Picea abies* (L.), Karst, *J. Plant Physiol.,* 128: 233-244.

Webber, J.M., 1940, Polyembryony, *Bot. Rev.,* 6: 575-598.

Yamakawa, K., 1985, Application of artificial seed and its potential, *Agr. Chem. Today,* 29: 68-72, (in Japanese).

7

BIOTECHNOLOGICAL APPLICATION OF PLANT TISSUE CULTURE TO FORESTRY IN INDIA

A.F. Mascarenhas, S.S. Khuspe,
R.S. Nadgauda, P.K. Gupta,
E.M. Muralidharan and B.M. Khan

ABSTRACT

Forest tree tissue culture in India is being carried out in several universities and research institutions. The main thrust of these activities is directed towards micropropagation, using explants from both adult and juvenile embryonic or seedling tissues. The tree species used in these studies cover a wide spectrum which include desert and arid zone trees, leguminous trees and also trees used for timber, fuel, paper and pulp and in rayon industries.

A.F. Mascarenhas, S.S. Khuspe, R.S. Nadgauda, P.K. Gupta, E.M. Muralidharan and *B.M. Khan* * Biochemical Sciences Division. National Chemical Laboratory. Pune - 411 008. India.

NCL Communication No. 4394.

The micropropagation of hardwood forest species is still a controversial issue because of the high cost of plantlet production, which is mainly a result of the labor-intensive stages in the process, and because of the uncertainties regarding the likely gains. This requires careful field evaluation of tissue culture-raised plants with respect to growth rates and wood quality, depending on the end-uses.

The results of small-scale outplantings at the National Chemical Laboratory with tissue culture-raised plants of *Dendrocalamus strictus, Eucalyptus* spp., *Salvadora persica,* and *Tamarindus indica* have been presented, together with results of the different approaches followed in making the micropropagation processes cost-effective and beneficial for use in forestry programs.

INTRODUCTION

Considerable progress has been made over the last two decades on the development of tissue culture methodologies for trees and their possible application to forestry (Bonga and Durzan, 1987). The major hurdles to large-scale production are:

* the high cost of plantlet production mainly as a result of the capital and labor-intensive stages, and
* the uncertainties regarding the likely gains.

This requires careful field evaluation of tissue culture-raised plants with respect to growth rates, wood quality and other traits depending on the end-uses.

This article is divided into two main sections. The first summarises the progress of tissue culture in India and is presented in a tabular form, sub-divided on the basis of the tree species (Table 1).

Table 1: Tissue culture of forest tree species in India

Species	Explant	Response	Reference
	Pulpwood trees		
Bamboos			
Bambusa arundinacea	Seed	C,SE.	Mehta et al., 1982
B. vulgaris	M. Nodes	MS.	NCL (Unpubl.)
Dendrocalamus strictus	Seedl.	MS,PL,S,F.	Nadgir et al., 1984
	Seeds	SE,PL,S,F.	Rao et al., 1985
	M. Nodes	MS.	NCL (Unpubl.)
	Seedl.	Flowering	Univ. Delhi (Unpubl.)

Species	Explant	Response	Reference
Eucalyptus			
E. camaldulensis	M. Nodes	MS,PL,S,F.	Gupta et al., 1983
	M.L. (in vitro)	C,ORG,PL,S.	Muralidharan and Mascarenhas, 1987a
E. grandis	M. Nodes	MS,PL,S.	Lakshmi Sita and Shobha Rani, 1985 Rao and Venketes- wara, 1985
E. tereticornis	M. Nodes	MS,PL,S,F.	Mascarenhas et al., 1982
E. torelliana	M. Nodes	MS,PL,S,F.	Gupta et al., 1983
Populus ciliata	M.L.	ORG,PL,S.	Mehra and Cheema, 1980
	M.Buds	C,ORG,PL,S.	Mehra and Cheema, 1980
	M. Stem	C,SUS,SE.	Cheema, 1987
Timber trees			
Albizzia lebbeck	Hyp.	SE,PL.	Gharyal and Maheshwari, 1981
	Seedl.	ORG,PL,S.	Upadhyaya and Chandra, 1983
	Anther	PL(HAP)	Gharyal et al., 1983
Cedrus deodara	Embryo	C,ORG.	Bhatnagar et al., 1983
Dalbergia lanceolaria	Seedl.	C,ORG,PL.	Anand and Bir, 1984
D. latifolia	Seedl.	MS,PL.	Nataraja and Sudhadevi, 1985
	Seedl.	C,ORG,PL.	Rao, 1986
	Roots, Nodes (5 yr.tree)	ORG,PL,S.	Mascarenhas et al., 1982
D. sissoo	Seedl. Root	ORG,PL.	Mukhopadhyaya and Mohan Ram, 1981
	M. Nodes	MS,PL.	Datta et al., 1983
Picea smithiana	Cot.	C,ORG.	Mehra and Verma, 1981
Pinus gerardiana	Hyp.	C,ORG.	Konar, 1975
P. roxburghii	Embryo	C,ORG.	Bhatnagar et al., 1983
P. wallichiana	Embryo	C,ORG,PL.	Konar and Singh, 1980

Species	Explant	Response	Reference
Santalum album	Hyp.	ORG,PL.	Rao and Bapat, 1978
		C,SE,PL,S,F.	Bapat and Rao, 1979
	M.Shoot tip	C,SE,PL,F.	Lakshmi Sita et al., 1979
	Endosperm	ORG,PL, (TRIP),S, F.	Lakshmi Sita et al., 1980
	Hyp, M.Stem	PROT,C, SE,PL.	Bapat et al., 1985
	Sus.	PROT,C, SE,PL.	Rao and Ozias-Akins, 1985
Tamarindus indica	Seedl.	MS,PL,A,F.	Kulkarni et al., 1981
	M. Nodes	MS,PL,S.	Kulkarni et al., 1981
Tectona grandis	Seedl.	MS,PL,S,F	Gupta et al., 1980
	M. Nodes	MS,PL,S,F	Gupta et al., 1980

Oil yielding trees

Species	Explant	Response	Reference
Eucalyptus citriodora	Lignotuber	C,PL.	Aneja and Atal, 1969
	Seedl. Node	MS,PL,S.	Gupta et al., 1978
	Cot.	C,ORG,PL.	Lakshmi Sita, 1979
	Shoot tip (2 yr.tree)	MS.	Grewal et al , 1980
	M. Nodes	MS,PL,S,F.	Gupta et al., 1981
	Seed	SE,PL,F.	Muralidharan and Mascarenhas, 1987a
E. globulus	M. Nodes	MS,PL,S.	Mascarenhas et al., 1982
Salvadora persica	M. Nodes	MS,PL,S,F.	Rao, 1987
Sapium sebiferum	M. Nodes	MS,PL,S,F.	Kotwal et al., 1983

Multipurpose and other trees

Species	Explant	Response	Reference
Acacia nilotica	M.Stem	ORG,PL.	Mathur and Chandra, 1983
	Cot.	SE.	NCL (Unpubl.)
Azadirachta indica	M.L.	C,PL.	Narayan and Jaiswal, 1985
	Cot.	ORG,PL	Muralidharan and Mascarenhas, 1987b
Biota orientalis	Embryo	SE.	Konar and Oberoi, 1965

Species	Explant	Response	Reference
Ficus religiosa	M.Stem	C,PL.	Jaiswal and Narayan, 1985
Leucaena leucocephala	M. Nodes	MS,OL,S.	Kulkarni et al., 1984 Datta and Datta, 1985 Dhawan and Bhojwani, 1985
Morus indica	M.L Nodes,	ORG,MS,PL.	Mhatre et al., 1985
Prosopis cineraria	Hyp.	ORG,PL.	Goyal and Arya, 1981
	M. Nodes	MS,PL,S.	Goyal and Arya, 1984
Putranjiva roxburghii	Endosperm	C,PL.	Srivastava, 1973
Salix babylonica	M. Nodes	MS,PL,S.	Dhir et al., 1984
Sesbania grandiflora	Hyp,Cot	C,ORG,PL.	Khattar and Mohan Ram, 1983

Abbreviations : C - Callus: Cot - Cotyledon: F - Field Trials: HAP - Haploid Plants: Hyp - Hypocotyl: L - Leaf: M - Mature: MS - Multiple Shoot: ORG - Organogenesis (Shoots): PL - Plantlet: PROT - Protoplast: S - Transfer to soil: SE - Somatic Embryogenesis: Seedl. - Seedling: Sus. - Suspension Culture: Tripl. - Triploid Plants: Unpubl. - Unpublished information.

References have been cited in the text mainly to published data where the studies have a biotechnological application. Apart from these, there are several laboratories engaged in tissue culture of forest trees, but the data is unpublished. The second section is restricted mainly to the results of our field evaluation studies carried out on sample plots using tissue culture plants raised from mature trees of *Eucalyptus tereticornis, E. torelliana* and *S. persica,* and from seedlings of *D.strictus* and *T. indica*. These studies have allowed us to assess general differences between tissue culture plantlets and seedlings. Based on indications regarding the likely benefits to be accrued from tissue culture, we have established better designed field tests spread over different locations. This research and the assurance that abnormalities will not occur, as has been observed in the case of oil palm (Corley et al., 1986), are essential before applying micropropagation for large-scale production.

In the latter part of the second section, refining of the techniques for *Eucalyptus* has been described. An analysis has been given to bring out the possibilities for reduction in the cost of plants using these methods. This article has emphasised the experimental results with regard to *Eucalyptus*.

MATERIALS AND METHODS

The experimental procedures for culture of explants from mature elite trees of *E. tereticornis* and *E. torelliana* and their transfer to the field have been described in earlier publications (Gupta et al., 1981, 1983; Gupta and Mascarenhas, 1987). The conditions for inducing somatic embryogenesis from *E. citriodora* are as cited earlier in Muralidharan and Mascarenhas (1987b). Plantlets produced were field-grown in Pune in 1983 in a non-replicated trial, as described by Khuspe et al. (1987), and comparisons were made with seedlings raised from the seeds of the same elite trees from which explants were collected. A preliminary rogueing was carried out so that all plants were of similar height and identical in appearance at the beginning of the trial.

The analysis of the organic carbon and available nitrogen (Walkely and Black, 1934), available phosphorus (Olson et al., 1954), potassium levels (Perur et al., 1973), pH and electrical conductivity (Jackson, 1967) of soil samples collected from zones 40 to 80 cm below the surface around the elite trees and also from different regions of the experimental plot in Pune, have been described earlier (Mascarenhas et al., 1988).

Height and girth measurements were mainly taken in the months of May and October. Biomass was estimated according to the methods described by Chaturvedi and Venkatraman (1973). These data for both the species, have been published earlier (Mascarenhas et al., 1988). The specific gravity was determined by the method developed by Smith (1955) with samples collected with a cork borer, at breast height from 40-month-old trees. The basic laboratory-scale procedures for the clonal multiplication of bamboo (Nadgir et al., 1984) and tamarind (Mascarenhas et al., 1987) using seedling explants have been reported earlier.

RESULTS

Growth Analysis and Biomass Yield

Tissue culture-grown plants of *E. tereticornis* showed greater height, diameter and biomass values than control plants. The biomass yields of tissue culture-grown plants at 12 and 34 months were respectively 200 and 34 percent more than controls. At 52 months, the increase was 16 percent.

On the other hand, the biomass increases in tissue culture-raised plants of *E. torelliana* were 700 percent and 100 percent higher, above the controls at the end of 12 and 34 months respectively. This increase did not diminish after 52 months and was still 100 percent higher than controls.

The results with both *E. tereticornis* and *E. torelliana* indicate very high early increase in growth rates of tissue culture plants as compared to controls.

Some of the wood-properties such as specific gravity and total pulp content were also tested by collecting random samples from 25 percent of the control and tissue culture stands. There was a high degree of variation between the maximum and minimum values, both in tissue culture-raised plants and in controls (Mascarenhas et al., 1988).

A similar variation was also observed in pulp yields (Table 2) for debarked wood samples analysed by the sulphate process (Rydholm, 1965), the values being 13 percent higher in *E. torelliana* tissue culture-raised plants.

Table 2: Percent (dry weight) pulp yield in debarked wood samples.*
Age of plants - 34 months

	E. tereticornis		E. torelliana	
	T.C.	Control	T.C.	Control
Minimum	39.25	35.75	50.50	42.25
Maximum	47.25	42.00	57.75	52.25
Average	42.75	40.00	54.10	47.58

The data have been computed based on measurements from 9 plants for each treatment.

* By Standard Sulphate Process

A similar observation was also noted for rooted cuttings of *E. grandis* in the Aracruz experiment (Brando, 1984). The density has an important influence on the structural and mechanical properties of paper. The density range of the more acceptable *Eucalyptus* species for paper pulp is 400 to 720 kg m^{-3}. The density of the tissue culture-raised plants even at 34 months falls within this range. This observation suggests the possibility of reducing the rotation cycles so that the trees are harvested at a younger age when biomass increases are maximum.

ECONOMICS OF CLONING USING TISSUE CULTURE PROCEDURES

In studies dealing with economic comparisons, it is often necessary to make certain assumptions within reasonable limits. In an attempt to obtain some fairly reliable comparisons of the yield of biomass per year in the case of *Eucalyptus*, it was necessary to calculate the costs of the plantlets. The details of capital and recurring expense have been published earlier (Mascarenhas et al., 1988). The cost per plantlet, based on capital and recurring expenses is approximately two rupees (1 rupee = $ 0.06) divided between 50 percent for capital and other 50 percent for recurring expenses. The cost per seedling is around 25 percent of the tissue culture-raised plantlet, but 50 percent when the recurring costs alone are considered. At the end of 34 months, the biomass yields of *E. tereticornis* and *E. torelliana* planted at 2m x 2m were calculated where the gains based on recurring costs were higher than with seedlings. The profits with *E. torelliana* were greater, being nearly double that obtained from seedlings. The cost of plants were calculated on the assumption that the capital expenditure on laboraotry, greenhouse, etc. has a capacity for producing one million plants and a 90 percent survival.

In an attempt to reduce recurring expenditure, different procedures were tested. These included the liquid nutrient replenishment system (LNRS; Aitken-Christie and Jones, 1987) wherein the cultured shoots were hedged at regular intervals and fresh liquid medium was added. A comparison of the costs was also made using a controlled system of somatic embryogenesis (Muralidharan and Mascarenhas, 1987a) that has been recently developed from seedlings of *E. citriodora*. A comparative break-up of the recurring expenditure through these two approaches and also by standard micropropagation is given in Table 3.

The costs vary from 0.83 rupee for the standard micropropagation to 0.66 rupee by the LNRS and 0.18 rupee through somatic embryogenesis, assuming a survival rate of 50 percent as compared to 90 percent in the first two procedures.

At present, the major constraint associated with mass propagation of plants from somatic cells is financial. It is reported that genetically improved seeds of certain species of pines are worth $ 2200 per kg (Brown, 1981). This can give rise to about 44,000 plantable seedlings or it will cost $ 50.00 per thousand for seeds alone and $ 50.00 per thousand for production costs. These costs fall within the range of those for micropropagated plants.

The costs could be further reduced when suitable methods for the production of somatic embryos and their encapsulation become a common practice with an increasing number of tree species. In *E. citriodora,* we have successfully encapsulated somatic embryos and obtained about three percent germination, using the method of Redenbaugh et al. (1986).

Hasnain and Cheliak (1986) and Hasnain et al. (1986) compared the economics of tissue culture-raised trees in New Zealand, United States and Canada. In Canada,

Table 3: Comparative break-up of recurring expenditure (through different approaches for one million *Eucalyptus* plants) at NCL[*] by all manual operations

	Micropropagation (90% survival) (in million rupees)	Hedging (90% survival) (in million rupees)	Somatic embryo-genesis (50% survival) (in million rupees)
Labor (including nursery practice)	0.70	0.54	0.130
Chemicals	0.03	0.02	0.006
Power	0.06	0.06	0.020
Breakage and Unforeseen expenses	0.04	0.04	0.020
	0.83	0.66	0.176

* National Chemical Laboratoy, Pune

where seedling costs are higher, micropropagation should be more feasible than in United States or New Zealand, where the differences in cost between seedlings and tissue culture plantlets are higher. Franclet and Boulay (1983) also observed that the cost of frost-resistant *Eucalyptus* clones raised by tissue culture was twice as much as that for seed-raised plants.

Salvadora persica L. (Tooth-Brush Tree)

Salvadora persica is a large, many-branched evergreen tree that grows in dry, arid, saline and coastal regions of India. Its seeds yield a pale yellow solid fat, the fatty acid composition of which is similar to coconut oil. Being rich in lauric acid, it is used in the soap industry as a substitute for coconut oil (Chadha, Y.R.).

A tissue culture process was developed to successfully multiply high-yielding trees and those growing near the sea shore (Rao, 1987). Small-scale plantations were raised in June 1984. The first flowering was observed in control plants after 20 months, whereas tissue culture-raised plants flowered after 31 months (Mascarenhas et al., 1988). In general, tissue culture-raised plants had a higher degree of uniformity even after 40 months. This could be an advantage in plants where seeds are the final product. An added advantage is the capacity of this species to grow on saline and alkaline soils, which occupy over 7.5 million hectares in India. If some of these wastelands can be covered with high-yielding *Salvadora* trees, it would be a major benefit, both in wasteland development and for a reduction in the quantum of imports of edible oils, a substantial portion of which is at present diverted towards the non-edible-oil industry.

Dendrocalamus strictus and *Tamarindus indica*

Very preliminary pilot-scale outplantings that were carried out using the tissue culture-raised plants of bamboo and tamarind have yielded useful and interesting information. In the case of bamboo, seven tissue culture-raised plants of identical age were planted side by side with seed-raised plants of the same age, height, etc. The tissue culture plants were raised from seedlings of the same batch. The results with tamarind were obtained from 14 tissue culture plantlets raised from seedlings. Comparisons have not been made with seed-raised plants. Some of the observations for bamboo are summarised below :

1. Culm formation occurred within 30 months in tissue culture-raised plants as compared to four years in seed-raised plants.

2. At 52 months, the height of the main culm, the number of culms per plant, the number of nodes of the main culm and the girth of the second internode were nearly double of those in the controls.

These results, though preliminary, could be indicative of the potential application of tissue culture for bamboo, where the rotation periods can be cut down, besides permitting a steady supply of plantlets all the year round.

In the outplantings with tamarind (raised through tissue culture), flowering and fruiting was observed within 42 months, as compared to 8 to 12 years in seedling-raised plants. This could shorten the period before the first harvest in tamarind. These results are thought-provoking, since plantlets have been developed from juvenile seedlings and not from mature trees, where such a response could be explained.

DISCUSSION

When we developed tissue culture methods in our laboratory for different species, we were interested in answering two questions: The first was whether there would be any advantages or benefits in using tissue culture-raised plants in forestry. The second was how to reduce costs. To answer these questions, we outplanted all the plantlets that could be produced from each species. The planting sizes thereby varied from a very small number to larger plots and were conducted in our laboratory estate. The techniques have since been greatly improved and we are now conducting larger multilocation replicated trials. This is necessary to confirm the preliminary results obtained on the likely benefits of using tissue culture.

In our studies with *Eucalyptus,* higher biomass yields were obtained within a shorter period with tissue culture plantlets rather than seedlings. Frampton and Isik (1986) studied differences in field performance between tissue culture plantlets and seedlings of loblolly pine. After 3.5 years, the tissue culture plants were 82 percent taller than the seedlings and appeared more mature. The *Eucalyptus* differs from loblolly pine in terms of accelerated growth rates in the first 34 months. The specific gravity and cellulose content were also of the correct specification for use in the paper and pulp industry. An additional advantage in using *Eucalyptus* is its coppicing ability.

In general, the benefits of tissue culture, if any, will be known only after outplanting of the propagules. For instance, in bamboo, early culm formation was observed after the second year, as compared to four years in case of seed. Similarly, in tamarind, flowering and fruiting took place after three years, as compared to 8-10 years in seed raised-plants. Early flowering six to twelve months of tissue culture plants was observed in pomegranate (unpublished) and also in teak (Mascarenhas et al., 1987). This could reduce the breeding cycles and could be used to raise seed orchard of elite trees. A large genetic base could be maintained to avoid deleterious effects of inbreeding. This would provide superior seed with the capacity for improved growth rates and other desired characteristics. Breeding of selected trees to combine the most favorable genes can give even higher returns (Mukhopadhyay and Mohan Ram, 1981).

According to Brown (1981), it appears that shoot tip cloning of some species is now a feasible procedure to use in establishing vigorously-growing clones for afforestation in short rotation forestry.

It seems certain that some of the current economic problems could be solved by refining of the labor-intensive stages of the tissue culture process.

ACKNOWLEDGEMENTS

We acknowledge the help and cooperation extended by Mr. S. Kondas, Mr. H.P. Vishwanathan, Mr. P.A. Mukhedkar and Mr. A.C. Gupta of the Forest Corporations of Tamil Nadu, Karnataka, Andhra Pradesh and Uttar Pradesh, respectively. We also express gratitude to Mr. S.C. Jain of Grasim Forest Institute, Karnataka, Mr.C.H. Laxmipathy, Bhadrachalam Paper Boards, Andhra Pradesh and Mr. R.J. Tilani, Aegis Chemicals Ltd., Bombay, for supplying plant material of the different tree species.

We are also thankful to the National Bank for Agriculture and Rural Development for financing this forestry project and to Miss Shobha Nimhan for typing this paper.

REFERENCES

Aitken-Christie, J. and Jones, C., 1987, Towards automation : Radiata pine shoots hedges *in vitro*, *Plant Cell Tissue Organ Cult.*, 8: 185-196.

Anand, M. and Bir, S.S., 1984, Organogenetic differentiation in tissue cultures of *Dalbergia latifolia*, *Curr. Sci.*, 53: 1305-1307.

Aneja, S. and Atal, C.K., 1969, Plantlet formation from tissue culture from lignotuber of *Eucalyptus citriodora*, *Curr. Sci.*, 38: 69.

Bapat, V.A. and Rao, P.S., 1979, Somatic embryogenesis and plantlet formation in tissue cultures of sandalwood *(Santalum album)*, *Ann. Bot. (Lond.)*, 44: 629-630.

Bapat, V.A., Gill, R. and Rao, P.S., 1985, Regeneration of somatic embryos and plantlet from stem callus protoplast of sandalwood tree *(Santalum album)*, *Curr. Sci.*, 54: 978-982.

Bhatnagar, S.P., Singh, M.N. and Kapur, N., 1983, Preliminary investigations on organ differentiation in tissue cultures of *Cedrus deodara* and *Pinus roxburghii*, *Indian J. Exp. Biol.*, 21: 524-526.

Bonga, J.M. and Durzan, J., 1987 (eds.), *Cell and Tissue Culture in Forestry*, Vols. *1,2 and 3*, Martinus Nijhoff, Dordrecht.

Brando, L.G., 1984, The new *Eucalyptus* forest, *The Marcus Wallenberg Foundation Symposia Proceedings*, Falun, Sweden, pp. 3-15.

Brown, C.L., 1981, Application of tissue culture technology to production of woody biomass. In: *Proc. Planning Gp. B.* pp. 1-18, Int. Energy Agency, Brighton.

Chadha, Y.R. (ed.). In: *Wealth of India, Vol. 9*, pp. 193-195, Publications and Information Directorate, CSIR, New Delhi.

Chaturvedi, A.N. and Venkatraman, K.G., 1973, Volume and weight table for *Eucalyptus* hybrid, *Indian Forester*, 99: 599-608.

Cheema, G.S., 1987, Tissue culture of poplars with particular reference to improvement of Poplars in India. In: *Proc. Symp. Genetic Manipulation of Woody Plants*, pp. 461, Michigan State Univ., Michigan.

Corley, R.H.V., Lee, C.H., Law, L.H. and Wong, C.Y., 1986, Abnormal flower development in oil palm clones, *Planter (Kuala Lumpur)*, 62: 233-240.

Datta, K. and Datta, S.K., 1985, Auxin + KNO_3 induced regeneration of leguminous tree *Leucaena leucocephala* through tissue culture, *Curr. Sci.*, 54: 248-250.

Datta, S.K., Datta, K. and Pramanik, T., 1983, *In vitro* clonal multiplication of mature trees of *Dalbergia sissoo, Plant Cell Tissue Organ Cult.,* 2: 15-20.

Dhawan, V. and Bhojwani, S.S., 1985, *In vitro* vegetative propagation of *Leucaena leucocephala, Plant Cell Rep.,* 4: 315-318.

Dhir, K.K., Angrish, R. and Bajaj M., 1984, Micropropagation of *Salix babylonica* through *in vitro* shoot proliferation, *Proc. Indian Acad. Sci. (Plant Sci.),* 93: 655-660.

Frampton, L.J. and Isik, K., 1986, Comparison of field growth among loblolly pine seedlings and three plant types produced *in vitro.* In: *Research and Development Conference,* TAPPI Press, Raleigh.

Franclet, A. and Boulay, M., 1983, Micropropagation of frost-resistant *Eucalyptus* clones, *Aust. For. Res.,* 13: 83-89.

Gharyal, P.K. and Maheshwari, S.C., 1981, *In vitro* differentiation of somatic embryoids in a leguminous tree *Albizzia lebbeck, Naturwissenschaften,* 68: 379-380.

Gharyal, P.K., Rashid, A. and Maheshwari, S.C., 1983, Production of haploid plants in anther cultures of *Albizzia lebbeck, Plant Cell Rep.,* 2: 308-309.

Goyal, Y. and Arya, H.C., 1981, Differentiation in cultures of *Prosopis cineraria, Curr. Sci.,* 50: 468-469.

Goyal, Y. and Arya, H.C., 1984, Tissue culture of desert trees : I Clonal multiplication of *Prosopis cineraria* by bud culture, *J. Plant Physiol.,* 115: 183-189.

Grewal, S., Ahuja, A. and Atal C.K., 1980, *In vitro* proliferation of shoot apices of *Eucalyptus citriodora, Indian J. Exp. Biol.,* 18: 775-776.

Gupta, P.K. and Mascarenhas, A.F., 1987, Eucalyptus. In : *Cell and Tissue Culture in Forestry* Vol. 3 (J.M. Bonga and D.J. Durzan, eds.), pp. 385-399, Martinus Nijhoff, Dordrecht.

Gupta, P.K., Mascarenhas, A.F. and Jagannathan, V., 1981, Tissue culture of forest trees - Clonal propagation of mature trees of *Eucalyptus citriodora* Hook by tissue culture, *Plant Sci. Lett.,* 20: 195-201.

Gupta, P.K., Mehta, U.J. and Mascarenhas, A.F., 1983, A tissue culture method for rapid multiplication of mature trees of *Eucalyptus torelliana* and *E. camaldulensis, Plant Cell Rep.,* 2: 296-299.

Gupta, P.K., Nadgir A.L., Mascarenhas A.F. and Jagannathan V. 1980, Tissue culture of forest trees - Clonal multiplication of *Tectona grandis* (Teak) by tissue culture, *Plant Sci. Lett.,* 17: 259-268.

Gupta, P.K., Nadgauda, R.S., Hendre, R.R., Mascarenhas, A.F. and Jagannathan, V., 1978. In : *Proc. All India Symp. and 3rd Conf. of Plant Tissue Cult.,* pp. 63-64, M.S. Univ., Baroda.

Hasnain, S. and Cheliak, W., 1986a, Tissue culture in forestry: economic and genetic potential, *The Forestry Chronicle* (August issue), 219-225.

Hasnain, S., Pigeon, R. and Overend, R.P., 1986b, Economic analysis of the use of tissue culture for rapid forest improvement, *The Forestry Chronicle* (August issue), 240-245.

Jackson, M.L., 1967, *Soil Chemical Analysis,* Prentice - Hall of India, New Delhi.

Jaiswal, V.S. and Narayan, P., 1985, Regeneration of plantlets from the callus of stem segments of adult plant of *Ficus religiosa, Plant Cell Rep.,* 4: 256-258.

Khattar, S. and Mohan Ram, H.Y., 1983, Organogenesis and plantlet formation *in vitro* in *Sesbania grandiflora, Indian. J. Exp. Biol.,* 21: 252-253.

Khuspe, S.S., Gupta, P.K., Kulkarni, D.K., Mehta, U.J. and Mascarenhas, A.F., 1987, Increased biomass production by tissue culture of *Eucalyptus., Can. J. For.,* 17.

Konar, R.N., 1975, *In vitro* studies on *Pinus* II. The growth and morphogenesis of cell cultures from *Pinus gerardiana, Phytomorphology,* 15: 137-140.

Konar, R.N. and Oberoi, Y.P., 1965, *In vitro* development of embryoids on cotyledons of *Biota orientalis, Phytomorphology,* 15: 137-140.

Konar, R.N. and Singh, M.N., 1980, Induction of shoot buds from tissue cultures of *Pinus wallichiana, Z. Pflanzenphysiol.,* 99: 173-177.

Kotwal, M., Gupta, P.K. and Mascarenhas, A.F., 1983, Rapid multiplication of *Sapium sebiferum* by tissue culture, *Plant Cell Tissue Organ Cult.,* 2: 133-139.

Kulkarni, D.K., Gupta, P.K. and Mascarenhas, A.F., 1984, Tissue culture studies on *Leucaena leucocephala, Leucaena Res. Rep.,* 5: 37-39.

Kulkarni, V.M., Gupta, P.K., Mehta, U.J. and Mascarenhas, A.F., 1981, Tissue culture of woody trees : clonal propagation of *Tamarindus indica* Linn. (Tamarind) by tissue culture, Abst. of the 6th All India Plant Tissue Culture Conference, Dept. of Bot., Pune University.

Lakshmi Sita, G., 1979, Morphogenesis and plant regeneration from cotyledonary cultures of *Eucalyptus, Plant Sci. Lett.,* 14: 63-68.

Lakshmi Sita, G. and Shobha Rani, B.,1985, *In vitro* propagation in *Eucalyptus grandis* by tissue culture, *Plant Cell Rep.,* 4: 78-80.

Lakshmi Sita, G., Raghav Ram, N.V. and Vaidyanathan C.S., 1979, Differentiation of embryoids and plantlets from shoot callus of sandalwood, *Plant Sci. Lett.,* 15: 265-270.

Lakshmi Sita, G., Raghav Ram, N.V. and Vaidyanathan, C.S., 1980, Triploid plants from endosperm cultures of sandalwood by experimental embryogenesis, *Plant Sci. Lett.,* 20: 63-69.

Mascarenhas, A.F., Hazra, S., Potdar, U., Kulkarni, D.K. and Gupta, P.K., 1982, Rapid clonal multiplication of mature forest trees through tissue culture. In: *Plant Tissue Culture.* (A. Fujiwara, ed.), pp. 719-720. Jpn. Assoc. Plant Tissue Cult., Tokyo.

Mascarenhas, A.F., Kendurkar, S.V., Gupta, P.K., Khuspe, S.S. and Agrawal, D.C., 1987, Teak. In: *Cell and Tissue Culture in Forestry, Vol. 3,* (J.M. Bonga and D.J. Durzan, eds.), pp. 300-315, Martinus Nijhoff, Dordrecht.

Mascarenhas, A.F., Nair, S., Kulkarni, V.M., Agrawal, D.C., Khuspe, S.S. and Mehta, U.J., 1987, Tamarind, In: *Cell and Tissue Culture in Forestry, Vol. 3* (J.M. Bonga and D.J. Durzan, eds.), pp. 316-330, Martinus Nijhoff, Dordrecht.

Mascarenhas, A.F., Khuspe, S.S., Nadguada R.S., Gupta, P.K. and Khan, B.M., 1988, Potential of cell culture in plantation forestry programmes. In: *Genetic Manipulation of Woody Plants,* pp. 391-412, Plenum Press, New York.

Mathur, I. and Chandra, N., 1983, Induced regeneration in stem explants of *Acacia nilotica, Curr. Sci.,* 52: 882-883.

Mehra, P.N. and Cheema G.S., 1980, Clonal multiplication *in vitro* of Himalayan Poplar *(Populus ciliata), Phytomorphology,* 30: 336-343.

Mehra, P.N. and Verma, V., 1981, Callus induction and shoot bud formation on cotyledons of West Himalayan spruce *(Picea smithiana), Phytomorphology,* 31: 60-69.

Mehta, U., Ramanuja Rao, I.V. and Mohan Ram, H.Y., 1982, Somatic embryogenesis in bamboo. In: *Plant Tissue Culture* (A. Fujiwara, ed.), pp. 109-110, Jpn. Assoc. Plant Tissue Cult., Tokyo.

Mahatre, M., Bapat, V.A. and Rao, P.S., 1985, Regeneration of plants from the culture of leaves and axillary buds in mulberry *(Morus indica* L.), *Plant Cell Rep.,* 4: 78-80.

Mukhopadhyay, A. and Mohan Ram, H.Y., 1981, Regeneration of plantlets from excised roots of *Dalbergia sissoo, Indian J. Exp. Biol.,* 19: 1113-1115.

Muralidharan, E.M. and Mascarenhas, A.F., 1987a, *In vitro* plantlet formation by organogenesis in *Eucalyptus camaldulensis* and by somatic embryogenesis in *E. citriodora, Plant Cell Rep.,* 6: 256-260.

Muralidharan, E.M. and Mascarenhas, A.F., 1987b, *In vitro* morphogenesis in *Azadirachta indica* and *Eucalyptus citriodora.* In: *Abst. Intl. Workshop on Tissue Culture and Biotechnology of Medicinal and Aromatic Plants,* CIMAP, Lucknow (in press).

Nadgir, A.L., Phadke, C.H., Gupta, P.K., Parasharami, V.A., Nair, S. and Mascarenhas, A.F., 1984, Rapid multiplication of bamboo by tissue culture, *Silvae Geneti.* 33: 221-223.

Narayan, P. and Jaiswal, V.S., 1985, Plantlet regeneration from leaflet callus of *Azadirachta indica, J. Tree Sci.,* 4: 65-68.

Nataraja, K. and Sudhadevi, A.M., 1985, Induction of plantlet from seedling explants of *Dalbergia latifolia in vitro, Bitr. Biol. Pflanz.,* 59: 341-350.

Olson, S.R., Cole, G.V., Watanabe, F.W. and Dean, L.A., 1954, Estimation of available phosphorus in soils by extraction with sodium bicarbonate, USDA, Circ., pp. 939.

Perur, N.G., Subramaniam, C.K., Muhr, G.R. and Ray, H.E., 1973, *Soil fertility evaluation to serve Indian farmers*, Univ. Agric. Sci., Bangalore, pp. 124.

Rao, I.U., Rao, I.V.R. and Narang, V., 1985, Somatic embryogenesis and regeneration of plants in the bamboo *Dendrocalamus strictus, Plant Cell Rep.*, 4: 191-194.

Rao, K.S., 1986, Plantlets from somatic callus tissue of East Indian Rosewood, *(Dalbergia latifolia), Plant Cell Rep.*, 5: 199-202.

Rao, K.S., and Venkateswara, R, 1985, Tissue culture of forest trees: Clonal multiplication of *Eucalyptus grandis, Plant Sci.*, 40: 51-55.

Rao, P.S. and Bapat, V.A., 1978, Vegetative propagation of sandalwood plants through tissue culture, *Can. J. Bot.*, 56: 1153-1156.

Rao, P.S. and Ozias-Akins, P., 1985, Plant regeneration through somatic embryogenesis in protoplast cultures of sandalwood *(Santalum album), Protoplasma*, 124: 80-86.

Rao, S.M., 1987, *In vitro* studies on *Salvadora persica* Linn. M.Sc. Thesis, Univ. Pune, India.

Redenbaugh, K., Paasch, B.D., Nichol, J.W., Kossler, M.E., Vess, P.R. and Walker, K.A., 1986, A gel encapsulation system for alfalfa somatic embryos, *Biotechnology*, 4: 797-801.

Rydholm, S.A., 1965. *Pulping Processes*, John Wiley and Sons, New York.

Smith, D.M., 1955, A comparison of two methods for determining the specific gravity of small samples of secondary growth of Douglas-fir, U.S.F.P.L. Ref. No. 2033, pp. 13-20.

Srivastava, P.S., 1973, Formation of triploid plantlets in endosperm culture of *Putranjiva roxburghii, Z. Pflanzenphysiol.*, 69: 270-273.

Upadhyay, S. and Chandra, N., 1983, Shoot and plantlet formation in organ and callus culture of *Albizzia lebbeck, Ann. Bot. (Lond.)*, 52: 421-424.

Walkely, A. and Black, T.A., 1934, Determination of organic carbon by rapid titration method, *Soil Sci.*, 37: 29-35.

8

TISSUE CULTURE OF PLANTATION CROPS

S. Bhaskaran and
V.R. Prabhudesai

ABSTRACT

Plantation crops constitute an important renewable resource for raw materials and contribute a great deal towards meeting human requirements for food, timber, medicines, spices and beverages. In a developing country like India, there is an urgent need to increase productivity of plantation crops in order to meet the growing demands for home consumption and export. This can be achieved by planting *en masse,* high yielding elite varieties produced through tissue culture. Tissue culture has become a valuable tool for rapid clonal multiplication of several economically important plantation crops. Ingenuity will lie in blending traditional breeding methods with new technologies. Tissue culture can act as an important bridge in this regard.

S. Bhaskaran and *V.R. Prabhudesai* * Hindustan Lever Research Centre, Andheri, Bombay - 400 099, India.

INTRODUCTION

Plantation crops are a perennial source of raw materials and also offer protection against natural calamities such as floods and help in soil and water conservation, thus maintaining the ecological and environmental balance. Plantation crops have a vital role in the Indian economy. About 41 percent of tea, 50 percent of coffee, 57 percent of cardamom and 77 percent of pepper produced in India is exported to other countries (Muliyar, 1983). The areas and production per annum for some of the important plantation crops in India are shown in Table 1.

Table 1: Area and production of some plantation crops in India

Plant species	Area '000 ha	Production per annum '000 tonnes
Camellia sinensis	400	645
Coffea arabica	220	190
Theobroma cacao	23	4
Elettaria cardamomum	100	3.5
Piper nigrum	110	30
Curcuma longa	88	200
Zingiber officinale	40	80
Anacardium occidentale	460	150
Hevea brasiliensis	350	235
Cocos nucifera	1100	5700 *

* million nuts

Source: Plantation Crops: Opportunities and Constraints, Vol. I & II, Oxford & IBH Publ. Co., New Delhi, 1986.

Because of the rising population and shrinking land and other resources, it is not possible to extend the area under plantation crops in India to a great extent; the only alternative then is to increase the output of plantation crops several fold per unit area.

Unlike the annuals, plantation crops present certain unique problems for the plant breeder in terms of their improvement. The long pre-bearing age and highly heterozygous cross-pollinating nature of these crops constitute a serious handicap for their rapid improvement and assessment of available variability. For want of rapid clonal propagation methods in many of these crops, breakthroughs in production have not been possible. Tissue culture can be of immense help in the multiplication of true-to-type high yielding plants on a large-scale for planting. Producing thousands of plants starting from a small portion of tissue taken from an elite plant is not at all unrealistic, as has been exemplified in banana, cardamom, eucalyptus, and teak.

The nature of problems associated with some economically important plantation crops and the approaches used to solve them through tissue culture are discussed in Table 2.

Table 2: Successful micropropagation in major plantation crops of India

Plant species	Response[*]	Multiplication/ Propagation	Reference
Beverage Crops			
Camellia sinensis	C,SH	Plantlets	Wu et al.,1982; Kato, 1985
	SH	Plantlets	Phukan and Mitra, 1984
C. canephora var. *Robusta* Arobusta	SE	Plantlets	Sondahl and Sharp, 1979
Spices & Condiments			
Elettaria cardamomum	SH	Plantlets	Srinivasa Rao et al., 1982; Nadgauda et al., 1983; Kumar et al., 1985
	SH	X 5000 Plantlets	Bhaskaran (Unpublished)
Curcuma longa	SH	X 200,000 Plantlets	Nadgauda et al., 1978
Zingiber officinale	SH	Plantlets	Hosaki and Sagawa, 1977
	SH	$X\ 15 \times 10^6$ Plantlets	Nadgauda et al., 1980
Palms			
Cocos nucifera	SE	Single/few Plantlets	Branton and Blake, 1983; Raju et al., 1984; Bhaskaran, 1985a,b
Elaeis guineensis	C,SE	Plantlets	Rabechault and Martin, 1976; Jones, 1983; Thomas and Rao, 1985
Phoenix dactylifera	SE	Plantlets	Reuveni et al., 1972; Tisserat, 1979; Sharma et al., 1984
Rubber			
Hevea brasiliensis	C	Haploid plantlets	Hu and Hao, 1980; Chen et al., 1981

Plant species	Response[*]	Multiplication/ Propagation	Reference
Fruit Crops			
Musa paradisiaca	SH	Plantlets	Berg and Bustamante, 1974; Ma et al., 1978; De Guzman et al., 1980; Doreswamy et al., 1983; Cronauer and Krikorian, 1984; Vuylsteke and De Langhe, 1985; Banerjee et al., 1986
	SH	X 5000	Vatsya and Bhaskaran Plantlets (Unpublished)
Mangifera indica	C,SE	Plantlets	Litz et al., 1984, 1985
Citrus spp.	C,SE	400 plantlets per 1 gm callus	Chaturvedi and mitra, 1974
	NE	Plantlets	Spiegel-Roy and Kochba, 1980 (please refer George and Sherrington, 1984 for an exhaustive list)
Carica papaya	SH	Plantlets	Litz and Conover, 1977
	C	Plantlets	Jordan et al., 1982,1983
	SH	X 25 plantlets per 3 weeks	Rajeevan and Pandey, 1986
Punica granatum	SH	X 20,000 Plantlets	Gupta et al., 1981

Abbreviations: C - Callus, NE - Nucellar embryos, SE - Somatic embryos, SH - Shootlets

Beverage Crops

Beverage crops such as tea and coffee are important cash crops, earning sizeable foreign exchange of about Rs. 10 billion for India, of which tea accounts for Rs. 7 billion. Tea (*Camellia sinensis* L.) is of paramount importance to India, which is the world's leading producer, consumer and exporter of tea and, therefore, plays an important role in the trade of this commodity. At present, tea is being grown on about 400,000 ha of land and the annual production is 645 M kg. In order to retain its 28 percent share of international trade and to meet the increasing domestic demand, India needs to improve the productivity of tea substantially. This objective can be realised by replanting the old, less productive tea plantations with improved plant material. Genetically improved cultivars bred through selection of bi- and poly- clonal material and vigorous triploids can be used for planting. Tea is an exclusively cross-pollinating, self-incompatible crop, polymorphic in origin, which are the characters that hamper both genetic studies and breeding as well. The intense heterogeneity in the seedling populations provide tremendous scope for clonal selection which is a widely accepted method of improvement in tea (Sharma and Ranganathan, 1985).

The advantages of culturing the tea plant could be two fold:

(i) Certain high yielding tea bushes are often identified in a plantation, which, if selectively propagated, could enhance yields considerably. However, the number of cuttings for planting from such a bush is limited to about 200. Through shoot tip or meristem culture, a perpetual process of producing an unlimited number of propagules can be developed.

(ii) The second avenue for improvement through tissue culture are the spontaneous changes occurring in the callus cells (somaclonal variation) and subsequent regeneration of plantlets showing variability with respect to desirable characters such as resistance to disease, pests, drought and with other desirable agronomical characters.

In the tea plant, the differentiation from callus has been obtained only in a limited number of cases (Doi, 1981; Wu et al., 1982). Pollen pro-embryoids, anther callus and roots from callus have been reported (Doi, 1981; Raina and Iyer, 1983) but no plantlets were obtained. Kato (1982, 1985) was able to induce some shootlets from calli originated from peeled epidermis and sub-epidermal tissues. They exhibited slow growth in culture and only a few could be established in the soil. Shootlet formation from nodal explants has been reported by Phukan and Mitra (1984) but complete plantlets could not be established. More intensive efforts are needed towards this end.

Coffee (*Coffea arabica* L.) is second only to oil in world trade in terms of global turnover (Anon., 1986a). India accounts for only about 2-2.5 percent of world production and exports about 50 percent of its produce worth about Rs. 3 billion annually. Both robusta and arabica types of coffee are cultivated in almost equal quantities, with the result that India gets premium prices in the international market. Area, production and productivity of coffee have shown a marked increase in the past few decades. However, cultivated robusta shows great variability, since it is a cross-pollinating species. Efforts through tissue culture should be geared up to reduce variation in the stands of robusta. There is a need to breed shorter robustas to facilitate easy picking. These objectives can be achieved through clonal propagation of desirable robusta material. Arabica is a tetraploid, self-pollinating species which can also be multiplied on a mass-scale through tissue culture.

Coffee has been amenable to tissue culture (Table 3), which is evident from a number of reports as reviewed by Monaco et al. (1977). Embryogenesis through callus was reported in the early 1970s (Staritsky, 1970). Callus cultures of several different varieties of coffee have been established without difficulty. Callus has been obtained from both seedling and mature leaf explants. High frequency somatic embryogenesis

Table 3: Tissue culture of *Coffea* spp.

Explant source	Response*	Reference
C. arabica		
Stem	C	Staritsky, 1970
Endosperm	C	Keller et al., 1972
Various plant parts	R,E	Sharp et al., 1973
Var. "Bourbon"		
Leaf	SE,	Sondahl and Sharp, 1977;
	plantlets	Sondahl and Sharp, 1979
C. canephora var. "Robusta"		
Stem	C,SE	Staritsky, 1970
Leaf	C,SE	Sondahl and Sharp, 1979
Anther culture		
C. arabica	C	Sharp et al., 1973
C. canephora	C, SE	Lanaud and Parvis, 1980
C. excelsa	C, SE	Lanaud and Parvis, 1980
C. liberica & *C. arabica*	Haploid C, pro-embryos	Sondahl et al., 1980

Abbreviations: C - Callus; E - Embryoid; R - Roots; SE - Somatic embryos

could be achieved and plantlets obtained (Sondahl and Sharp, 1977; Sondahl and Sharp, 1979).

Shoot apices of *Coffea arabica* seedlings were cultured to produce multiple shootlets and plants (cited in George and Sherrington, 1984). Undifferentiated haploid callus tissue was obtained in *C. arabica* and pro-embryos in cultured anthers have been reported in some species of coffee (Sharp et al., 1973; Lanaud and Parvis, 1980). Therefore, tissue culture appears to have the scope for large-scale multiplication and would be helpful in yield stabilisation. A combined effort involving the use of new varieties (e.g., var. *Cauvery*), rapid clonal multiplication, mycorrhiza for better phosphorus utilisation and integrated pest management will be helpful in increasing productivity (Menon, 1986).

Cacao (*Theobroma cacao* L.) is a crop that does not require heavy capital expenditure and can be produced on any scale. It can be grown as an intercrop in coconut and arecanut gardens (Anon., 1986b). It has been shown that it does not hamper the growth of these crops; on the contrary, it ensures the producer a reasonable additional income.

Traditionally, cacao plants are propagated by seedlings, budding and rooted stem cuttings. Seedlings are genetically variable, while stem cuttings have to be taken

Table 4: Tissue culture of *Theobroma cacao*

Explant source	Response[*]	Reference
Cambium	C,R	Archibald, 1954
Seedling tissue	C,R	Hall and Collin, 1975
	C,SC,SE	Lee and Rao, 1982
Leaf, Cotyledons	C,R	Pence et al., 1979
Dormant shoot apices	SH	Orchard et al., 1979
Zygotic embryos	SE	Pence et al., 1980
Immature cotyledons	C,SC	Tsai and Kinsella, 1981
Hypocotyls	SE,C-SE	Kononowicz et al., 1984
Zygotic embryos & cotyledons	SE	Rao and Lee, 1986

[*] *Abbreviations:* C - Callus; R - Roots; SC - Suspension culture; SE -Somatic embryos; SH - Shootlets

from upright (orthotropic) shoots. If horizontal (plagiotropic) branches are used for cuttings, they result in trees with a low spreading habit which makes harvesting difficult. As orthotropic shoots are limited in number, it is suggested that tissue culture could be used for the multiplication of cacao cultivars (George and Sherrington, 1984). This would enable new varieties to be introduced rapidly and virus-free stocks can be produced.

Shoot tip culture (Orchard et al., 1979) has not yet been successful in cacao (Table 4). Callus and in some cases, callus with roots, has been successfully obtained from various tissues including cambium from mature stems (Archibald, 1954), seedling root, stem, hypocotyl, cotyledon (Hall and Collin, 1975), embryo and somatic tissue of anthers (Prior, 1977). Proliferation of asexual embryos was induced from immature zygotic embryos (Pence et al., 1979) and hypocotyls (Kononowicz et al., 1984). Organogenesis and formation of embryoids was studied in seedling tissues and immature seed embryos cultured on liquid medium (Lee and Rao, 1982; Rao and Lee, 1986). As yet, no regeneration of plantlets on a large-scale has been reported, that can be of practical use.

With cacao, as with other plantation crops, the tissue culture technique can be adapted as an aid to increase the F_1 generation stocks for cloning recognised high yielders. In cocoa, the method has been shown to have a propagation potential but the asexual embryos must be induced to germinate if this technique is to be adapted commercially. It would be more advantageous, however, if embryos could be induced from vegetative tissues. In order to meet the growing demands of the mushrooming chocolate industry in India, research should be aimed at increasing productivity and improving bean quality.

Spices and Condiments

Spices have been an important component of India's trade since ancient times. Spices and condiments include cardamom, pepper, ginger, turmeric, nutmeg and mace, clove, cinnamon and vanilla. Unlike other economic crops, spices continue to be cultivated in the same way today as they were thousands of years ago. Tissue culture has considerable scope in some of these for large-scale multiplication of the elite types so that productivity of these can be increased.

Cardamom (*Elettaria cardamomum* Maton) is an important plantation crop in India, earning sizeable foreign exchange in the order of about Rs. 650 million (1986). However, the productivity has stagnated at around 63 kg ha^{-1} compared to 250 kg ha^{-1} in Guatemala, the major competitor for Indian cardamom in the world market. The productive life span of a cardamom plant is about 7-8 years, but in India, the majority of plantations are about 20 years old. The crop has also suffered extensive damage due to a disease called *Katte*. Tissue culture provides an efficient and rapid method of propagation of high yielding clones and virus-free plants can be produced through meristem culture. Callus culture and regeneration in cardamom provides a good system for selecting variants with characters such as early flowering, bold capsules and resistance to diseases.

Callus initiation from embryos and root stocks was successfully achieved and regeneration of plantlets was studied (Srinivasa Rao et al., 1982). Apical shoot tip culture and panicle culture to produce multiple shoots and plantlets have been reported (Bhaskaran, Unpublished; Nadgauda et al., 1983; Kumar et al., 1985; Anon., 1986c). In our laboratory, a detailed procedure of scale up has been worked out. Shoot buds and immature panicles were selected as starting material from elite clones identified in planters' fields, that had traits such as compound panicles, bold capsules and high yield potential. The rate of multiplication achieved is about 5,000 plantlets per shoot explant in a year. The plantlets have been transferred to the field and are being evaluated in plantations side by side with original elite plants.

Pepper (*Piper nigrum* L.) is the most important foreign exchange earner among spices as Indian black pepper is considered to be superior in quality. The export earnings are reported to be about Rs. 2 billion in the current year. However, there is no organised cultivation of pepper on a large-scale, as is the case in Brazil and Indonesia. The present average yield is about 275 kg ha^{-1} as against 3,400 kg ha^{-1} in Brazil (Dineshkumar et al., 1986). Newly developed hybrids such as *Panniyur* yields over 5 kg per vine which can be used for planting. Tissue culture can be of use in rapid multiplication of such high yielding material. Shoot tip culture and callus culture from various parts of the pepper plant have been established (Mathews and Rao, 1984). However, large-scale multiplication through tissue culture needs to be worked out.

Ginger and turmeric are traditional condiments which together earn between Rs. 350-400 million in foreign exchange for India annually (Nair, 1982). They are also sources of essential oil, oleoresin and curcumin which find various uses in industry. Turmeric (*Curcuma longa* L.) is grown in an area of about 88,000 ha and is produced to the extent of 170 to 200 thousand tonnes per annum. India is the world's largest producer and exporter of turmeric. The agro-climatic conditions in most parts of India are suited for this crop.

The conventional method of propagation through rhizome is a slow process, whereby only five to ten plants are obtained from one rhizome in a year. Faster multiplication can be brought about using the shoot tip culture method. A process to produce 200,000 plantlets in a year, starting with a single shoot, has been reported (Nadgauda et al., 1978). A high curcumin containing variant was isolated through tissue culture (Anon., 1983).

India is the leading ginger (*Zingiber officinale* Roscoe) producing country and accounts for about 50 percent of dry ginger output in the world (Menon, 1986). The ginger produced in Kerala is considered to be one of the best types in the world. It is grown in about 40,000 ha and the total yield is about 80,000 tonnes. It is exported in three forms: fresh, pickled or processed and dry. The average yield ha^{-1} is about 1.9 tonnes which can be substantially improved through clonal propagation of high yielding types employing tissue culture techniques. Multiplication rates of over 15×10^6 plantlets per year from an initial bud of ginger have been estimated (Nadgauda et al., 1980).

Perpetually vegetatively propagated crops such as ginger and turmeric probably carry certain viruses in them persistently. Meristem cultured plants show luxuriant growth under field conditions (Bhaskaran and Prabhudesai, 1987).

Plantation Crops

Cashew and rubber are two important plantation crops. Of these, Indian cashew (*Anacardium occidentale* L.) accounts for 27 percent in the world market. However, India imports about 300,000 tonnes of raw nuts from East African countries in order to meet the demands of the processing industry (Bavappa, 1982). Cashew has been traditionally neglected as a wasteland crop with a very low average yield of less than one kg per tree. In a regularly bearing garden the yield is about 5 kg per tree. There are trees which are known to give very high yields of 20 kg per tree. Seed progeny shows marked variation in fruit and nut characters. It can also be propagated by bud grafting, side grafting or soft wood grafting. However, the rate of multiplication is very slow. If this can be doubled or trebled through meristem culture or somatic embryogenesis in a shorter period of time, the rate of multiplication of elite trees can be greatly accelerated.

In cashew, the serious pest, *tea mosquito* (*Helopeltis antonii* S.) causes total loss of the inflorescence and secretes some toxin which blackens and withers the flowering shoots. If tissues can be screened for the onslaught of this toxin *in vitro,* it will be a valuable tool for the breeder to screen a large number of accessions in a short period of time, without having to expose large plantations.

Apart from a brief abstract on callus formation (Ninan et al., 1983) no work has been reported in cashew. There is ample scope to develop technology for rapid multiplication of this important earner of foreign exchange.

Table 5: Tissue culture of *Hevea brasiliensis*

Explant source	Response*	Authors
Zygotic embryos	Plantlets	Toruan and Suryatmana, 1977
Shoot apex	SH	M.P. Asokan, personal communication
Axillary buds	R	Carron and Enjalric, 1982
Stem	C,E	Wilson and Street, 1975
	SC	Wilson et al., 1976
Leaf	C,SE	Carron and Enjalric, 1982
Anthers	C - SE plantlets	Paranjothy, 1974 but no survival
	C - SE, 100 plantlets	Wang et al., 1980
	Haploid and Aneuploid E, haploid plantlets	Hu and Hao, 1980

* *Abbreviations:* C - Callus: E - Embryos: R - Roots; SC - Suspension culture: SE -Somatic embryos; SH - Shootlets

Rubber (*Hevea brasiliensis* Muell. Arg.) is the principal source of natural rubber covering over 350,000 ha in India and producing about 200,000 tonnes of natural rubber. About 50,000 tonnes of rubber, natural and synthetic, are being imported to meet indigenous requirements (George, 1986). The gap between demand and supply is widening and this is likely to continue unless adequate and immediate action is taken to increase the internal production of natural rubber (Anon., 1987). In the traditional rubber-growing areas, the scope for further expansion is extremely limited. Clones suited to non-traditional areas will have to be selected and identified high yielders will have to be selectively propagated.

Rubber plantations are established from clones of high yielding trees that are produced by grafting shoots onto seedling root stocks which are of variable genotypes. The grafted cultivars are variable, sometimes the coefficient of variation is above 50 percent. If clonal root stocks could be produced by shoot tip/ meristem culture, the need for seedling stocks could be reduced or eliminated. Scion or budwood can also be propagated *in vitro* (Table 5). This would result in greater uniformity in plantings and their performance. Rubber has a prolonged pre-bearing age of seven or more years. Breeding and clone evaluation in rubber take over 30 years. Tissue culture can play an important role in the multiplication of identified high yielding material.

As the genetic base available in *Hevea* is rather limited, induced tetraploids, triploids and genetic dwarfs have been used in the breeding program. Rapid multiplication of such types is possible through tissue culture. Callus cultures, cell suspension cultures and subsequent somatic embryogenesis would provide a method of multiplying self rooted trees. Besides, it also ensures the direct development of plantlets with a tap root, thus obviating a rooting phase. Callus and organogenesis would be useful in looking for variants in order to enhance the crop's genetic base. As the rubber tree is cross-pollinating and highly heterozygous, it is difficult to obtain trees with a high percentage of homozygous loci for breeding. Anther culture has been exploited in China to produce haploid and aneuploid plants. On the whole, attempts to exploit tissue culture for mass production of clonal material for improvement are still in their infancy. Work along these lines has been initiated at the Rubber Research Institute, India and initial success with shoot tip culture is reported (M.P. Asokan, personal communication).

Palms

Palms are perennial woody monocots which constitute an important group among plantation crops. Tissue culture techniques have been successfully used in the vegetative propagation of both oil palm and date palm and the commercial development of the method has commenced for oil palm (Choo et al., 1981).

Coconut (*Cocos nucifera* L.), which is next only to oil palm in productivity per unit area, is an important oil crop of India. India is the third largest coconut-producing country in the world. Compared to the annual oilseed crops, its yield is four to five times higher per ha and its oil yield is up to 65 percent of the dried kernel. It is a perennial source of oil. Apart from oil, all parts of the plant are utilised for various purposes, thus providing employment to a substantial section of the population. However, the production of coconut in the country has been showing a declining trend during the last decade. The reasons for this are:

* existence of a large number of senile and unproductive palms;
* exhaustion of soil caused by continuous cultivation of the crop without proper agronomic care;
* widespread occurrence of debilitating diseases such as root wilt, stem bleeding, etc.;
* inadequate availability of superior planting material, particularly hybrids; and
* lack of a regular replanting program (Bavappa, 1985).

In order to overcome the acute vegetable oil shortage in the country, it is imperative that the declining trend be reversed immediately. Seedling progeny is highly heterozygous and there is no known method of vegetative propagation. Crop improvement will be facilitated only by producing clones from selected trees and from newly developed F_1 hybrids.

Although the average yield of widely cultivated *West Coast Tall* variety is 40 nuts per tree per year, there are single elite palms yielding 200 to 400 nuts and Dwarf x Tall hybrid palms yielding 180 nuts per annum. There is scope for increasing productivity by rapidly multiplying such proven high yielders using the tissue culture technique.

Various explant sources have been used by workers to obtain somatic embryos which have the potential to grow into clonal plantlets (Table 6). Despite efforts from a large number of laboratories, coconut has remained a highly recalcitrant crop.

The first success in producing a clonal plantlet was reported in 1983 by Branton and Blake from Wye College, London, through somatic embryogenesis from root callus. Sporadic reports on the regeneration of plantlets have appeared subsequently (Raju et al., 1984; Bhaskaran, 1985a). In our laboratory, somatic embryos were produced directly from leaf explants or through a callus mass (Bhaskaran, 1985b). Two types of embryos are produced:

* tripolar with haustorium, and
* bipolar without haustorium.

However, the major problem is that a vast majority of the embryos are non-functional. In spite of intensive efforts, not much success has been achieved in the production of functional embryos in sufficient numbers, which is essential for scale up and further commercialisation. The frequency of successful growth into plants is very low. Oil palm plantations have been started in India on about 5000 ha of land in Kerala and in the Andaman and Nicobar Islands by Oil Palm India Ltd. and the Forest and Plantation Development Corporation. The results have been encouraging. It is the highest yielding oil crop and average yields of about 3.5 tonnes per ha have been reported (Abraham, 1986), which can be further increased with the use of high yielding clonal material.

India spends about Rs. 150 million in foreign exchange for the import of dates. Although there are areas in the country suitable for date palm cultivation, there are no commercial plantations of good quality dates. There is a potential area of 0.3 million ha in the arid North-West which is proposed to be covered by date palm groves (Sharma and Chowdhury, 1986). This would need about 30 million offshoots for plantation. Only a limited number of offshoots are formed in the early life of the palm.

Seed progeny is variable and since the plant is dioecious, half the plants turn out to be unproductive males that can be distinguished only after several years. Date palm is amenable for mass propagation through tissue culture, as is evident from a number of reports. Details of work on tissue culture in oil palm and date palm are covered elsewhere in this volume.

There is no better method immediately available in palms for breaking the yield barrier or producing propagules on a mass scale, than adopting the technique of tissue culture.

Fruit Crops

India has a varied climate and, therefore, temperate as well as tropical fruits are grown here. The problem here is, once again, the want of suitable planting material for the improvement of yield and quality.

Table 6: Tissue culture of *Cocos nucifera*

Explant source	Response*	Reference
Seed embryos	R	De Guzman, 1969
	Seedlings	Sajise and De Guzman, 1972
	C,E,R	De Guzman et al., 1978
	C,SH	D'Souza, 1982
Cotyledons	C,R	Jagdeesan and Padmanabhan, 1982
Endosperm	C	Euvens and Blake, 1977
Shoot tip	R	Blake and Euvens, 1982
Leaf	C	Euvens, 1976
	C,SE	Pannetier and Bufford-Morel, 1982
	SE, Plantlets	Raju et al., 1984
	C, SE	Gupta et al., 1984
	SE or C - SE, Plantlets	Bhaskaran, 1985a,b
Stem	C	Euvens, 1976
	C	Apavatjrut and Blake, 1977
	C,E	Branton and Blake, 1983
	C,SE	Gupta et al., 1984
Roots	C, Single plantlet	Branton and Blake, 1983
Inflorescence	C	Euvens, 1976
	SH	Euvens and Blake, 1977
	C,R	Euvens, 1978
	SH,R	Blake and Euvens, 1982
	C,E	Branton and Blake, 1983
Anthers	Pollen embryos	Thanh-Tuyen and De Guzman, 1983
	Pollen embryos	Monfort, 1985

* *Abbreviations:* C - Callus; E - Embryos; R - Roots; SH - Shootlets; SE - Somatic embryos

Banana (*Musa paradisiaca* L.), which is referred to as the common man's fruit in India, grows in about 270,000 ha of land. Numerous varieties exist with yields ranging from 26,000 to 55,000 kg per ha. Most of the commercial cultivars are triploid and thus seed sterile. Conventional vegetative methods are slow; therefore, attention has been turned to *in vitro* techniques for clonal propagation (Berg and Bustamante, 1974; Ma et al., 1978; De Guzman et al., 1980; Doreswamy et al., 1983; Cronauer and Krikorian, 1984; Vuylsteke and De Langhe, 1985; Banerjee et al., 1986; Vatsya and Bhaskaran, Unpublished).

For *in vitro* propagation, all possible meristem containing plant parts, e.g., suckers, peepers, dormant eyes and the base of the parental pseudostem, have been used. The rate of multiplication could be correlated with the type of genome. Somatic embryogenesis has been reported but the embryos failed to germinate into plantlets (Litz, 1985). The process for scale up of clonal plantlets through meristem culture was developed in our laboratory and it is possible to produce 5,000 plantlets from a single shoot tip in a year. Meristem culture of banana enables:

* rapid multiplication, and
* elimination of *bunchy top* virus, thus resulting in healthy, high yielding plants.

Mango (*Mangifera indica* L.), the king of fruits, has been domesticated in India for several thousand years. It accounts for about half the total area under fruit plantations in India (Menon, 1986). Both ancient and modern cultivars have been derived from seedling trees that resulted from uncontrolled pollination. Indian cultivars are monoembryonic and are propagated vegetatively by grafting. There is an urgent need to multiply elite varieties at a faster rate to cover new areas, as well as to replace old plantations.

Regeneration of somatic embryos from nucellus could be more readily induced from polyembryonic mango cultivars than the monoembryonic types (Litz, 1984, 1985). Mature somatic embryos attained height of five to six cm prior to germination. The Indian varieties such as *Dusehri* and *Alfonso* have also been amenable to somatic embryogenesis (Litz, personal communication). Callus tissue and rooting has been reported from cotyledons (Rao et al., 1982).

Mango is of vital importance, as its plantation potentialities can be realised in unused common lands, marginally saline soils and other wastelands. Numerous varieties exist in India which can be selectively propagated through somatic embryogenesis and planted in suitable areas. Some varieties have great export potential.

The total area under *Citrus* fruit in India exceeds 68,000 ha. At present, yields are on the decline as a result of the plantations being old and diseased. Propagation can be done from mature root cuttings or seed. Commercial cultivars of *Citrus* are all clones and are normally propagated by budding or grafting scions on to new root stocks. *Citrus* varieties are characterised by a long juvenile phase, heterozygosity, and, in most species, nucellar embryony. Spontaneous mutations are frequent and the world's most important varieties have arisen by spontaneous mutations. Nucellar embryony plays an important role in natural and artificial selection in the evolution of *Citrus*. Nucellar embryos are virus-free and a genetically uniform source for root stocks.

Explanted nucellus tissue of both naturally monoembryonic and polyembryonic *Citrus* varieties can directly form adventitious embryos in culture (Rangan et al., 1968; Esan, 1973). Somatic embryos are also formed directly on cultured embryos (Rangaswamy, 1959, 1961; Sabharwal, 1963). Embryogenic callus has been obtained from unfertilised ovules and nucelli of *C. sinensis*. There are a few reports of organogenesis from callus or shoot tip culture. Proliferative shoot cultures were obtained from nodal explants of mature and young tissues of Rangpur lime by Barlass and Skene (1982).

Chaturvedi and Mitra (1974) obtained callus from stem and leaf segments of *C. maxima* (Pomelo) shoots in culture. It had the potential to produce 400 plants from each gram in five to six months. After a prolonged period of culture, this callus became embryogenic (Chaturvedi and Mitra, 1975).

At the National Botanical Research Institute, Lucknow, complete plants obtained through tissue culture in several *Citrus* species have been established successfully in soil.

Adventive nucellar embryos of *Citrus* are of importance because they are genetically uniform and reproduce the characters of the maternal parent alone. Moreover, many of the desirable characteristics such as plant vigour and fruiting associated with juvenility, are restored in trees which are newly established from nucellar seedlings. This advantage has been made use of on a commercial scale for propagating desirable varieties of *Citrus* (Rangan, 1982). The work on tissue culture of *Citrus* has been exhaustively reviewed (Button and Kochba, 1977; Spiegel-Roy and Kochba, 1980) and documented (George and Sherrington, 1984).

Carica papaya L. is an important fruit crop and a source for extraction of papain. It suffers heavily due to leaf mosaic virus and its plantations are decreasing day by day. Nagai (1974) and Khuspe et al. (1980) have reported a method of growing immature hybrid embryos of *Carica papaya* x *C. cauliflora* with a view to transferring virus resistance from the latter to the former.

The improvement of papaya is hindered by its heterozygosity, dioecious habit and susceptibility to viruses. Clonally propagated selected genotypes would be of great value in breeding programs and in commercial cultivation. Conventional grafting or multiplication by cutting is not possible in papaya. Tissue culture techniques may provide the only method to clonally propagate selected genotypes on a large-scale. The attempts to propagate papaya *in vitro* have been through regeneration from callus (Yie and Liaw, 1977) or from multiple shoot cultures (Litz and Conover, 1977; Rajeevan and Pandey, 1983). A rapid method of multiplication has been developed for cultivars *Coorg Honey Dew* and *Pusa Dwarf* wherein 25 fold multiplication from shoot tips was possible in three weeks. The process is continuous, self perpetuating and has reached the stage of commercial application (Rajeevan and Pandey, 1986). A very high multiplication rate reduces the number of subcultures and labor charges during multiplication are only about 13 percent of laboratory costs. Labor intensive rooting and hardening stages could be further reduced by *in vivo* rooting.

Pomegranate (*Punica granatum* L.) is grown in India on about 1,200 ha, of which Maharashtra accounts for two-thirds of the area (Purohit, 1986). Its drought hardy nature, low maintenance cost, steady good yields, fine table and therapeutic value, keeping quality and its resting capacity at low irrigation potential, make it ideally suited for the hot semi-arid and desert regions of India. Its suitability for plantations remains underexplored. It is a hardy plant which can be cultivated in poor soils. Tissue culture can be helpful in two respects:

* Clonal selection in indigenous cultivars and their rapid multiplication.

* The superior exotic varieties such as *Muskat, Kabul Bedana* and *Kandhari* from Afghanistan failed in India since they require chilling temperature (7°C). Hybridisation between indigenous and exotic varieties combine desirable characters and clonal multiplication of the hybrids through tissue culture.

At National Chemical Laboratory, Pune, a rapid method of clonal multiplication through shoot tip culture from mature trees has been established and it has been estimated that 20,000 plantlets can be produced within a year, starting with a single shoot tip (Gupta et al., 1981).

RESEARCH NEEDS

The major emphasis of tissue culture application to plantation crops on a short-term basis has to be in rapid clonal multiplication of identified high yielding varieties. The improved types can also be multiplied and used for large-scale plantation in proper areas.

Location specific research would be beneficial in rapid application of the technique and transfer of technology. Scale up procedures and hardening aspects need to be developed for most of the crops. Most important of all, field evaluation is a prerequisite for wider application.

A much better understanding of somatic embryogenesis from cell cultures and the phenomena underlying them is needed so that it can pave the way for mass multiplication at a faster rate. Especially, where meristem culture is not possible, somatic embryogenesis is the route of choice.

On a long-term basis, the use of protoplast culture technology in basic understanding of plantation crops is desirable. Somaclonal variation can be employed to select plants with desired characters, such as tolerance to environmental variables and resistance to diseases and pests.

REFERENCES

Abraham, V.K., 1986, Oil palm cultivation in India. In: *Plantation Crops - Opportunities and Constraints,* Vol. II, (H.C. Srivastava, B. Vatsya and K.K.G. Menon, eds.), pp. 97-105, Oxford & IBH Publ. Co., New Delhi.

Anonymous, 1983, Report on Biotechnology Research at National Chemical Laboratory, Pune.

Anonymous, 1986a, Report of the Evaluation Committee, September 1986, Coffee Board, Bangalore.

Anonymous, 1986b, Annual Report 1985, Central Plantation Crops Research Institute, Kasaragod.

Anonymous, 1986c, Research Highlights 1986, Central Plantation Crops Research Institute, Kasaragod.

Anonymous, 1987, A Comprehensive report on the activities of the Rubber Research Institute of India.

Apavatjrut, P. and Blake, J., 1977, Tissue culture of stem explants of coconut (*Cocos nucifera* L.), *Oleagineux,* 32: 267-271.

Archibald, J.I., 1954, Culture *in vitro* of cambial tissue of cacao, *Nature, (Lond.)* 173: 351-352.

Banerjee, N., Vuylsteke, D. and De Langhe E.A.L., 1986, Meristem tip culture of *Musa* : Histo-morphological studies of shoot bud proliferation. In: *Plant Tissue Culture and Its Agricultural Applications,* (L.A. Withers and P.G. Alderson, eds.), pp. 139-147, Butterworths, London.

Barlass, M. and Skene, K.G.M., 1982, *In vitro* plantlet formation from *Citrus* species and hybrids, *Sci. Hortic.,* 17: 333-341.

Bavappa, K.V.A., 1982, *Cashew package of practices,* Pamphlet No. 8, ICAR, Central Plantation Crops Research Institute, Kasaragod.

Bavappa, K.V.A., 1985, Plantation crops research - 2000 A.D., *J. Plant. Crops,* 13: 1-10.

Berg, L.A. and Bustamante, M., 1974, Heat treatment and meristem culture for the production of virus free bananas, *Phytopathology,* 64: 320-322.

Bhaskaran, S., 1985a, Clonal propagation *in vitro* from coconut leaves, NFI Bull. (Bull. Nutrition Foundation of India) No. 6.

Bhaskaran, S., 1985b, Tissue culture technology for higher vegetable oil production. In: *Oilseed Production - Constraints and Opportunities* (H.C. Srivastava, S. Bhaskaran, B. Vatsya and K.K.G. Menon, eds.), pp. 537-544, Oxford & IBH Publ. Co., New Delhi.

Bhaskaran, S. and Prabhudesai, V.R., 1987, Pathogen eradication in crop plants through cell and tissue culture, *National Symposium on Role of Biotechnology in Crop Protection* held at Bidhan Chandra Krishi Vidyapeeth, Kalyani, W. Bengal, (In Press).

Blake, J. and Euvens, C.J., 1982, Culture of coconut palm tissues with a view to vegetative propagation. In: *Tissue Culture of Economically Important Plants* (A.N. Rao, ed.), pp. 145-148, COSTED and ANBS, Nat. Univ., Singapore.

Branton R.L. and Blake, J., 1983, Development of organised structures in callus derived from explants of *Cocos nucifera* L., *Ann. Bot. (Lond.),* 52: 673-678.

Button, J. and Kochba, J., 1977, Tissue culture in the citrus industry. In: *Applied and Fundamental Aspects of Plant, Cell, Tissue and Organ Culture* (J. Reinert and Y.P.S. Bajaj, ed.), pp. 70-92, Springer-Verlag, Berlin.

Carron, M.P. and Enjalric, F., 1982, Studies on vegetative micropropagation of *Hevea brasiliensis* by somatic embryogenesis and *in vitro* microcutting. In: *Plant Tissue Culture 1982* (A. Fujiwara, ed.), pp. 751-752, Jpn. Assoc. Plant Tissue Cult., Tokyo

Chaturvedi, H.C. and Mitra, G.C., 1974, Clonal propagation of *Citrus* from somatic callus cultures, *Hortscience,* 9: 118-120.

Chaturvedi, H.C. and Mitra, G.C., 1975, A shift in morphogenetic pattern in *Citrus* callus tissue during prolonged culture, *Ann. Bot. (Lond.),* 39: 683-687.

Chen, Z., Huang, J. and Chen, Z., 1981, Techniques of anther culture of *Hevea brasiliensis,* Muell. Arg. 9p. *Int. Trg. Course on Somatic cell genetics applied to cereals,* Beijing.

Choo, W.K., Yew, W.C. and Corley, R.H.V., 1981, Tissue culture of palms - A review. In: *Tissue Culture of Economically Important Plants* (A.N. Rao, ed.), pp. 138-144, COSTED and ANBS Nat. Univ., Singapore.

Cronauer, S.S. and Krikorian A.D., 1984, Multiplication of *Musa* from excised stem tips, *Ann. Bot. (Lond.),* 53: 321-328.

De Guzman, E.V., 1969, The growth and development of coconut *Makapuno* embryo *in vitro* 1. The induction of rooting, *Philip. Agric.* 53:65-78.

De Guzman, E.V., Decena, A.C. and Ubalde, E.M., 1980, Plantlet regeneration from unirradiated and irradiated banana shoot tip tissues cultured *in vitro. Philip. Agric.,* 63: 140-146.

De Guzman, E.V., Del Rosario, A.G. and Ubalde, E.M., 1978, Proliferative growths and organogenesis in coconut embryo and tissue culture, *Philip J. Coconut Stud.,* 3: 1-10

Dineshkumar, M.A., Kundapurkar, A. and Vatsya, B., 1986, Scope and potential of pepper (*Piper nigrum* L.) cultivation in India. In: *Plantation Crops - Opportunities and Constraints, Vol. I,* (H.C. Srivastava, B. Vatsya and K.K.G. Menon, eds.), pp. 269-273, Oxford & IBH Publ. Co., New Delhi.

Doi, Y., 1981, Frequency of root differentiation in anther callus of tea, *Study of Tea,* 60: 1-3.

Doreswamy, R., Srinivasa Rao, N.K. and Chacko, E.K., 1983, Tissue culture propagation of banana, *Sci. Hortic.,* 18: 247-252.

D'Souza, L., 1982, Organogenesis in coconut embryo callus. In: *Plant Tissue Culture 1982* (A. Fujiwara, ed.), pp. 179-180, Jpn. Assoc. Plant Tissue Cult., Tokyo.

Esan, E.B., 1973, A detailed study of adventive embryogenesis in the Rutaceae, Ph.D. Thesis, Riverside, Univ. Calif., USA.

Euvens, C.J., 1976, Mineral requirements for growth and callus initiation of tissue explants excised from mature coconut palms *(Cocos nucifera)* and cultured *in vitro, Physiol. Plant.,* 36: 23-38.

Euvens, C.J., 1978, Effects of organic nutrients and hormones on growth and development of tissue explants from coconut *(Cocos nucifera)* and date *(Phoenix dactylifera)* palms cultured *in vitro, Physiol. Plant.,* 42: 173-178.

Euvens, C.J. and Blake, J., 1977, Culture of coconut and date palm tissue with a view to vegetative propagation, *Acta Hortic.,* 78: 277-286.

George, C.M., 1986, Role of rubber plantation industry in socio-economic development. In: *Plantation Crops - Opportunities and Constraints Vol.II,* (H.C. Srivastava, B. Vatsya and K.K.G. Menon, eds.), pp. 227-232, Oxford & IBH Publ. Co., New Delhi.

George, E.F. and Sherrington, P.D., 1984, Plantation Crops. In: *Plant Propagation by Tissue Culture, Handbook and Directory of Commercial Laboratories,* pp. 470-471, Exegetics Ltd., Hants.

Gupta, P.K., Iyer, R. and Mascarenhas, A.F., 1981, Tissue culture of fruit trees: Rapid clonal multiplication of *Punica granatum* Linn. (Pomegranate) from mature trees by tissue culture, Abs. In: *The 6th All India Plant Tissue Culture Conference,* Univ. Pune.

Gupta, P.K., Kendurkar, S.V., Kulkarni, V.M., Shirgurkar, M.V. and Mascarenhas, A.F., 1984, Somatic embryogenesis and plants from zygotic embryos of coconut (*Cocos nucifera* L.) *in vitro, Plant Cell Rep.,* 3: 222-225.

Hall, T.R.H. and Collin, H.A., 1975, Initiation and growth of tissue cultures of *Theobroma cacao, Ann. Bot. (Lond.),* 39: 555-570.

Hosaki, T. and Sagawa, Y., 1977, Clonal propagation of ginger (*Zingiber officinale* Roscoe) through tissue culture, *Hortscience,* 12: 451-452.

Hu, H. and Hao, S., 1980, The present status of investigations of plant tissue and cell culture in China. In: *Plant Cell Cultures - Results and Perspectives* (F. Sala, B. Parisi, R. Cella and O. Ciferri, eds.), pp. 93, Elsevier/North-Holland Biomedical Press, Amsterdam.

Jagdeesan, M. and Padmanabhan, D., 1982, Induction of rooting in cotyledon callus of coconut, *Curr. Sci.,* 51: 567.

Jones, L.H., 1983, The oil palm and its clonal propagation by tissue culture, *Biologist,* 30: 181-188.

Jordan, M., Cortes, I. and Montenegro, G., 1983, Regeneration of plantlets by embryogenesis from callus cultures of *Carica candamarcensis, Plant Sci. Lett.,* 28: 321-326.

Kato, M., 1982, Results of organ culture in *Camellia japonica* and *C. sinensis, Jpn. J. Breed.,* 32: 276-277.

Kato, M., 1985, Regeneration of plantlets from tea stem callus, *Jpn. J. Breed.,* 35: 317-322.

Keller, H., Wanner, H. and Baumann, T.W., 1972, Kaffein synthese in fruchten und gewebekulturen von *Coffea arabica, Planta,* 108: 339-350.

Khuspe, S.S., Hendre, R.R., Mascarenhas, A.F., Jagannathan, V., Thombre, M.V. and Joshi, A.B., 1980, Utilization of tissue culture to isolate interspecific hybrids in *Carica.* In: *Plant Tissue Culture, Genetic Manipulation and Somatic Hybridization of Plant Cells* (P.S. Rao, M.R. Heble and M.S. Chadha, eds.), pp. 198-205, BARC, Department of Atomic Energy, Bombay.

Kononowicz, H., Kononowicz, A.K. and Janick, J., 1984, Asexual embryogenesis via callus of *Theobroma cacao* L., *Z. Pflanzenphysiol.,* 113: 347-358.

Kumar, K.B., Prakash Kumar, P., Balachandran, S.M. and Iyer, R.D., 1985, Development of clonal plantlets from immature panicles of cardamom, *J. Plant. Crops,* 13: 31-34.

Lanaud, C. and Parvis, J-P., 1980, Observations, avant mise en culture, des divisions anormales des noyaux de grains de pollen de cafeier induits a froid, influence du stade de development lors, de l'induction, *Cafe Cacao The,* 24: 305-312.

Lee, S.K. and Rao, A.N., 1982, Induction of callus and organogenesis in cocoa tissues. In: *Tissue Culture of Economically Important Plants* (A.N. Rao, ed.), pp. 107-112, COSTED and ANBS, Nat. Univ. Singapore.

Litz, R.E., 1984, *In vitro* somatic embryogenesis from nucellar callus of monoembryonic *Mangifera indica* L., *Hortscience.,* 19: 715-717.

Litz, R.E., 1985, Somatic embryogenesis in tropical fruit trees. In: *Tissue Culture in Forestry and Agriculture,* (R.R. Henke, K.V. Hughes, M.J. Constantin and A. Hollaender, eds.), pp. 179-193, Plenum Press, New York.

Litz, R.E. and Conover, R.A., 1977, Tissue culture propagation of papaya, *Proc. Florida. State Hortic. Soc.,* 90: 245-246.

Litz, R.E., Knight, R.J. and Gazit, S., 1984, *In vitro* somatic embryogenesis from *Mangifera indica* L. callus, *Sci. Hortic.,* 22: 233-240.

Litz, R.E., Moore, G.A. and Srinivasan, C., 1985, *In vitro* systems for propagation and improvement of tropical fruits and palms, *Hortic. Rev.,* 7: 157-200.

Ma, S., Shii, C. and Wang, S.O., 1978, Regeneration of banana plants from shoot meristem tips and inflorescence sections *in vitro.* In: *Abstracts XXth International Horticultural Congress,* No. 1639, Sydney.

Mathews H.V. and Rao, P.S., 1984, *In vitro* response of black pepper *(Piper nigrum),* Curr. Sci., 53: 183-186.

Menon, K.K.G., 1986, Plantation opportunities - Need for long term planning. In: *Plantation Crops, Opportunities and Constraints,* Vol.I (H.C. Srivastava, B. Vatsya and K.K.G. Menon, eds.), pp. 11-28, Oxford & IBH Publ. Co., New Delhi.

Monaco, L.C., Sondahl, M.R., Carvalho, A., Crocomo, O.J. and Sharp, W.R., 1977, Applications of tissue culture in the improvement of coffee. In: *Applied and Fundamental Aspects of Plant Cell, Tissue and Organ Culture* (J. Reinert and Y.P.S. Bajaj, eds.), pp. 102-109, Springer-Verlag, Berlin.

Monfort, S., 1985, Androgenesis of coconut: Embryos from anther culture, Z. *Pflanzenzucht.,* 94: 251-254.

Muliyar, M.K., 1983, Transfer of technology in plantation crops, *J. Plantation Crops,* 11: 1-12.

Nadgauda, R.S., Kulkarni, D.D., Mascarenhas, A.F. and Jagannathan, V., 1980, Development of plants from cultured tissues of ginger *(Zingiber officinale* Roscoe). In: *Plant Tissue Culture, Genetic Manipulation and Somatic Hybridization of Plant Cells* (P.S. Rao, M.R. Heble and M.S. Chadha, eds.), pp. 358-365, BARC, Department of Atomic Energy, Bombay.

Nadgauda, R.S., Mascarenhas, A.F., Hendre, R.R. and Jagannathan, V., 1978, Rapid multiplication of turmeric *(Curcuma longa* Linn.) plants by tissue culture, *Indian J. Exp. Biol.,* 16: 120-122.

Nadgauda, R.S., Mascarenhas, A.F. and Madhusoodhanan, K.J., 1983, Clonal multiplication of cardamom *(Elettaria cardamomum* Maton) by tissue culture, *J. Plantation Crops,* 11: 60-64.

Nagai, H., 1974, *In vitro* culture of hybrid embryos of papaya and squash. In: *3rd Intern. Congr. Plant Tissue and Cell Culture,* Abst. 118, Univ. Leicester, Leicester.

Nair, M.K., 1982, In: *Proc. Nat. Seminar on Ginger and Turmeric,* Central Plantation Crops Research Institute, Kasaragod.

Ninan, C.A., Mohankumar, P. and Thomas, J., 1983, Tissue culture studies in coconut, cashewnut and tapioca, *Abstracts of Contributed Papers Part I,* Abstr. No. 40, XV International Congress of Genetics, pp. 418, Oxford & IBH Publ. Co., New Delhi.

Orchard, J.E., Collin, H.A. and Hardwick, K., 1979, Culture of shoot apices of *Theobroma cacao, Physiol. Plant.,* 47: 207-210.

Pannetier, C. and Bufford-Morel, J., 1982, Production of somatic embryos from leaf tissue of coconut *(Cocos nucifera* L.). In: *Plant Tissue Culture 1982* (A. Fujiwara, ed.), pp. 755-756, Jpn. Assoc. Plant Tissue Cult., Tokyo.

Paranjothy, K., 1974, Induced root and embryoid differentiation in *Hevea* tissue cultures, *3rd Int. Cong. Plant Tissue and Cell Culture.* Abst. No. 67, Univ. of Leicester, Leicester.

Pence, V.C., Hasegawa, P.M. and Janick, J., 1979, Asexual embryogenesis in *Theobroma cacao, J. Am. Soc. Hortic. Sci.,* 104: 145-148.

Pence, V.C., Hasegawa, P.M. and Janick, J., 1980, Initiation and development of asexual embryos of *Theobroma cacao* L. *in vitro, Z. Pflanzenphysiol.,* 68: 1-14.

Phukan Mina, K. and Mitra, G.C., 1984, Regeneration of tea shoots from nodal explants in tissue culture, *Curr. Sci.,* 53: 874-876.

Prior, C., 1977, Growth of *Oncobasidium theobromae* in dual culture with callus tissue of *Theobroma cacao, J. Gen. Microbiol.,* 99: 219-222.

Purohit, A.G., 1986, Pomegranate. In: *Plantation Crops - Opportunities and Constraints, Vol.II* (H.C. Srivastava, B. Vatsya and K.K.G. Menon, eds.), pp. 273-281, Oxford & IBH Publ. Co., New Delhi.

Rabechault, H. and Martin, J.P., 1976, Multiplication vegetative du palmier a huile (*Elaeis guineensis* Jacq.) a l'aide de cultures de tissues foliares, *C.R. Acad. Sci.,* 283: 1735-1737.

Raina, S.K. and Iyer, R.D., 1983, Multicelled pollen pro-embryoids and callus formation in tea anther cultures, *J. Plant. Crops* (Suppl.) 11: 63-67.

Rajeevan, M.S. and Pandey, R.M., 1983, Propagation of papaya through tissue culture, *Acta Hortic.,* 131: 131-139.

Rajeevan, M.S. and Pandey, R.M., 1986, Economics of mass propagation of papaya through tissue culture. In: *Plant Tissue Culture and its Agricultural Applications* (L.A. Withers and P.G. Alderson, eds.), pp. 211-215, Butterworths, London.

Raju, C.R., Prakash Kumar, P., Chandramohan, M. and Iyer, R.D., 1984, Coconut plantlets from leaf tissue cultures, *J. Plantation Crops,* 12: 75-91.

Rangan, T.S., 1982, Ovary, ovule and nucellus culture. In: *Experimental Embryology of Vascular Plants* (B.M. Johri, ed.), pp. 105-129, Springer-Verlag, Heidelberg.

Rangan, T.S., Murashige, T. and Bitters, W.P., 1968, *In vitro* initiation of nucellar embryos in monoembryonic *Citrus, Hortscience,* 3: 226-227.

Rangaswamy, N.S., 1959, Morphogenetic response of *Citrus* ovules to growth adjuvants in culture, *Nature (Lond.),* 183: 735-736.

Rangaswamy, N.S., 1961, Experimental studies on female reproductive structures of *Citrus microcarpa* Bunge., *Phytomorphology,* 11: 109-127.

Rao, A.N. and Lee, S.K., 1986, An overview of the *in vitro* propagation of woody plants and plantation crops. In: *Plant Tissue Culture and its Agricultural Applications* (L.A. Withers and P.G. Alderson, eds.), pp. 123-138, Butterworths, London.

Rao, A.N., Sin, Y.M., Kothagoda, N. and Hutchinson, J., 1982, Cotyledon culture of some tropical fruits. In: *Tissue Culture of Economically Important Plants* (A.N. Rao, ed.), pp. 124-137, COSTED and ANBS, Nat. Univ. Singapore.

Reuveni, O., Adato, Y. and Lilien-Kipnis, H., 1972, A study of new and rapid methods for vegetative propagation of date palms, pp. 17-24. In: 49th Ann. Rep. Date Growers' Institute. Indio. California

Sabharwal, P.S., 1963, *In vitro* culture of ovules, nucelli and embryos of *Citrus reticulata* Blanco var. Nagpuri. In: *Plant Tissue and Organ Culture - A Symp. Intn. Soc. Plant Morphologists,* Univ. Delhi, Delhi, pp. 265-274.

Sajise, J.U. and De Guzman, E.V., 1972, Formation of adventitious roots in coconut macapuno seedlings grown in medium supplemented with naphthalene acetic acid, *Kalikasan Philip. J. Biol.,* 1: 197-206.

Sharma, D.R. and Chowdhury, J.B., 1986, Date palm (*Phoenix dactylifera* L.) tissue culture - Future propagation and research needs. In: *Plantation Crops - Opportunities and Constraints, Vol. I* (H.C. Srivastava, B. Vatsya and K.K.G. Menon, eds.), pp. 207-213, Oxford & IBH Publ. Co., New Delhi.

Sharma, D.R., Dawra, S. and Chowdhury, J.B., 1984, Somatic embryogenesis and plant regeneration in date palm (*Phoenix dactylifera* Linn.) cv. Khadravi through tissue culture, *Indian J. Exp. Biol.,* 22: 596-598.

Sharma, V.S. and Ranganathan, V., 1985, The world of tea today. *Outlook on Agric.,* 14: 35-40.

Sharp, W.R., Caldas, L.S., Crocomo, O.J., Monaco, L.C. and Carvalho, A., 1973, Production of *Coffea arabica* callus of three ploidy levels and subsequent morphogenesis, *Phyton (B. Aores)*, 31: 67-74.

Sondahl, M.R. and Sharp, W.R., 1977, High frequency induction of somatic embryos in cultured leaf explants of *Coffea arabica*, L., *Z. Pflanzenphysiol.*, 81: 395-408.

Sondahl, M.R. and Sharp, W.R., 1979, Research in coffee species and application of tissue culture methods. In: *Plant Cell and Tissue Culture - Principles and Applications* (W.R. Sharp, P.O. Larsen, E.F. Paddock and Raghavan, V., eds.), pp. 527-584, Ohio State Univ. Press, Columbus.

Sondahl, M.R., Evans, D.A. and Sharp, W.R., 1980, Coffee cell culture, *I.A.P.T.C. Newslett.*, 30: 2-7.

Spiegel-Roy, P. and Kochba, J., 1980, Embryogenesis in Citrus tissue cultures. In: *Advances in Biochemical Engineering* (A. Fiechter, ed.), pp. 27-48, Springer-Verlag, Heidelberg.

Srinivasa Rao, N.K., Narayanaswamy, S., Chacko, E.K. and Doreswamy, R., 1982, Regeneration of plantlets from callus of *Elettaria cardamomum* Maton. In: *Proc. Indian Acad. Sci., Plant Sci.* 91: 37-41.

Staritsky, G., 1970, Embryoid formation in callus tissues of coffee, *Acta. Bot. Neerl.*, 19: 509-514.

Tisserat, B., 1979, Tissue culture of the date palm, *J. Hered.*, 70: 221-222.

Thanh-Tuyen, N.T. and De Guzman, E.V., 1983, Formation of pollen embryos in cultured anthers of coconut (*Cocos nucifera* L.), *Plant Sci. Lett.*, 29: 81-88.

Thomas, V. and Rao, P.S., 1985, *In vitro* propagation of oil palm (*Elaeis guineensis* Jacq. var. *Tenera*) through somatic embryogenesis in leaf-derived callus, *Curr. Sci.*, 54: 184-185.

Toruan N.L. and Suryatmana, N., 1977, Kultur jaringan *Hevea brasiliensis*, Muell. Arg, *Menara Perkebunan*, 45: 17-21.

Tsai, C.H. and Kinsella, J.E., 1981, Initiation and growth of callus and cell suspensions of *Theobroma cacao* L., *Ann. Bot. (Lond.)*, 48: 549-558.

Vuylsteke, D. and De Langhe, E., 1985, Feasibility of *in vitro* propagation of bananas and plantains, *Trop. Agric. (Trin.)*, 62: 323-328.

Wang, Z., Zeng, X., Chen, C., Wu, H., Li, Q., Fan, G. and Lu, W., 1980, Induction of rubber plantlets from anthers of *Hevea brasiliensis* Muell. Arg. *in vitro.*, *Chinese J. Tropic. Crops*, 1: 16-26.

Wilson, H.M. and Street, H.E., 1975, The growth, anatomy and morphogenetic potential of callus and cell suspension culture of *Hevea brasiliensis*, *Ann. Bot. (Lond.)*, 39: 671-682.

Wilson, H.M., Eisa, M.Z. and Irwin, S.W.B., 1976, The effects of agitated liquid medium on *in vitro* culture of *Hevea brasiliensis*, *Physiol. Plant.*, 36: 399-402.

Wu, C., Huang, T., Chen, G. and Chen, S., 1982, A review on the tissue culture of economically important plants. In: *Tissue Culture of Economically Important Plants* (A.N. Rao, ed.), pp. 104-106, COSTED and ANBS, Nat. Univ. Singapore.

Yie, S.T. and Liaw, S.I., 1977, Plant regeneration from shoot tips and callus of papaya, *In Vitro*, 13: 564-568.

9

IN VITRO STRATEGIES FOR TROPICAL FRUIT TREE IMPROVEMENT

Richard E. Litz

ABSTRACT

Vegetatively propagated tropical fruit trees can be highly sensitive to plant disease epidemics. These trees are usually very heterozygous and have long generation cycles (e.g., 7-20 years). Some tropical fruit species such as mangosteen, which is obligately apomictic (nucellar embryony), are recalcitrant to conventional plant breeding approaches. Strategies for the recovery of horticulturally useful plants from cell and tissue cultures have great potential for the improvement of tropical fruit species. *De novo* regeneration pathways have been described from explants of selected mature trees of perennial tropical fruit species in Anacardiaceae, Euphorbiaceae, Moraceae, Myrtaceae, Oxalidaceae, Rosaceae, Rubiaceae, Rutaceae, Sapindaceae and

Richard E. Litz * Tropical Research and Education Center, University of Florida, 18905 S.W. 280 St., Homestead, Florida, 33031 U.S.A.

Florida Agricultural Experimental Station Journal Series No. 8677.

Sterculiaceae. The regeneration of tropical fruit trees and the application of the approaches for cultivar improvement are the subjects of this discussion.

INTRODUCTION

Perennial tropical fruit species are important in the Third World as additional sources of revenue through export and as important diet supplements. However, there has been scarcely any conscious effort directed towards cultivar improvement through classical plant breeding. Consequently, outstanding fruit trees that have occurred by chance have been vegetatively propagated and maintained in orchards of the same genotype. In the ancient civilisations of south and south-east Asia, some of these tree selections are several hundred or more than one thousand years old. Vegetative propagation of these selections has preserved the unique genetic composition that confers outstanding horticultural characteristics.

The limitations of monoculture in tropical agriculture have become painfully evident during the past century, with successive disease epidemics affecting the production of banana and plantain *(Musa)*, cacao *(Theobroma cacao)*, coffee *(Coffea arabica)*, papaya *(Carica papaya)* and coconut *(Cocos nucifera)* in the neotropics. In response to these challenges, the genetic improvement of tropical fruit trees by conventional plant breeding approaches has been confounded by the long generation cycle of these plants, the heterogeneity of superior tree selections, the absence of continuing breeding and genetic studies and the inaccessibility or inadequacy of germplasm repositories.

There are certain common cultivar improvement or breeding objectives that can be designated for many tropical fruit tree species. Among these, disease and pest resistance have traditionally held high priority. The control of virus diseases of *Citrus* has been accomplished by the use of micro-grafting. (Navarro, 1981); however, the control of fungal and bacterial pathogens has continued to be dependent on chemical applications or more drastic measures such as quarantine and plant destruction, in order to prevent the spread of threatening pathogenic micro-organisms. The occurrence and spread of crop-threatening diseases being extremely rapid, conventional breeding strategies are inadequate to meet the challenge of these epidemics involving tropical fruit trees. Consequently, highly esteemed tropical fruit cultivars that lack the necessary genetic protection against pathogens are usually removed from cultivation and are replaced by horticulturally inferior selections.

Other breeding goals must include the development of improved rootstocks that can tolerate soil-borne fungi, saline conditions and/or those that will confer a dwarf habit to tropical fruit trees, many of which originated in the tropical rain forests as tall canopy trees. Many other breeding imperatives remain, e.g., overcoming alternate bearing, improving fruit quality, extending the geographical range by selection for cold tolerance, etc. However, these objectives cannot be achieved readily by using either traditional approaches or currently available biotechnology strategies. It is anticipated that procedures for producing transgenic plants will be greatly improved in the next few years, together with identification and cloning of horticulturally useful genes.

The development and application of *in vitro* strategies for tropical fruit tree cultivar improvement has been reviewed previously by Litz et al. (1985) and by Litz (1985; 1987). The potential use of cell and tissue culture techniques as integral components in tropical fruit tree breeding certainly has great appeal. The ability to manipulate the genome of well-established fruit tree cultivars, and to direct the breeding effort toward specific goals such as disease or pest resistance, would obviate the sexual process and the long (7-20 years) juvenile cycles that would ensue. The basic nature of the cultivar could be maintained, although certain traits could be altered in a discrete manner. The use of *in vitro* approaches for tropical fruit cultivar improvement is, however, predicated on the assumption that regeneration from somatic tissues of the mature tree is possible.

PERMISSIVE PATTERNS OF REGENERATION

Regeneration from callus derived from somatic tissues of mature tropical fruit trees has, in fact, been the most limiting factor in the application of biotechnology to these species. The morphogenetic potential of explants from juvenile or embryonic tissues of angiosperm trees has been well-characterised. However, until recently, relatively few tree species have been regenerated from cell or callus cultures derived from mature trees. Ironically, somatic embryogenesis from cultured ovules of polyembryonic *Citrus* sp. was described some thirty years ago by Stevenson (1956). In subsequent studies by Maheshwari and Rangaswamy (1958), Rangaswamy, (1961) and Sabharwal (1962), this regeneration pathway was described in greater detail. Somatic embryogenesis or embryogenic pseudobulbil callus appeared to have developed directly from adventitious embryos already present in the nucellar explant. This pattern of direct somatic embryogenesis from nucellar explants was confirmed by Rangan et al. (1968) with monoembryonic *Citrus* sp. Button et al. (1974) showed that the *Citrus* pseudobulbil callus was entirely composed of small globular proembryos at various stages of development. Hence, a true unorganised embryogenic *Citrus* callus may not occur, and even if it does, it will have a brief existence. Other studies have indicated that, unlike somatic embryogenesis in *Daucus carota*, the process involving *Citrus* nucellar tissue is not dependent on the presence of an auxin in the medium (Murashige and Tucker, 1969; Kochba and Spiegel-Roy, 1977; Tisserat and Murashige, 1977). Although the regeneration of *Citrus* by somatic embryogenesis from nucellar tissues has been well-characterised for many years, it did not serve as a model for other tropical trees until recently. Tropical tree species have been regarded to be somewhat recalcitrant *in vitro*. Using approaches adapted from previously reported *Citrus in vitro* studies, several tropical fruit tree species representing several plant families have been regenerated from tissue cultures derived from cultured nucelli (Table 1). Somatic

Table 1: Permissive somatic embryogenesis from the nucellus of tropical fruit trees

Species	Common name	Family	Reference
Citrus sp.	citrus	Rutaceae	Stevenson (1956)
Eriobotrya japonica	loquat	Rosaceae	Litz (1985)
Eugenia sp.	rose and malay apples	Myrtaceae	Litz (1984b)
Mangifera indica	mango	Anacardiaceae	Litz et al. (1982)
Myrciaria cauliflora	jaboticaba	Myrtaceae	Litz (1984c)

embryogenesis in all of these species is not dependent on exogenous growth regulators and occurs directly from the nucellar explant or by secondary budding from globular proembryos.

Nucellar Cells of many species evidently possess the ability to produce somatic embryos without the influence of an external stimulus. This regeneration pathway has been described to involve pre-embryogenic determined (PED) cells (Sharp et al., 1980). The PED cells only require release from the inhibitory environment of the ovule to become embryogenic. Ammirato (1987) has defined somatic embryogenesis from PED cells as being permissive, because the nucellar cells are already fully competent to produce somatic embryos.

Is the nucellus the explant of choice for regenerating tropical fruit trees? Permissive somatic embryogenesis from the nucellus of tropical trees is generally restricted to species with relatively large seeds, genera in which the incidence of nucellar polyembryony is notable and species within a narrow range of plant families. It is probable that many tropical fruit tree species could be regenerated by this pathway. Ultimately, this must be determined for each species and cannot be predicted.

Cultivar or genotype can influence somatic embryogenesis from the nucellus. This has been observed not only in citrus (Moore, 1986), but also in mango (Litz, 1984a; 1987). There are significant differences in the responses among cultivars and between polyembryonic and monoembryonic cultivars of the same species. Even within a cultivar, there are different morphogenetic responses from cultured nucelli that are dependent on the stage of development of the ovule at the time of explanting (Fig. 1). In mango, the optimum stage of development of the ovule for establishing embryogenic cultures corresponds to that period during which the embryo mass occupies approximately one half of the embryo sac. This period also corresponds with high endogenous concentrations of the polyamines (putrescine and spermidine) in the nucellus (Litz, 1987). The interaction between the growth medium and the nucellar explant is also very important. Although somatic embryogenesis can occur directly from the excised nucellus on medium without growth regulators, the presence of 2,4-D can stimulate the process of regeneration, presumably by cloning the PED cells (Litz, 1987).

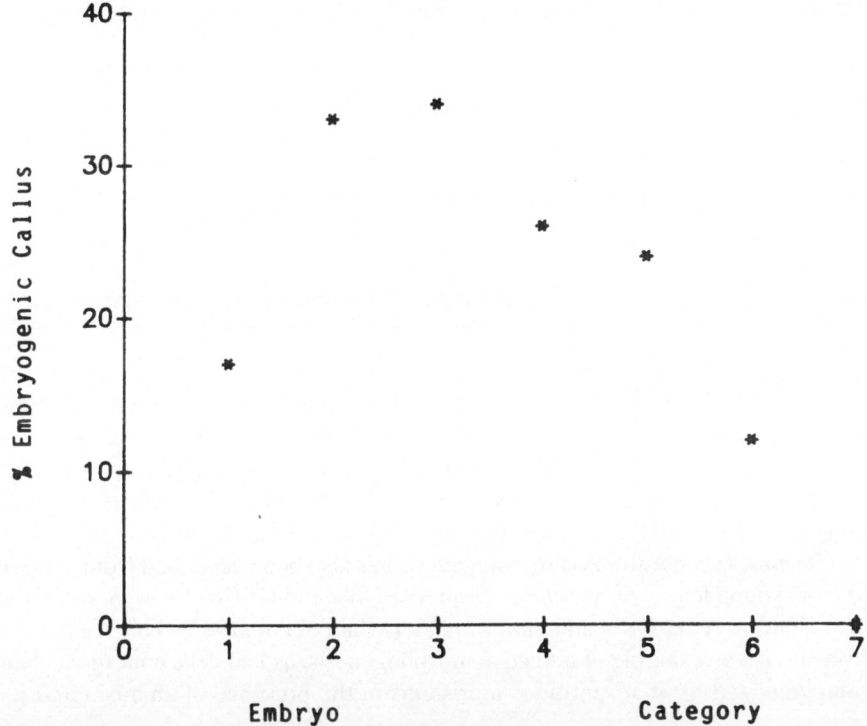

Fig. 1: Effect of ovule development on somatic embryogenesis from nucellar explants of mango. Embryo categories refer to the ratio of embryo length: ovule length. Category 1-ratio 0.15; Category 2-0.2; Category 3-0.4; Category 4-0.6; Category 5-0.8; Category 6: 1.0.

INDUCTIVE PATTERNS OF REGENERATION

The *de novo* regeneration of tropical fruit tree species from tissues other than the nucellus has only recently been described (Table 2). Young leaves in fresh vegetative flushes of certain fruit species can be induced to produce an unorganised regenerative callus on growth media containing 2,4-D and a cytokinin.

The induction of embryogenic callus in *Euphoria longan* is strongly cytokinin-dependent; regeneration can only occur if kinetin is present in the induction medium. Similar conditions promote the formation of embryogenic callus in the tropical forest trees *Sapindus trifoliatus* (Desai et al., 1986) and in *Coffea arabica* (Sondahl and Sharp, 1977). Interestingly, the media that induce morphogenesis in leaf callus of *E. longan* and *S. trifoliatus* have no such effect on leaf explants of the closely related *Litchi chinensis* (Litz, unpublished data). Genotype, in addition to medium, consequently has an important role in realising the morphogenetic potential of cultured leaves of tree species in the Sapindaceae.

Table 2 : Inductive Morphogenesis from somatic tissues of mature tropical fruit trees

Species	Common name	Family	Regeneration pathway	Reference
Averrhoa carambola	carambola	Oxalidaceae	organogenesis	Litz & Griffis (1988)
Euphoria longan	longan	Sapindaceae	somatic embryogenesis	Litz (1988)
Morus indica	mulberry	Moraceae	organogenesis	Mhatre et al. (1985)
Solanum quitoense	lulo	Solanaceae	organogenesis	Hendrix et al. (1987)

Regeneration from leaf callus of *E. longan* is consistent with the description of inductive somatic embryogenesis (Ammirato, 1987) because a change in competency of leaf cells has occurred in response to the stimulus of an auxin and a cytokinin. This regeneration pathway is distinct from that described for nucellar explants. Implicit in this response is the induction of a subculturable, unorganised callus.

De novo regeneration via organogenesis has also been described from callus derived from young leaves of *Averrhoa carambola* (Litz and Griffis, 1988), *Solanum quitoense* (Hendrix et al., 1987) and *Morus indica* (Mhatre et al., 1985). This regeneration pathway is also an example of inductive morphogenesis, as leaf cells undergo a change of competency to form adventitious meristems in the presence of an auxin and a cytokinin.

APPLICATION OF *IN VITRO* SYSTEMS

With the exception of *Citrus* spp., there have been relatively few attempts to utilise *in vitro* systems to address important cultivar improvement problems. This is largely due to the fact that regeneration protocols have been described only recently. However, there are also considerable problems associated with stimulating of normal maturation and germination of somatic embryos of tropical fruit tree species having large seeds. Mature somatic embryos of mango can attain length of 4.0 cm at the time of germination. *In vitro* conditions for controlling maturation and germination of somatic embryos in mango (DeWald, 1987) and longan (Litz, 1988) have recently been described. It is necessary to prevent precocious germination of these somatic embryos by altering the physical conditions required for growth and by manipulating the growth medium during maturation (DeWald, 1987; Litz, 1988; Fig. 2).

Research involving *Citrus* has attempted to address certain breeding problems using two approaches:

* *in vitro* selection of somaclonal variants for salt tolerance, 2,4-D resistance and resistance to *mal secco* toxin; and
* protoplast fusion between *Citrus* sp. and related genera in order to produce disease resistant, frost resistant, dwarfing rootstocks.

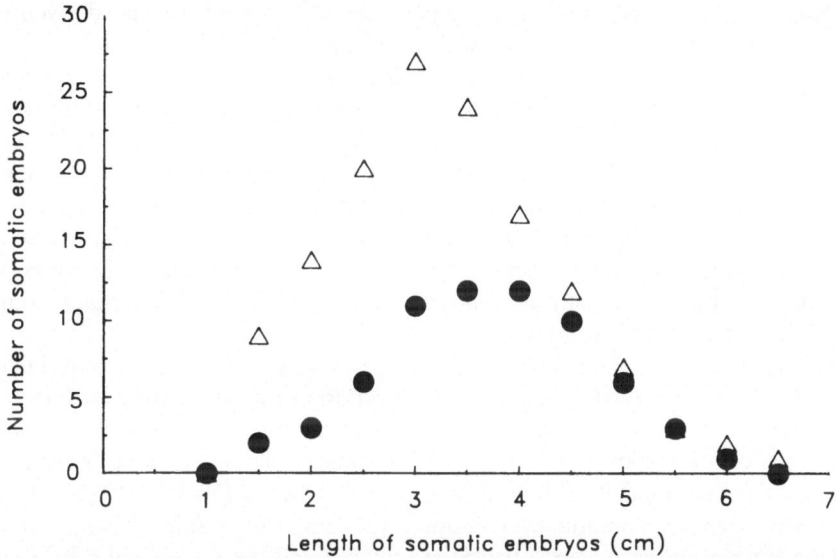

Fig. 2: Relationship between precocious maturation of mango somatic embryos and abnormal development. Somatic embryos with roots and shoots; somatic embryos with roots only. (Sample size = 202 somatic embryos).

Exploiting the genetic instability that is known to exist in cell and tissue cultures, i.e., somaclonal variation, Kochba et al. (1980; 1982) were able to recover embryogenic cell lines of 'Shamouti' and sour orange calli that were apparently resistant to sodium chloride concentrations up to 10.0 gl^{-1}. None of the resistant lines, however, could be stimulated to produce embryos. A similar approach was adopted to select for heightened resistance to the herbicide 2,4-D in callus of 'Shamouti' orange (Kochba et al., 1980; Speigel-Roy et al., 1983). Callus with resistance to 10M 2,4-D was selected; however, plants were not regenerated.

The recovery of disease resistant tropical fruit trees derived from somaclonal variants in the selection medium has been a high priority. Efforts have been made to select for specific disease resistant *Citrus* by exposing embryogenic callus to the toxin associated with *mal secco* disease (Nachmias et al., 1977). Another approach has been adopted for mango. *Anthracnose,* a fruit and foliage disease of mango, is caused by the fungus *Colletotrichum gloeosporiodes.* The disease is particularly severe in the humid tropics. The monoembryonic Indian and Florida cultivars are generally highly susceptible to *anthracnose,* whereas the polyembryonic south-east Asian cultivars have considerable resistance to this pathogen. It is evident that resistance to *anthracnose* is present within the species, but transfer of resistance to monoembryonic mango cultivars with superior horticultural quality has not been undertaken. The infection of plant tissues by *Colletotrichum* sp. is accompanied by the release of a polysaccharide elicitor from the fungal walls, which triggers a hypersensitive response within the host tissue (Anderson-Prouty and Albersheim, 1975). The hypersensitive response is characterised by the accumulation of phytoalexins in the tissue and is always accompanied by the browning of the tissue. Partially purified extracts derived from culture filtrates of *Colletotrichum* sp. will also elicit the same response. It has recently been demonstrated that the fungal culture filtrate can be incorporated into the tissue culture medium, in which it can trigger a similar hypersensitive response as reported in cell cultures of the forage legume *Stylosanthes* sp. (Lopez et al., 1987). Plants regenerated from the cells that survived this selection pressure were found to be more resistant to *anthracnose* (Lopez et al., 1987). Preliminary studies in which mango callus was exposed to the partially purified culture filtrate of *Colletotrichum gloeosporiodes* have demonstrated that a hypersensitive response is elicited within 48-72 hours (Litz, unpublished). It is hoped that this approach can be utilised for cultivar improvement among the monoembryonic mangos.

The application of protoplast technology to tropical fruit trees has only been reported for *Citrus* sp. Vardi et al. (1975) first reported the successful recovery of somatic embryos from protoplasts derived from *Citrus* nucellar callus. In subsequent studies, the conditions for improving *Citrus* protoplast plating efficiency were greatly improved and the range of cultivar responses was measured (Vardi, 1981; Vardi et al., 1982). Inter-generic somatic hybridisation between *Citrus sinensis* and *Poncirus trifoliata* (Ohgawara et al., 1985) has been achieved and the parasexual hybrid plants have been regenerated. Using similar protocols, Grosser et al. (1988a,b) have created artificial hybrids between several *Citrus* spp. and species of other related genera, thereby anticipating the development of suitable germplasm for use as dwarfing rootstocks.

CONCLUSION

Considerable progress has been made in overcoming the problems associated with morphogenesis in callus derived from explants from mature tropical fruit trees. This should facilitate the application of modern genetics to the improvement of this important group of plants.

ACKNOWLDEGEMENTS

The author acknowledges with gratitude the assistance of Ms Roe C. Hendrix and Callie Sullivan.

REFERENCES

Ammirato, P.V., 1987, Organization events during somatic embryogenesis. In: *Plant Tissue and Cell Culture* (C.E. Green, D.A. Somers, W.P. Hackett and D.D. Biesboer, eds.), pp. 57-82, Alan R. Liss, New York.

Anderson-Prouty, A.J. and Albersheim, P., 1975, Host-pathogen interactions, VIII. Isolation of a pathogen-synthesised fraction rich in glucan that elicits a defense response in the pathogen's host, *Plant Physiol.,* 56: 286-291.

Button, J., Kochba, J. and Bornman, C.H., 1974, Fine structure of embryoid development from embryogenic ovular callus of 'Shamouti' orange (*Citrus sinensis* Osb.), *J. Exp. Bot.,* 25: 446-457.

Desai, H.V., Bhatt, P.N. and Mehta, A.R., 1986, Plant regeneration of *Sapindus trifoliatus* L. (soapnut) through somatic embryogenesis, *Plant Cell Rep.,* 3: 190-191.

DeWald, S.G., 1987, *In vitro* somatic embryogenesis and plant regeneration from mango (*Mangifera indica* L.) nucellar callus, Ph.D. Dissertation, Univ. of Florida, Gainesville, USA.

Grosser, J.W., Gmitter, F.G. and Chandler, J.F., 1988a, Intergeneric somatic hybrid plants of *Citrus sinensis* cv. 'Hamlin' and *Poncirus trifoliata* cv. 'Flying Dragon', *Plant Cell Rep.* (in press).

Grosser, J.W., Gmitter, F.G. and Chandler, J.F., 1988b, Intergeneric somatic hybrid plants from sexually incompatible woody species, *Citrus sinensis* and *Severinia disticha, Theor. Appl. Genet.* (In Press).

Hendrix, R.C., Litz, R.E. and Kirchoff, B.K., 1987, *In vitro* organogenesis and plant regeneration from leaves of *Solanum candidum* Lindl., *S. quitoense* Lam. (naranjilla) and *S. sessiliflorum* Dunal, *Plant Cell Tissue Organ Cult.,* 11: 67-73.

Kochba, J. and Spiegel-Roy, P., 1977, The effects of auxins, cytokinins and inhibitors on embryogenesis in habituated ovular callus of the 'Shamouti' orange *(Citrus sinensis), Z. Pflanzenphysiol.,* 81: 283-288.

Kochba, J., Spiegel-Roy, P. and Saad, P., 1980, Selection for tolerance to sodium chloride (NaCl) and 2,4-dichlorophenoxyacetic acid (2,4-D) in ovular callus lines of *Citrus sinensis.* In: *Plant Cell Cultures: Results and Perspectives* (F. Sala, B. Parisi, R. Cella and O. Ciferri, eds.), pp. 187-192, Elsevier/North Holland Biomedical Press, Amsterdam.

Kochba, J., Ben-Hayyim, G., Spiegel-Roy, P., Saad, S. and Neumann, H., 1982, Effect of carbohydrates on somatic embryogenesis in sub-cultured nucellar callus of *Citrus* cultivars, *Z. Pflanzenphysiol.,* 105: 359-368.

Litz, R.E., 1984a, *In vitro* somatic embryogenesis from nucellar callus of monoembryonic *Mangifera indica* L., *Hortscience,* 19: 715-717.

Litz, R.E., 1984b, *In vitro* responses of adventitious embryos of two polyembryonic *Eugenia* species, *Hortscience,* 19: 720-722.

Litz, R.E., 1984c, *In vitro* somatic embryogenesis from callus of jaboticaba, *Myrciaria cauliflora, Hortscience,* 19: 62-64.

Litz, R.E., 1985, Somatic embryogenesis in tropical fruit trees. In: *Tissue Culture in Forestry and Agriculture* (R.R. Henke, K.W. Hughes, M.P. Constantin and A. Hollaender, eds.), pp. 179-193, Plenum Press New York.

Litz, R.E., 1987, Application of tissue culture to tropical fruits. In: *Plant Tissue and Cell Culture* (C.E. Green, D.A. Sommers, W.P. Hackett and D.D. Biesboer, eds.), pp. 407-418, Alan R. Liss, New York.

Litz, R.E., 1988, Somatic embryogenesis from cultured leaf explants of the tropical tree *Euphoria longan* Stend, *J. Plant Physiol.* (In Press).

Litz, R.E. and Griffis, J.L. Jr., 1988, Carambola (*Averrhoa carambola* L.). In: *Biotechnology in Agriculture and Forestry Trees II* (Y.P.S. Bajaj, ed.), Springer-Verlag, Heidelberg (In Press).

Litz, R.E., Knight, R.J. and S. Gazit, 1982, Somatic embryos from cultured ovules of polyembryonic *Mangifera indica* L., *Plant Cell Rep.,* 1: 264-266.

Litz, R.E., Moore, G.A., and Srinivasan, C., 1985, *In vitro* systems for propagation and improvement of tropical fruits and palms, *Hortic. Rev.,* 7: 157-200.

Lopez, P., Largnelet, A., Szabados, L., Roca, W.M. and Lenne, J., 1987, Effect of *Colletotrichum* culture filtrates on growth of *Stylosanthes* sp. cell cultures, In : *Proc. Intl. Cong. Plant Tissue Culture,* Colombia, pp. 52. Bogota (Abs.).

Maheshwari, P. and Rangaswamy, N.S., 1958, Polyembryony and *in vitro* culture of embryos of *Citrus* and *Mangifera, Indian J. Hortic.,* 15: 275-283.

Mhatre, M., Bapat, V.A. and Rao, P.S., 1985, Regeneration of plants from the culture of leaves and axillary buds in mulberry (*Morus indica* L.), *Plant Cell Rep.,* 4: 78-80.

Moore, G.A., 1986, Factors affecting *in vitro* embryogenesis from undeveloped ovules of mature *Citrus* fruit, *J. Am. Soc. Hortic. Sci.,* 110: 66-70.

Murashige, T. and Tucker, D.P.H., 1969, Growth factor requirements of citrus tissue culture. In: *Proc. 1st Int. Citrus Symp. Vol. 3* (H.D. Chapman, ed.), pp. 1115-1161, Univ. of California, Riverside.

Nachmias, A., Barash, I., Solel, Z. and Strobel, G.A., 1977, Translocation of *mal secco* toxin in lemons and its effect on electrolyte leakage, transpiration and citrus callus growth, *Phytoparasitica,* 5: 94-103.

Navarro, L., 1981, Citrus shoot tip grafting *in vitro* (STG) and its applications: a review. In: *Proc. Int. Soc. Citriculture,* pp. 452-456.

Ohgawara, T., Kobayashi, S., Ohgawara, E., Uchimija, H. and Ishii, S., 1985, Somatic hybrid plants obtained by protoplast fusion between *Citrus sinensis* and *Poncirus trifoliata., Theor. Appl. Genet.,* 71: 1-4.

Rangan, T.S., Murashige, T. and Bitters, W.P., 1968, *In vitro* initiation of nucellar embryos in monoembryonic *Citrus, Hortscience,* 3: 226-227.

Rangaswamy, N.S., 1961, Experimental studies on female reproductive structures of *Citrus microcarpa* Bunge, *Phytomorphology,* 11: 109-127.

Sabharwal, P.S., 1962, *In vitro* culture of nucelli and embryos of *Citrus aurantiifolia* Swingle. In: *Plant Embryology - A Symposium* (P. Maheshwari, ed.), pp. 239-243, CSIR, New Delhi.

Sharp, W.R., Sondahl, M.R., Caldas, L.S. and Maraffa, S.B., 1980, The physiology of *in vitro* asexual embryogenesis, *Hortic. Rev.,* 2: 268-310.

Sondahl, M.R. and Sharp, W.R., 1977, High frequency induction of somatic embryos in cultured leaf explants of *Coffea arabica* L., *Z. Pflanzenphysiol.,* 81: 395-408.

Spiegel-Roy, P., Kochba, J. and Saad, S., 1983, Selection for tolerance to 2,4-dichlorophenoxyacetic acid in ovular callus of orange *(Citrus sinensis), Z. Pflanzenphysiol.,* 109: 41-48.

Stevenson, F.F., 1956, The behaviour of *Citrus* tissues and embryos *in vitro,* Ph.D. Dissertation, Univ. of Michigan, Ann Arbor, Michigan.

Tisserat, B. and Murashige, T., 1977, Effects of ethephon, ethylene, and 2,4-dichlorophenoxy acetic acid on asexual embryogenesis *in vitro, Plant Physiol.,* 60: 437-439.

Vardi, A., 1981, Protoplast-derived plants from different *Citrus* species and cultivars. In: *Proc. Int. Soc. Citriculture,* (1981) 1 : 149-152

Vardi, A., Spiegel-Roy, P. and Galun, E., 1975, Citrus cell culture: Isolation of protoplasts, plating densities, effects of mutagens, and regeneration of embryos, *Plant Sci. Lett.,* 4: 231-236.

Vardi, A., Spiegel-Roy P. and Galun, E., 1982, Plant regeneration from *Citrus* protoplasts: Variability in methodological requirements among cultivars and species, *Theor. Appl. Genet.,* 62: 171-176.

10

THE CONTEXT AND STRATEGIES FOR TISSUE CULTURE OF DATE, AFRICAN OIL AND COCONUT PALMS

Abraham D. Krikorian

ABSTRACT

The benefits that can accrue from the application of aseptic culture techniques to palms such as African oil palm, date palm, coconut palm, etc. are considerable. Here, the challenge is, as in other instances, in developing methods to such a degree that they reflect a high level of reliability. In addition to effective tissue and cell culture methods, this means that sooner or later, *in vitro* derived or manipulated materials must be grown to maturity under field conditions, evaluated agronomically or horticulturally and shown to be acceptable. From that point on, predominantly economic considerations will determine the implementation of production programs. Information from preliminary field trials is just beginning to emerge. It is unclear

Abraham D. Krikorian * Department of Biochemistry, Division of Biological Sciences, State University of New York, Stony Brook, New York 11794-5215, U.S.A.

where the subject now stands and will remain so until more data accrue. An attempt is made to provide a brief overview of the range of available tissue culture techniques for palms and their potential for production. Special emphasis has been placed on culture strategies as they reflect cell biological principles. Suggestions are made for a rethinking of approaches towards achieving the required reliability.

INTRODUCTION

The aseptic culture of palms is of special significance because the Arecaceae represents a major family that has wide and far-reaching economic importance (Pesce, 1985; Uhl and Dransfield, 1987). Palm tissue culture is also of importance because it typifies an activity which has been among the earliest research attempts directed by industry-based or related research laboratories to develop and use tissue culture towards a practical end. It is frequently stated that plant biotechnology is the application of knowledge, tools and skills to solve practical problems and to extend human capabilities in the broad realm of plant science, horticulture, agriculture and forestry (Krikorian, 1988 and references cited therein). The pioneer work on the so-called coconut sport *makapuno* by Emerita de Guzman and her associates at the University of the Philippines at Los Banos (de Guzman and Del Rosario, 1964; de Guzman et al., 1983); African oil palm by the group headed by Henri Rabechault at ORSTOM (Office de la Recherche Scientifique et Technique Outre-Mer) particularly at the Institut de la Recherches pour les Huiles et Oleagineux (IRHO) in Bondy, France (Rabechault, 1962; Rabechault et al., 1970, 1972; Rabechault and Martin, 1976); the Wye College (University of London) group, especially on coconut (Blake, 1983) and the Unilever oil palm group at Colworth House, Bedford, England (Smith and Jones, 1970; Smith and Thomas, 1973; Jones, 1974a,b, 1984) and the work done by various groups on date palm, *Phoenix dactylifera* (Schroeder, 1970; Reuveni et al., 1972; Tisserat, 1979; Rhiss, 1980), should provide not only a historical background but a realistic view, a *model* if you will, of what progress can be made in a given period of time if one is, in essence, starting *de novo*.

The perspective from which one can seek to evaluate whether these palms can provide a model for future tissue culture investigations can, of course, be any of several. Following will be examined in this paper:

1. the objective(s) and background from which the tissue culture activity was approached and initiated;

2. the strategy(ies) that were and/or might be adopted to achieve the objective(s);

3. the progress made, if any;

4. the problems encountered, if any; and

5. the lessons learned.

OBJECTIVES

This author had earlier reviewed the progress of tissue culture activities on tropical and sub-tropical plants (Table 1). In such activities, the terms *research* and *facilitate* are emphasised. The most apparent of the activities with potential benefits in the short term were and still are those involving rapid multiplication of select specimens. Since African oil palm, *Elaeis guineensis,* does not normally produce vegetative branches because the inflorescences are derived from axillary buds and also the coconut palm, *Cocos nucifera,* which has a similar growth habit, an obvious goal is to use tissue culture to multiply elite germplasm. In the case of date palm, *Phoenix dactylifera,* vegetative multiplication can occur by off-shoot production but the number of off-shoots is limited and very few of them survive upon transplantation (Nixon, 1966): we need to develop methods to produce large populations of clonal materials. Infestation with insects and contamination with soil borne and root transmitted vascular pathogens with diseases such as *Bayoud* caused by *Fusarium oxysporum* f. sp. *albedinis* can be particularly troublesome, and multiplication through off-shoots from infested date palm parent materials and their transport to and planting in new areas is to be avoided at all costs (Elmer et al., 1968; Hussain, 1974; Djerbi 1982). Diseases of coconuts such as *cadang-cadang* caused by RNA viroids (Haseloff et al., 1982) and various lethal yellowings caused by mycoplasma like organisms (MLOs) or wilts (Tsai, 1980; Maramorosch and Hunt, 1981; Tsai and Thomas, 1981; Maramorosch, 1988; Tsai and Maramorosch, 1988), necessitate new planting, and raise problems in germplasm exchange and transport (Harries, 1977). The need is especially acute when one seeks to replant large plantation areas where the plants have grown too tall and/or yields have declined, with clean, high performance materials. The identification or production through breeding of coconut palms that have special qualities (Le Saint and de Nuce de Lamothe, 1987) has drawn further attention to the desirability of cloning certain individuals. In the case of African oil palm, and more recently, with their hybrids with the American oil palm *(Elaeis melanococca),* elite palms have been produced through breeding (e.g., Meunier, 1975; Meunier et al., 1976; Hartley, 1977; Hardon et al., 1985). Here too, problems of disease (Turner, 1977, 1981), the inconvenience and expense of producing African oil palm hybrid seed *(tenera)* by crossing *Elaeis guineensis* var. *dura* with *E. guineensis* var. *piscifera,* and its subsequent storage and germination (Rees, 1962), have drawn attention to developing methods of aseptic cloning.

The potential value of tissue culture for a host of activities associated with various specialised needs in breeding, management and production of various economically important plants, has been reiterated (Bonga, 1987). This applies to palms as well. Other spin-off benefits are sure to emerge, as additional emphasis is given to the possibilities, and as experience emerges. For instance, in the case of *E. guineensis* var. *pisifera,* in the course of establishing a readily available supply of material for tissue culture studies, we discovered at Stony Brook that sanitised, i.e., disinfested, seeds of *pisifera* could be retained in a healthy, viable state for extended periods merely by aseptic storage under water. By being able to do so, *pisifera* seeds that would ordinarily perish, or at best, germinate at a very low level due to microbial degradation, could be preserved better and made available for assessment in breeding activities (Nwankwo and Krikorian, 1982, 1983a). There was no intention at the outset of our studies to develop the means with which one could retain the viability of seeds over very long periods; it was a chance outcame.

Table 1: Tissue culture for the tropics and sub-tropics: Possible benefits for agriculture, forestry, floriculture, etc.

Research with potential for near term impact

Culture techniques might facilitate

* Rapid multiplication of select specimens
* Elimination of virus and specific pathogens
* Virus indexing
* Germplasm introduction and evaluation
* Germplasm collection, preservation and management
* Production of polyploids, haploids, somaclonal variants for new crop production and use in breeding, etc.
* Elimination of certain breeding barriers
    - *in vitro* fertilisation *in ovulo*
    - embryo rescue and/or storage
    - androgenesis
    - gynogenesis

Research with potential for intermediate impact

* All the above in recalcitrant species
* Selection for complex traits such as tolerance to stress
    - biotic: diseases and pests
    - abiotic: temperature, salt and herbicides
* *In vitro* mutation breeding
* Cryopreservation

Research with potential for long range implications

* Extending it further to recalcitrant species and adult tissue
* Genetic Engineering
    - transformation by selectable genes, etc.
    - organelle transfer
    - wide crosses: somatic hybridisation
* Understanding controls in development and physiological processes

---

Louis Pasteur stated that chance favours only the prepared mind. No doubt, the projected benefits for agriculture, forestry, floriculture, etc., will include those that cannot now be envisioned, or will even be dominated by them. From my perspective, virtually all the activities on palms should be seen in the context of *research*, not *development*, and certainly not yet from the viewpoint of large-scale commercial production (Krikorian, 1988 and references cited therein).

If the objective is to multiply materials rapidly, then the means whereby this may be done assume paramount importance (Table 2). So much has been said and written about the alleged difficulties in working with palms (Staritsky, 1970) that

Table 2: Strategies for multiplication of higher plants *in vitro* (Krikorian et al., 1986)

Shoots from terminal, axillary or lateral buds

* shoot apical meristems (no leaf primordia present)
* shoot tips (leaf primordia or young leaves present)
* buds
* nodes
* shoot buds on roots

Direct organogenesis

* adventitious shoot and/or root formation on an organ or tissue explant with minimal or no intervening callus

Indirect organogenesis

* adventititous shoot and/or root formation on a callus

Somatic embryogenesis

* direct formation on a primary explant with minimal or no intervening callus
* indirect formation from cells grown in suspension or semi-solid media
* secondary somatic embryogenesis. Budding from a somatic embryo generated by either of the above routes

Direct plantlet formation via an organ of perennation formed *in vitro*

* Micrografting
* Ovule culture
* Embryo rescue
* Mega-and microspore culture
* Infection with a crown gall plasmid genetically altered to give teratoma-like tumors
* Infection with *Agrobacterium rhizogenes* to induce roots on *traditional* cuttings

prospective investigators are often intimidated. At the other extreme, there is the perception that palm multiplication strategies are well defined operations nowadays - biotechnologies if you will - that need merely be instituted, scaled up and put into full *commercial operation*. As is the case in many different kinds of activities, the truth lies somewhere in between.

Investigators seeking to work on species traditionally perceived as recalcitrant should test their skills and laboratory facilities using more tractable materials. If one is forced later to conclude that the system in question is more difficult to deal with than had been anticipated, one can at least feel confident that the failure to test the laboratory's capabilities to sustain high quality plant tissue culture work was not the reason. After all, if one cannot deal effectively with easier materials, why should it be possible with those that are more difficult? Curiously, quite a few laboratories that have not even initiated, much less undergone, *shakedown* operations, are engaged in working on the *commercialisation* of systems that are not facile.

In the case of initial and preliminary attempts to deal with the species of choice, it can be very helpful to work with juvenile material. Excised embryos at various stages of development (Hodel, 1977), various parts of aseptically raised seedlings and even explants from the nucellar region of more mature flowering specimens should be tried (Srinivasan et al., 1985). For reasons that are not all that easy to characterise, one frequently observes explants from immature sources in date palm (Sharma et al., 1980). If consistent responses can be obtained from juvenile specimens, this provides a firm base from which to depart for work on mature plants. At least it shows responsiveness and provides valuable clues and learning experience. Criticisms that one should not work on juvenile or seedling material must be appreciated for their merit, but one should not, at least at the outset, slavishly adhere to the warnings. Even in the case of date palm, where sex cannot be determined before flowering (Anonymous, 1914), seeds can be used as planting material (Trabut, 1921; Nixon, 1966). In the case of African oil palm, they always provide, in the context of traditional technologies, the primary propagule. In coconut, likewise, seeds are the only propagules. In this latter instance, since elite seedling material is generally in short supply, any ability to multiply embryos, in a clonally faithful manner, would be a major advance. In fact, the relatively simple operation of splitting an embryo longitudinally to bisect the growing point, can yield two plantlets (Balaga, 1975; Fisher and Tsai, 1979). Twinning in coconut is encountered in the field but it is a relatively rare phenomenon (Furtado, 1926-29). It goes without saying that a reliable, inexpensive means of rearing zygotic embryos of any palm to maturity could provide an alternative strategy for germplasm exchange and storage (Harries, 1977; Hodel, 1977; Nwankwo, 1981; Zaid and Tisserat, 1984; Assy Bah, 1986).

Roots from mature oil palms (Purvis, 1956; Ruer, 1968), while requiring some attention to obtain them clean (Krikorian and Kann, 1986), can provide responsive explants; it should be a logical extension to evaluate the same from mature coconut and date (Zaid and Tisserat, 1983).

Since bulbil shoot formation and development of vegetative shoots from inflorescences in certain palms such as *Cocos* has been noted (Davis, 1967, 1969; Sudasrip et al., 1978; Davis et al., 1981), explants of these may provide good starting material as well (Drira and Benbadis, 1985, for *Phoenix*). However, it must be recalled that in coconut, there as yet have been no major success in controlling these bulbil shoots so as to produce normal inflorescences. In *Borassus,* spontaneous reversal has been encountered (Davis and Basu, 1969). When all is said and done, the investigator, almost inevitably, is first forced to adopt the attitude of *going with the system.* This means that one essentially takes what one can get. As progress is made, one can take conscious decisions as to how to make the most effective changes in the approach taken.

Historically, the main approach with palms has been to initiate cultures that can yield somatic embryos in profusion (Rabechault et al., 1970; Reynolds and Murashige, 1979). As in all cases involving plant tissue culture, a primary and guiding principle should be that all living cells of the higher plant body are totipotent. This means they can yield callus masses which can adventitiously produce shoots and roots.

The generation of precociously branching cultures from lateral buds of date palm and their significance for multiplication should be assessed from as wide a range of germplasm as possible (Zaid and Tisserat, 1984). At Stony Brook, such cultures have been obtained in oil palm but the practical implications for clonal multiplication have not so far been addressed (Krikorian and Kann, 1986).

The rationale behind the manipulation and/or selection of cells and tissues for somatic embryo production is perforce focused in such a way as to institute or elicit their competence, impose various controls on them and so obtain the desired form of organised development. In theory, an investigator has a host of cells and tissues to work with. There are many conventional and unconventional strategies that can be brought to bear on a given set of tissues. There is extensive summarised literature on palms now available, which gives details of and guidance on what has been done. This includes methods for inducing cultures that are capable of yielding somatic embryos (Kovoor, 1981; Reynolds, 1982; Blake, 1983; Paranjothy, 1982, 1984; Krikorian and Kann, 1986; Tisserat, 1981, 1983, 1984 a,b,c, 1987). Much of this literature has, however, major flaws. This is particularly true of the work from *industry* which deals with work on oil palm and some of the more recent work on coconut palm (Pannetier and Buffard-Morel, 1982, 1985).

The details of methodology have become more and more obscure; even less is disclosed about the complement of growth regulators employed, though there are exceptions (Hanower and Hanower, 1984). Those who have sought to repeat such work are, of course, at a disadvantage but the lack of detail is not nearly as counterproductive as inexperienced workers generally think. In the first instance, one should take the attitude that all methods published reflect generic approaches and are, in fact, experimental in nature and not geared to commercial production (despite some claims and intimations to the contrary); therefore, one reasonable approach ought to be as good as another. In tissue culture, there are many different ways to achieve the same general objectives (Krikorian, 1982). The use of hormones and nutrient media in plant tissue culture is, at this point of time, and as practised by the majority, largely an empirical activity, it ought not to be (Krikorian et al., 1987). Many monocotyledonous systems seem to be able to tolerate relatively high amounts of auxin, especially in the presence of adsorbants such as activated charcoal (Fridborg and Eriksson, 1975; Kohlenbach and Wernicke, 1977; Fridborg et al., 1978; Johansson, 1983) and high molecular weight polyvinylpyrrolidone, PVP (Loomis, 1974). Simultaneously, they seem to profit by the addition of very low levels of cytokinins and these seem to permit the growth to be sustained in a relatively disorganised mode. Lowering of the auxin, or even its elimination, while retaining the low level of cytokinins, can, in many cases, lead to organised growth. The mineral nutrient medium components can be quite important (Del Rosario and de Guzman, 1978; Eeuwens and Blake, 1978). This matter has been insufficiently considered in palm tissue culture.

Most plant tissue culturists often fail to recognise fully that the explants they are working with comprise numerous cell types. This biological fact becomes even more apparent in monocotyledons where the shoot-growing regions are comprised of zones of cells that contribute for limited periods to growth in girth (the primary thickening meristem), basal meristems and even intercalary meristems (Ball, 1941; Tomlinson, 1970). The number and kind of cells responding to *in vitro* manipulation will vary according to a given temporal and spatial setting.

One can think of this situation in terms of a callus or tissue or explant mass. Some cells are responsive to the culture conditions and these can and do divide (Fig. 1A). Some cells are non-dividing cells (1B) but these can revert or be induced to the dividing state. These may or may not be dividing because they are or are not (for whatever reason) appropriately sensitive to the culture medium or other growth medium or environmental components, as supplied. The key point here is that they can be turned on and off according to the situation (Bryant and Francis, 1985; Gray and Darzynkiewicz, 1987). Lastly, in monocotyledons in particular and in their localised growing zones, one usually has a large number of non-dividing cells (1C). These are either to be thought of as differentiated or capable of becoming terminally differentiated. They do not revert to status. B. *In situ,* these comprise cells that are pushed aside, outwards and/or upwards to take their places among the differentiated or relatively quiescent cell lineages. *In vitro,* the situation may be such that they may not only survive, but may grow. If they did not grow, that would be best, since we would not have to deal with them, but as they grow, they can ruin a culture that one is trying to elevate to the competent state.

(In Fig. 1C, it says *of little concern except for physical presence).* This must be qualified by our understanding that the physical presence of these is an obstacle to success unless they are selected out and discarded. If a culture becomes populated with cells of the wrong kind, this is counter-productive, to say the least. One has to be able to produce and maintain through serial culture, populations of cells with a preponderance of cells, as at A. Also, too frequently, investigators seem to be hesitant to discard cells or tissues which are unresponsive. They seem to adhere doggedly to the view that they can become responsive. However, based on the results from a vast literature and the author's experience at the Stony Brook laboratory, it is suggested that one should select, as early as possible, the cells that can respond, and discard those that do not. To select the proper ones from the mixture in a growing culture is not easy. The pioneers of the French school of tissue culture (founded by Gautheret and others - Gautheret,

1959) enunciated this principle in their *repiquage* or sub-culture (literally translated as *pricking out*) strategies. Over time, and as our scientific *corporate memories* have seemingly become more and more dulled, too many have apparently lost sight of what has long been known.

The challenge is to identify responsive cells, and to encourage their multiplication and growth in an unorganised state, with a view to then impose on them various nutrient or environmental changes and thus stimulate their organisation and further growth. If the maintenance culture mode can be achieved with minimal or preferably no organisation, this is to be preferred (Krikorian et al., 1986). It provides a good method of management: grow in a maintenance mode, and distribute or sub-divide on or in a medium that permits expression of organised development. Alternatively, somatic embryos can be stimulated to form (either induced or permitted to develop) from totipotent cells. If these competent ones can be identified at the pro-embryo or globular embryo stage (they are frequently *pearly white* in appearance) and husbanded so as to increase their numbers and then cultured and increased in a maintenance mode, this provides opportunities as well. One is, in essence, increasing (preferably in suspension culture) pro-embryos in such a way that they do not go through later stages of embryogenesis unless they are provided an appropriate environment. This generally means a lowering or total elimination of auxins such as 2,4-D, NAA, etc. Upon lowering of the auxin, and very frequently associated with a shift in conditions from liquid medium to semi-solid (agar, Gelrite etc.), the somatic embryos proceed to pass through the appropriate stages and eventually will grow out and produce plantlets. This latter step may or may not entail yet another medium or substrate change. A major consideration in all this is, again, that the maintenance mode can be sustained by virtue of the ability of the pro-embryos or globular stage embryos to yield, by detachment of cells and cell clusters, additional pro-embryonic cell clusters that can be sub-cultured. If this was not possible, the globular embryos would proceed to develop through the subsequent stages characteristic for the species mature, and sooner or the later, germinate. In doing so, means whereby the suspension is maintained would be

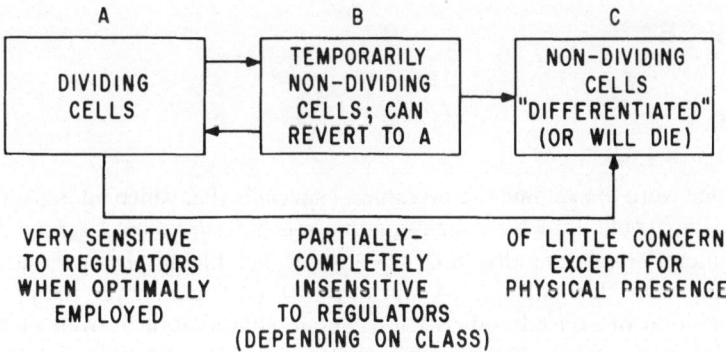

Fig 1. Relationships among categories or population of cells in a meristematic region or callus mass. See text for details.

exhausted due to organised development of the stem line. For ideal management potential, no *precocious germination* of the somatic embryo should occur prior to the time one wants it to sprout. Much emphasis has been placed on understanding the factors that influence this (Obendorf and Wettlaufer, 1984).

An examination of the literature on palms and related species discloses the viewpoint that frequently, insufficient management principles are imposed on *in vitro* systems that are potentially exclusively embryogenic and as a result, one is left to deal with cultures that are mixed in origin, composition as to cell and tissue type, and hence, in responsiveness. The variation in responsiveness can include cells that yield somatic embryos and cells which develop into adventitious shoots or roots as well. Such cultures are messy, to say the least (Wicart et al., 1984; Bornman, 1987).

In all these activities, there is a pressing need to be aware of the basic morphology and biology of the systems. Unfortunately, despite their importance as crop plants, the literature on palms is more sparse than might be suspected. Details of embryo structure and seed germination are known but tend to be in older or less accessible literature (Gatin, 1906; Guerin, 1949; Sento, 1972,1974). Details on embryological anatomy of coconut, lacking for a long time (Selvaratnam, 1952), is now well worked out (Haccius and Philip, 1979), and the same may be said of date palm (Lloyd, 1910; DeMason and Thomson, 1981; DeMason, 1985; DeMason et al., 1985; Chandrasekhar and DeMason, 1988) and oil palm (Vallade, 1966a,b, 1969). The detailed anatomy and histology of growing points of the vast majority of palms is available only in the most general way [for example, wild palm (Ball, 1941); date palm (Hilgeman, 1954); coconut and oil palm (Henry, 1955, 1957,1961)]. Too little has been done in recent years on these systems and the needs are great (Padmanabhan, 1976 on axillary shoots of *Phoenix sylvestris;* Bouguedoura, 1980 on *Phoenix dactylifera;* DeMason and Tisserat, 1980 and DeMason et al., 1982 on *Phoenix dactylifera* floral buds, etc.). It is to be regretted that tissue culturists have, thus far, avoided this morphological and anatomical aspect. Also, much is buried in unpublished theses (see Hilgeman, 1951 for a good coverage of date palm; also see Johnson, 1983).

PROGRESS

Oil Palm

If one were to examine the literature, especially that which adopts a pragmatic approach, one would draw the conclusion that much progress has been made (Lioret and Ollagnier, 1981; Paranjothy, 1982, 1984). A recent bibliography on tissue culture of *Elaeis* is impressive for its length (Anonymous, 1987). There is even the possibility of cryopreservation of excised embryos for genetic conservation (Grout et al., 1983). More conservative stands (Blake, 1983; Krikorian and Kann, 1986) emphasise the problems, even as modest achievements have been described in detail. At Stony Brook, the procedures reported for *pisifera* oil palm (Nwankwo and Krikorian, 1983b) worked well for *dura* and *tenera*. The title of a paper implying that somatic embryogenesis in *Elaeis guineensis* and *E. guineensis* x *E. melanococca* hybrids had been achieved in liquid culture (Malaurie, 1987), is somewhat misleading. The published work discloses that one has essentially repeated the work normally carried out on semi-solid media and that suspension cultures were not involved. For most experienced investigators, *liquid culture* connotes suspension culture and not merely growing large callus masses in liquid (see Smith and Thomas, 1973 for early work on cell suspensions).

Date Palm

The work on date palm carried out by Tisserat and co-workers (Tisserat, 1979, 1981, 1983, 1984 a,b,c, 1987; Zaid and Tisserat, 1983, 1984; Gabr and Tisserat, 1985) has been reported in considerable detail. This includes anatomical investigations carried out on somatic embryogenesis (Tisserat and De Mason, 1985). A small handbook delineating procedural details for date palm culture will be helpful to those seeking to initiate studies of their own (Tisserat, 1981).

Coconut Palm

Culture of excised embryos of *makapuno* coconut which would otherwise not germinate due to rotting of the endosperm (Zuniga, 1959; Abraham et al., 1965), provided means of *rescue* as far back as the late 1960s and early 1970s (de Guzman and Del Rosario, 1964, 1974; de Guzman et al., 1971). Attempts to elevate these research activities to a commercial production mode have been partly successful but some problems still remain. These include, on the biological level, inability to fully control the level of shoot and root development (de Guzman, 1970; Balaga and de Guzman, 1971; Sajise and de Guzman, 1972; de Guzman and Manuel, 1977) and hence, the failure of the transition from *in vitro* to *ex vitro* conditions. Economic considerations further limit the commercialisation. Embryo culture of regular forms of *Cocos nucifera* for purposes of generating clean planting materials, etc. is showing promise (Assy Bah, 1986).

Perceived pressures to produce positive results *vis-a-vis* somatic embryogenesis in coconut have resulted, over the years, in a spate of lay press publications that do little more than emphasise that it is within the realm of reason to anticipate that one day one can work effectively with coconut (Pannetier and Buffard-Morel, 1982; Gupta et al., 1984). Some of the claims in India and Sri Lanka for cloning coconuts via somatic embryos are unjustified (Iyer, 1981) and suffer from a lack of rigorous studies on establishing that somatic embryos rather than adventitious growths, are, indeed, being produced. Cronauer-Mitra and Krikorian (1988a) have shown in banana the structures that could easily be mistaken for somatic embryos but which are, in fact, adventitious shoots! They similarly cited an example in *Musa* where the generation of somatic embryos is beyond all doubt (1988b).

Why coconut has proved so difficult to work with is not clear (Apavatjrut and Blake, 1977; Eeuwens and Blake, 1978; Blake, 1983; D'Souza, 1980, 1982; Thanh-Tuyen and de Guzman, 1983; Kumar et al., 1985).

PROBLEMS ENCOUNTERED

In virtually all the cases, the successful callus culture of the palm tissue in question has entailed adding an auxin such as 2,4-D or NAA to initiate callus and then lowering its concentration or removing it to elicit somatic embryos. The level of responsiveness has, on the whole, been low. The recovery of plantlets, once a system such as that of oil palm or date palm gets going, has, however, been impressive. At the same time when one considers (as in the case of oil palm) the large numbers of personnel involved in many of the operations attempting to deal with these predominantly plantation species, the production levels become less impressive. Clearly, innovative approaches, besides tissue culture, are in order to improve production levels (for date palm, see Oppenheimer and Reuveni, 1972).

The ultimate success of any tissue culture operation purporting to have as its prime objective the large-scale, clonal multiplication of elite specimens, will naturally be determined by one's ability to generate materials with clonal fidelity in a timely and cost effective manner.

Makapuno coconut trees which have a capacity to yield very high numbers of makapuno nuts, have been generated from aseptic-culture-rescued embryos for some time (de Guzman et al., 1983). In Indonesia, where the *makapuno* is called *kopyor,* the Philippine embryo culture experience has been repeated (Tahardi and Warda-Dalem, 1982). The recovery rate of rescued embryos, although adequate, is, however, not very high. Decisions as to the desirability of implementing production programs in the case of *makapuno* are essentially tied to economics. It seems that the idea of inducing vegetative branches in *Cocos* (Balaga, 1975; Fisher and Tsai, 1979) has not been pursued as a realistic approach to clonal multiplication of coconuts.

In the case of date palm, the field results of clonal multiplication activities, initiated at the laboratory level in 1960s, are beginning to emerge. A personal communication from Dr. Mostafa Abo El-Nil, of the Date Palm Research Centre, Al-Hassa, Saudi Arabia, states that somatic embryo production is routine and managed on a large-scale. Flowering of few specimens has yielded normal plants of uniform growth habits with improved yields. The full expectation is that the system can be used if there is need but because of the expense involved per plantlet produced, financial considerations dictate when or if advantage is taken of the capabilities. The author believes that if the methods of production were totally controllable, reliable and efficient, the costs would not be prohibitive. A conservative stance would include the view, then, that we still have considerable progress to make in the field of date palm.

Shoot tip culture (Tisserat, 1984a) and axillary bud break methods as well as embryogenesis from callus grown in semi-solid media have been used in date palm.

The date palm, therefore, presents us with a situation which is definitely more tractable than oil palm or coconut palm. Since the *biology* of date palm is such that it generally yields some axillary branches (Popenoe, 1973; Dowson, 1976), perhaps tissue culture may be viewed as merely increasing or enhancing the tendency.

The reliability from a clonal perspective of the generation of somatic embryos from African oil palm and the subsequent performance of plants derived from them has recently been questioned (Corley et al., 1986). For some years, the Unilever and French groups have been quite aggressive in projecting the image of being at the cutting edge of research and development activities associated with plant tissue culture. Others have rushed to emulate their success and laboratories devoted primarily to tissue culture of oil palm have sprung up all over Southeast Asia (especially Malaysia) and elsewhere. Attractive brochures and articles, albeit without any details, have drawn attention to progress (Jones, 1980; Ahee et al., 1981). As it turns out, the substance underlying these advertisements seems to have been fragile. Early reports projected *cloned* plants (Jones et al., 1982; Wooi et al., 1982) as having improved qualities and yields (Corley et al., 1977; Jones, 1984); rarely was a conservative evaluation presented (Hardon et al., 1985). These have been replaced by reports of plants showing abnormalities in flowering and fruit-set (Corley et al., 1986). Since palm oil production is intimately tied to flowering and fruiting, this has unfortunately generated skepticism among those polarised against or unsympathetic towards a tissue culture approach. The experience of the French group with oil palm, especially in Cote d'Ivoire, has supposedly not disclosed any *problems* (presentation by J.M. Noiret at International Congress of Plant Tissue Culture of Tropical Species, Bogota, Colombia, October, 1987). Even the oil palm plants showing flowering off-types in Malaysia might revert back to normal type after some time.

A number of recent studies have shown that *in vitro* plantlets of a large number of species multiplied via various tissue culture techniques may, give rise to plants under field conditions, that are considerably different from the *parent* plant from which they were initiated. These aseptic culture-associated variants, *somaclones,* were initially said to offer considerable opportunity for crop improvement (Scowcroft, 1985) but very few, if any, somaclones can yet be pointed to as an example of an economically significant or agronomically superior cultivar. At the same time, the author believes, one should not expect improvement over a parent ortet directly from tissue culture.

Before any serious work involving controlled mutation breeding or tissue culture improvement can be carried out, it is important to have an adequate store of baseline data. One can project that benefits, if any, will be most noticeable only after the flowering generation. In the case of those species that are fertile and can be bred, the *somaclones* would then enter into a conventional breeding scheme. If they are infertile and are normally vegetatively multiplied, then, just as in mutation breeding, selection would be carried out after the first year in the field. This has not been the case in most studies carried out so far, and certainly not in palms. Improvement via somaclones will rarely be detectable in the material directly derived from tissue culture. To allow meaningful interpretations to be made, cultures ought to:

1. represent material of genetic background which is well characterised;
2. derive from *in vitro* procedures that are very carefully monitored from the perspective of
   a) precise botanical and morphological origin,
   b) precise duration under *in vitro* conditions - i.e., the exact date of initiation and sub cultures,
   c) precise and preferably different tissue culture protocols which are used to generate plantlets *in vitro* i.e., a number of potentially crucial variations ought to be tested to generate plants.

In short, one needs an accurate *in vitro* pedigree of the system. If, from this, we are able to correlate an adopted tissue culture protocol with a tendency towards stability, or the reverse, we will have made a beginning.

The author does not believe that *clonal multiplication* by tissue culture is a *technology* that is straightforward, simple to understand and necessarily easy to achieve (Krikorian, 1982, 1988 and references cited therein). On the other hand, however, too many have field-planted what are essentially *traumatised* plants, and have been disappointed with their performance. One can almost expect *off-types* to be the normal case under such circumstances!

Table 3 lists some possible causes of culture-associated changes. The list is by no means complete, but it does provide sufficient suggestions that mutation related events *in vitro* are not much different from what might occur under natural conditions. What is encountered *in vitro* may be a matter of degree rather than of kind.

Moreover, as we learn more about the plant genome and how genetics *works* in development, our appreciation of what the clonal state means is sure to vary. Clearly, there is much plasticity in the higher plant genome (McClintock, 1978; Nevers et al., 1986; Schlichting, 1986). It is, perhaps, most accurate to think of a clonal population as a mixture of slightly differing individuals.

Repeated sub-culturing *in vitro* may well seem, at first glance, to increase the likelihood of genetic changes or variations taking place (Krikorian et al., 1983 and references cited therein). This generalisation is too broad, however, and has many exceptions. Even the term *duration in culture* needs to be more precisely stated. If one were to use chronological time measurement exclusively, it would imply that the number of calendar days is of paramount importance. But it is not always true and, in fact, the number of requisite sub-cultures as a measure of the rate of cell division is of greater significance (Bryant and Francis, 1985).

The situation involving aberrations in the flowering process in the oil palm plantings referred to earlier, emphasises that one is still essentially at the basic research stage and not yet at the commercialisation stage (Corley et al., 1986). It is not known what exactly *went wrong* in the tissue culture process but it has been said that field progeny of embryogenic cultures of oil palm that was still rather new in terms of the extent or their duration *in vitro* has, in fact, been acceptably stable. That is, tissue culture *per se* does not invariably lead to aberrations. However, as the length of the aseptic culture period during which somatic embryos were replicated or multiplied was

Table 3: Possible causes of somaclonal variation

---

* Specific karyotype selection from mosaic, chimeric and polysomatic tissues and plants
* Karyotype changes due to differential response to culture procedures (media composition and/or environment)
    - non-disjunctional aneuploidies in culture
    - mitotic arrest leading to polyploid lines
* Somatic gene re-arrangements or mutations of the karyotype
* Gene amplification or diminution
* Virus elimination from selected lines of a culture
* Re-arrangements or mutations in organellar genomes
* Altered nucleo-cytoplasmic interactions resulting in regulation changes
* Sudden reorganisation of the genome by transposable elements
* Variegated position effects of chromosomal re-arrangements (inversions, translocations, etc.)
* Late replication of heterochromatin

---

extended, the regenerates from these *old* cultures deteriorated in their flowering performance (personal communications to author in Malaysia).

    The reasons for this deterioration are still to be worked out. Visual examination of *tenera* oil palm embryogenic cultures in at least two commercial tissue culture laboratories in Malaysia has suggested to the author that the highly mixed nature of the cultures is at least partially responsible for the off-type difficulties. This is tied to or associated with *tissue culture management.* Cultures did, indeed, comprise embryogenic masses, but callus, calloid (epidermis present), adventitious roots and shoots were discernible as well. Insufficient attention was paid to *repiquage* during the initial culture establishment stage. Also, little attention seems to be placed on the details of initiation and on subsequent maintenance of morphogenetically competent cultures. Too high a concentration of a growth regulator such as 2,4-D, media which is too *rich* in reduced nitrogen (as in MS medium, Murashige and Skoog, 1962), etc. can favour changes such as habituation (*anergie* in French, Gautheret, 1942). The implications of all this for differential gene action etc. are not well understood. The precocious flowering *in vitro* of oil palm (Krikorian and Kann, 1986) certainly indicated some important physiological changes (Scorza, 1982). Some of the changes in growth substance requirement or selected-for traits or characteristics in culture are heritable and can give rise to cell populations with substantial stability for these traits (White, 1949). One view expressed by F.C. Steward was that plants that were less domesticated *(more wild),* and have emerged as *newer crop plants* such as oil palm, were likely to be more heterogenous in their *in vitro* responsiveness and thus, one could more readily elicit morphogenesis from them (personal communication to the author). However, the heterogeneity can favor change, including improvement.

The fact that off-types in oil palm derived from seed exist in nature (Bernard, 1964; Courtois, 1968; Rajanaidu, 1980; Corley et al., 1986; Paranjothy, 1987), suggests that extreme heterogeneity of cell types grown *in vitro* could be responsible for abnormal flowering behavior. Also, one is not sure as to which characteristics are genetically or environmentally determined (Baudouin et al., 1987). If the encountered tendency towards androgyny in the inflorescences of tissue-cultured palms reverts to a normal flowering mode, the problem might be attributed to hormonal imbalances. Unfortunately, too little is known about hormonal balances or effects in any palm (Corley, 1976 on gibberellins in oil palm and Mohammed, 1985 on gibberellins in date palm). In one of the sugar palms, *Borassus flabellifer,* the tendency for bulbil shoot production was seen to revert to the normal flowering mode after some time (Davis and Basu, 1969). Also, there is evidence that what may be termed as position effects can influence *in vitro* response (Raman et al., 1980). Therefore, we are unable at this time to come up with a firm answer.

Chromosomes of palms are not easy to work with (Sato, 1946; Sarkar, 1957; also see Ninan and Raveendranath, 1965; Raveendranath and Ninan, 1974; Imam, 1982). It would, however, be worth attempting high resolution chromosome mapping (Lawrence et al., 1988). Torres and Tisserat (1980) reported on isozyme analysis as genetic markers in date palms.

LESSONS LEARNED

Like many monocotyledons (Hunault, 1979), palms are not particularly easy to work with but they can be grown *in vitro* and provide opportunities to learn much about culturing cells. So long as the activities are viewed as experimental and not commercial production related, progress will be inevitable. Clearly, *plant tissue culture,* especially as it relates to palms, *is a process, not an event.* Premature attempts at commercial exploitation will be risky. There will be a continuous and on-going situation of trying to sort out perception *vs* reality. It is tempting to capitalise financially on and to protect one's perceived achievements. Patents confuse rather than elucidate the *real state-of-the-art.* Rarely do plant tissue culture patents include anything but the barest of details. They are too diffuse and fall generally into the category of *use patents* (Martin and Rabechault, 1978).

From a technical perspective, like all plants that must fit into rigorous agronomic schemes, palms require a considered approach with adequate commitment of funds and expertise (Table 4). Blackening or darkening of explants due to phenolics (Ibrahim, 1987) can be dealt with (Poulain et al., 1979), or is not necessarily a problem (Krikorian and Kann, 1986; Compton and Preece, 1988; Krikorian, 1988 and references cited therein on how to deal with darkening). Temperature considerations for incubating cultures should not be overlooked (Mason, 1925 for date palm; Dhawan and Bhojwani, 1985 for *Leucaena*).

Table 4: Some limitations and constraints to progress in the use of plant tissue culture

---

* Investigators and support personnel with adequate training and scientific background
* Financial resources
* A bureaucracy sympathetic to research with a long *lead-time* to *development* and administrators able to facilitate implementation of programs
* Adequate release time in universities from undergraduate teaching duties so that principle investigators can work in the laboratory
* Recognition not only of *capabilities* but *limitations* of tissue culture techniques
* Teamwork at individual, institutional, local, national and international levels
* Maximum integration with other disciplines
* Frequent communication is in process but open communication is a limited commodity

---

The preparation of protoplasts from oil palm (Eeuwens and Blake, 1978; Vouyouklis, 1981; Bass and Hughes, 1984) and coconut palm (Eeuwens and Blake, 1978; Haibou and Kovoor, 1981) has been successful. Whether the protoplast methods can do much for us remains to be demonstrated but there is much to recommend protoplast research (Fitter and Krikorian, 1982, 1988).

The finding that anti-fungal principles exist in *Bayoud*-resistant date palms (Assef et al., 1986) is encouraging. Tissue culture-raised plantlets could certainly be used for screening (Saaidi et al., 1981) to learn more about the infection process (Belarbi-Hall and Mangenot, 1986).

Clearly, much remains to be done. If there is a place in a given country's economy for a palm species [e.g. *Bactris* in Costa Rica (Arias and Huete, 1983); *Jubaea* in Chile (Yuri, 1987); *Elaeis guineensis* var. *idolatrica* in Ghana (Addae-Kagyah et al., 1988)], then it behooves workers to investigate the species from as many perspectives as possible, including tissue culture.

ACKNOWLEDGEMENTS

Travel to the workshop was made possible through the support of the U.S. National Science Foundation (NSF). Years of funding for tissue culture activities by the U.S. National Aeronautics and Space Administration is gratefully acknowledged, for it has provided the mechanism whereby a broad base of experience has been gained. Various grants over the years from N.S.F., U.S. Agency for International Development, Weyerhaeuser Company, United AgriSeeds, Phyto Resource Research, etc., are also recognised.

REFERENCES

Abraham, A., Ninan, C.A. and Gopinath, P., 1965, Cytology of development of abnormal endosperm in Philippine-makapuno coconuts, *Caryologia*, 18: 395-408.

Addae-Kagyah, K.A., Osafo, D.M., Olympio, N.S. and Atubra, O.K., 1988, Effect of seed storage, heat pretreatment and its duration on germination and growth of nursery stock of the *idolatrica* palm, *Elaeis guineensis* var. *idolatrica* (Chevalier), *Trop. Agric. (Trinidad)*, 65: 77-83.

Ahee, J. et al., 1981, La multiplication vegetative *in vitro* du palmier a huile par embryogenese somatique, *Oleagineux*, 36: 113-118.

Ammar, S., Benbadis, A. and Tripathi, B.K., 1987, Floral induction in date palm seedlings (*Phoenix dactylifera* var. *Deglet nour*) cultured *in vitro*, *Can. J. Bot.*, 65: 137-142.

Anonymous, 1914, The sex of date palm seedlings, *Kew Bulletin of Miscellaneous Information*, pp. 159-162.

Anonymous, 1987, La culture *in vitro* du palmier a huile (*Elaeis guineensis* Jacq.) Bibliographie analytique, *Oleagineux* 42: 101-112.

Apavatjrut, P. and Blake, J., 1977, Tissue culture of stem explants of coconut (*Cocos nucifera* L.), *Oleagineux*, 32: 267-271.

Arias, M.O. and Huete, F., 1983, Propagacion vegetativa *in vitro* de pejibaye (*Bactris gasipaes* HBK), *Turrialba*, 38: 102-108.

Assef, G.M., Assari, K. and Vincent, E.J., 1986, Occurrence of an anti-fungal principle in the root extract of a Bayoud-resistant date palm cultivar, *Neth. J. Plant Pathol.*, 92: 43-47.

Assy Bah, B., 1986, Culture *in vitro* d'embryons zygotiques de cocotiers, *Oleagineux*, 41: 321-328.

Balaga, H.Y., 1975, Induction of branching in coconut, Kalikasan, *Philip. J. Biol.*, 4: 135-140.

Balaga, H.Y. and de Guzman, E.V., 1971, The growth and development of coconut *Makapuno* embryos *in vitro*. II. Increased root incidence and growth in response to media composition and to sequential culture from liquid to solid medium, *Philip. Agric.*, 53: 551-565.

Ball, E., 1941, The development of the shoot apex and of the primary thickening meristem in *Phoenix canariensis* Chaub., with comparisons to *Washingtonia filifera* Wats. and *Trachycarpus excelsa* Wendl. *Am. J. Bot.*, 28: 820-832.

Bass, A. and Huges, W., 1984, Conditions for isolation and regeneration of viable protoplasts of oil palm (*Elaeis guineensis*), *Plant Cell Rep.*, 3: 169-171.

Baudouin, L., and Noiret, J.M., 1987, Importance of environmental factors in the choice of oil palm ortets, *Oleagineux*, 42: 268-269.

Belarbi-Hall, R. and Mangenot, F., 1986, Bayoud disease of date palm: Ultrastructure of root infection through nematodes, *Can. J. Bot.*, 64: 1703-1711.

Bernard, G., 1964, Palmier a huile: Selection en pepiniere, *Oleagineux*, 19: 243-246.

Blake, J., 1983, Tissue culture propagation of coconut, date and oil palm. In: *Tissue Culture of Trees* (J.H. Dodds, ed.), pp. 29-50, AVI Publishing, Westport.

Bonga, J.M., 1987, Tree tissue culture applications, *Adv. in Cell Cult.*, 5: 209-239.

Bornman, Ch., 1987, *Picea abies*. In: *Cell and Tissue Culture in Forestry Vol. 3* (J.M. Bonga and D. J. Durzan, eds.) pp. 2-29, Martinus Nijhoff, Dordrecht.

Bouguedoura, N., 1980, Morphologie et ontogenese des productions axillaires du palmier-dattier, *Phoenix dactylifera* L., *C. R. A. Sci. (Ser. D)*, 291: 857-860.

Bryant, J.A. and Francis, D. (eds.), 1985. *The Cell Division Sycle in Plants*, Society for Experimental Biology 26. Cambridge University Press, Cambridge.

Chandra Sekhar, K.N. and DeMason, D., 1988, Quantitative ultrastructure and protein composition of date palm (*Phoenix dactylifera*) seeds: A comparative study of endosperm vs. embryo, *Am. J. Bot.*, 75: 323-329.

Compton, M.E. and Preece, J.E., 1988, Response of tobacco callus to shoot tip exudation from five species, *Hortic. Sci.*, 23: 208-210.

Corley, R.H.V., 1976, Sex differentiation in oil palm: Effects of growth regulators, *J. Exp. Bot.*, 27: 553-558.

Corley, R.H.V., Barrett, J.N. and Jones, L.H., 1977, Vegetative propagation of the oil palm via tissue culture. In: *International Developments in Oil Palm* (D.A. Earp and W. Newall, eds.), pp. 1-8, Proc. Malaysian Int. Agric. Oil Palm Conf., The Incorporated Society of Planters, Kuala Lumpur.

Corley, R.H.V., Lee, C.H., Law, J.H. and Wong, C.Y., 1986, Abnormal flower development in oil palm clones, *Planter*, 62: 233-240.

Courtois, G., 1968, Arbres anormaux chez *Elaeis guineensis, Oleagineux*, 23: 641-644.

Cronauer-Mitra, S.S. and Krikorian, A.D., 1988a, Adventitious shoot production from calloid cultures of banana, *Plant Cell Rep.*, 6: 443-445.

Cronauer-Mitra, S.S. and Krikorian, A.D., 1988b, Plant regeneration via somatic embryogenesis in the seeded diploid banana *Musa ornata* Roxb., *Plant Cell Rep.*, 7: 23-25.

Davis, T.A., 1967, Foliation of coconut spadices and flowers, *Oleagineux*, 22: 19-23.

Davis, T.A., 1969, Clonal propagation of the coconut, *World Crops*, 21: 253-255.

Davis, T.A. and Basu, K., 1969, Two cases of bulbil-bearing *Borassus flabellifer* Linn., *J. Indian Bot. Soc.*, 48: 198-201.

Davis, T.A., Sudasrip, H. and Azio, H., 1981, Bulbil-shoot production from clonally propagated coconuts, *Principes*, 25: 124-129.

de Guzman, E.V., 1969, The growth and development of coconut *Makapuno* embryo *in vitro*, 1. Induction of rooting, *Philip. Agric.*, 53: 65-78.

de Guzman, E.V. and Del Rosario, A.G., 1964, The growth and development of *Cocos nucifera* L. *Makapuno* embryo *in vitro, Philip. Agric.*, 48: 82-94.

de Guzman, E.V. and Del Rosario, A.G., 1974, The growth and development in soil of *Makapuno* seedlings cultures *in vitro*, National Research Council of the Philippines, *Res. Bull.*, 29: 1-16.

de Guzman, E.V. and Manuel, G.C., 1977, Improved root growth in embryo and seedling cultures of coconut *Makapuno* by the incorporation of charcoal in the growth medium, *Philip. J. Coconut Studies*, 2: 35-39.

de Guzman, E.V., Del Rosario, A.G. and Eusebio, E.C., 1971, The growth and development of coconut *Makapuno* embryo *in vitro* 3. Resumption of root growth in high sugar media, *Philip. Agric.*, 53: 566-579.

de Guzman, E.V., Rafols, A.G. and Del Rosario, A.G., 1983, Preliminary observations on floral biology and fruiting of *in vitro* coconut palms. In: *Coconut Research and Development* (N.M. Nayar, ed.), pp. 316-321, Proceedings of the International Symposium on Coconut Research and Development, Wiley Eastern, New Delhi.

Del Rosario, A.G. and de Guzman, E.V., 1978, The growth of coconut *Makapuno* embryos *in vitro* as affected by mineral composition and sugar levels of the medium during the liquid and solid cultures, *Philip. J. Sci.*, 105: 215-222.

DeMason, D.A., 1985, Histochemical and ultrastructural changes in the haustorium of date (*Phoenix dactylifera* L.), *Protoplasma*, 126: 168-177.

DeMason, D.A. and Thomson, W.W., 1981, Structure and ultrastructure of the cotyledon of date palm (*Phoenix dactylifera* L.), *Bot. Gaz.*, 142: 320-328.

DeMason, D.A. and Tisserat, B., 1980, The occurrence and structure of apparently bisexual flowers in the date palm, *Phoenix dactylifera* L. (Arecaceae), *Bot. J. Linnean Soc.* 81: 283-292.

DeMason, D.A., Stolte, K.W. and Tisserat, B., 1982, Floral development in *Phoenix dactylifera, Can. J. Bot.*, 60: 1437-1446.

DeMason, D.A., Sexton, R., Gorman, M. and Reid, J.S.G., 1985, Structure and biochemistry of endosperm breakdown in date palm (*Phoenix dactylifera* L.) seeds, *Protoplasma*, 126: 159-167.

Dhawan, V. and Bhojwani, S.S., 1985, *In vitro* vegetative propagation of *Leucaena leucocephala* (Lam.) de Wit, *Plant Cell Rep.*, 4: 315-318.

Djerbi, M., 1982, Bayoud disease in North Africa: History, distribution, diagnosis and control. Baghdad, Iraq: (FAO Regional Project for Palm and Dates Research Centre in the Near East and North Africa) *Date Palm J.*, 1: 153-198.

Dowson, V.H.W., 1976, Bibliography of The Date Palm (H. Field and W.T. Gillis, eds.), Field Research Projects, Coconut Grove, Miami, Florida.

Drira, N. and Benbadis, A., 1985, Multiplication vegetative du palmier dattier (*Phoenix dactylifera* L.) par reversion, en culture *in vitro*, de'ebauches florales de pieds femelles, *J. Plant Physiol.*, 119: 227-235.

D'Souza, L., 1980, Hypocotyl budding in cultured coconut embryos. In: *Proceedings of a National Symposium on Plant Tissue Culture, Genetic Manipulation and Somatic Hybridization of Plant Cells* (P.S. Rao, M.R. Heble and M.S. Chadha, eds.), pp. 190-294, Bhabha Atomic Research Centre, Bombay.

D'Souza, L., 1982, Organogenesis in coconut embryo callus. In: *Plant Tissue Culture 1982* (A. Fujiwara, ed.), pp. 179-180, Jpn. Assoc. Plant Tissue Cult., Tokyo.

Eeuwens, C.J. and Blake, J., 1978, Culture of coconut and date palm tissue with a view to vegetative propagation, *Acta Hortic.,* 78: 277-286.

Elmer, H.S., Carpenter, J.B. and Klotz, L.J., 1968, Pests and diseases of the date palm, *FAO Plant Protection Bull.,* 16: 1-32.

Fisher, J. and Tsai, J.H., 1979, A branched coconut seedling in tissue culture, *Principes,* 23: 128-131.

Fitter, M.S. and Krikorian, A.D., 1982, Plant protoplasts: Some guidelines for their preparation and manipulation in culture, CalBiochem Behring, Division of American Hoechst Corporation, La Jolla.

Fitter, M.S. and Krikorian, A.D., 1988, Variation among plants regenerated from protoplasts of diploid day lily. In: *Progress and Prospects in Forest and Crop Biotechnology* (F.A. Valentine, ed.), pp. 242-256, Springer-Verlag, New York.

Fridborg, G. and Eriksson, T., 1975, Effects of activated charcoal on growth and morphogenesis in cell cultures, *Physiol. Plant.,* 34: 306-308.

Fridborg, G., Pederson, P., Landstrom, L.E. and Eriksson, T., 1978, The effect of activated charcoal on tissue culture; adsorption of metabolite inhibiting morphogenesis, *Physiol. Plant.,* 43: 104-106.

Furtado, C.X., 1926-29, Abnormalities in coconut palms, *Gardens' Bulletin Straits Settlements,* Singapore (Ser. 3,4): 78-84.

Gábr, M.F. and Tisserat, B., 1985, Propagating palms *in vitro* with special emphasis on the date palm (*Phoenix dactylifera* L.), *Sci. Hortic.,* 25: 255-262.

Gatin, C.L., 1906, Recherches sur la germination des palmiers, *Annales des Sciences Naturelles, Botanique,* (Ser. 9,3): 191-314.

Gautheret, R.J., 1942, Hetero-auxines et cultures de tissue vegetaux, *Bull. de la Soc. de Chimie Biol.,* 24: 13-46.

Gautheret, R.J., 1959, La culture des tissus vegetaux, Masson et Cie, Paris.

Gray, J.W. and Darzynkiewicz, Z., (eds.), 1987, *Techniques in Cell Sycle Analysis,* Humana Press, Clifton, New Jersey.

Grout, B.W.W., Shelton, K. and Pritchard, H.W., 1983, Orthodox behaviour of oil palm seed and cryopreservation of the excised embryo for genetic conservation, *Ann. Bot. (Lond),* 52: 381-384.

Guerin, H.P., 1949, Contribution of l'etude du fruit et de la graine des palmiers, *Annales des Sciences Naturelles, Botanique,* (Ser. 2, 10): 21-69.

Gupta, P.K., Kendurkar, S.V., Kulkarni, V.M., Shirgurkar, M.V. and Mascarenhas, A.F., 1984, Somatic embryogenesis and plants from zygotic embryos of coconut (*Cocos nucifera* L.) *in vitro, Plant Cell Rep.,* 3: 222-225.

Haccius, B. and Philip, V.J., 1979, Embryo development in *Cocos nucifera* L.: A critical contribution to a general understanding of palm embryogenesis, *Plant Systematics* and *Evolution,* 132: 91-106.

Haibou, T.K. and Kovoor, A., 1981, Regeneration of callus from coconut protoplasts. In: *Tissue Culture of Economically Important Plants* (A.N. Rao, ed.), pp. 149-151, COSTED and ANBS Natn. Univ. Singapore.

Hanower, J. and Hanower, P., 1984, Inhibition et stimulaiton, en culture *in vitro,* de l'embryogenese des souches issues de'explants foliaires de Palmier a Huile, *C. R. A. Sci., Ser. D,* 298: 45-48.

Hardon, J.J., Rao, V. and Rajanaidu, W., 1985, A review of oil-palm breeding. In: *Progress in Plant Breeding* (G.E. Russell, ed.), pp. 139-163, Butterworths, London.

Harries, H.C., 1977, Coconut (*Cocos nucifera* L.). In: *Plant Health and Quarantine in International Transfer of Genetic Resources* (W.B. Hewitt and L. Chiarappa, eds.), pp. 125-136. CRC Press, Boca Raton, Florida.

Hartley, C.W.S., 1977, *The oil palm,* 2nd ed., Longman, London.

Haseloff, J., Mohamed, N.A. and Symons, R.H., 1982, Viroid RNAs of *cadang-cadang* disease of coconuts, *Nature (Lond.),* 299: 316-321.

Henry, P., 1955, Sur le development des feuilles chez la palmier a huile, *Rev. Gen. Bot.,* 62: 231-237.

Henry, P., 1957, Recherches sur la croissance et le developpement chez *Elaeis guineensis* Jacq. et chez *Cocos nucifera* L. comparaison avec d'autres palmiers, Universite de Paris.

Henry, P., 1961, Recherches cytologiques sur l'appareil floral et la graine chez *Elaeis guineensis* et *Cocos nucifera.* II, Les fleurs et la graine, *Rev. Gen. Bot.,* 68: 164-198.

Hilgeman, R.H., 1951, The differentiation, growth and anatomy of the axis, leaf, axillary bud, inflorescence and offshoot in *Phoenix dactylifera* L., Ph.D. Thesis, University of California, Los Angeles.

Hilgeman, R.H., 1954, The differentiation, development and anatomy of the axillary bud, inflorescence and offshoot in the date palm, pp. 6-10. In: *31st Ann. Rep. Date Growers' Institute,* Indio, California.

Hodel, D., 1977, Notes on embryo culture of palms, *Principes,* 21: 103-108.

Hunault, G., 1979, Rescherches sur le comportement des fragments d'organes et des tissus de monocotyledones cultivees *in vitro, Rev. Cytol., Biol. Vegetale Bot.,* 2: 259-287.

Hussain, A.A., 1974, Date Palms and Dates and their Pests, Baghdad.

Ibrahim, R.K., 1987, Regulation of synthesis of phenolics. In: *Cell Culture and Somatic Cell Genetics of Plants, Cell Culture and Phytochemistry,* Vol. 4 (I.K. Vasil, ed.), pp.77-95, Academic Press, Orlando.

Imam, M.M., 1982, Mitosis in *Elaeis guineensis* Jacq. race Deli Dura, *Palm Oil Res. Inst. of Malaysia (PORIM) Bull.,* 5: 18-27.

Iyer, R.D., 1981, Embryo and tissue culture for crop improvement, especially of perennials, germplasm conservation and exchange. In: *Tissue Culture of Economically Important Plants* (A.N. Rao, ed.), pp. 229-230, COSTED and ANBS, Natn. Univ. Singapore.

Johansson, L., 1983, Effects of activated charcoal in anther culture, *Physiol. Plant.,* 59: 397-403.

Johnson, D., 1983, A bibliography of graduate theses on palms, *Principes,* 27: 85-88.

Jones, L.H., 1974a, Plant cell culture and biochemistry: Studies for improved vegetable oil production. In: *Industrial Aspects of Biochemistry,* Vol. 30 (II) (B. Spencer, ed.), pp. 813-833, North Holland/American Elsevier.

Jones, L.H., 1974b, Propagation of clonal oil palms by tissue culture, *Oil Palm News,* 17: 1-8.

Jones, L.H., 1980, *Clonal oil palm propagation by tissue culture,* Unilever Research, Colworth Laboratory, Sharnbrook.

Jones, L.H., 1984, Novel palm oils from cloned palms, *J. Am. Oil Chem. Soc.,* 61: 648.

Jones, L.H., Barfield, D., Barrett, J., Flook, A., Pollock, R. and Robinson, P., 1982, Cytology of *oil palm* cultures and regenerant plants. In: *Plant Tissue Culture 1982* (A. Fujiwara, ed.), pp. 727-728, The Jpn. Assoc. Plant Tissue Cult., Tokyo.

Kohlenbach, H.W. and Wernicke, W., 1977, Investigation of the inhibitory effect of agar and the function of activated carbon in anther culture, *Physiol. Plant.,* 86: 463-472.

Kovoor, A., 1981, *Palm Tissue Culture: State of the Art and its Application to the Coconut,* Food and Agriculture Organization of the United Nations.

Krikorian, A.D., 1982, Cloning higher plants from aseptically cultured tissues and cells, *Biol. Rev.,* 57: 151-218.

Krikorian, A.D., 1988, Tissue culture for tropical woody plants: The context, prospects and problems. In: *Tissue Culture of Forest Species,* Forest Research Institute Malaysia, Kuala Lumpur (in press).

Krikorian, A.D. and Kann, R.P., 1986, Oil palm improvement via tissue culture, *Plant Breed. Rev.,* 4: 175-202.

Krikorian, A.D., Kelly, K. and Smith, D.C., 1987, Hormones in tissue culture and micropropagation. In: *Plant Hormones and their Role in Plant Growth and Development* (P.J. Davies, ed.), pp. 593-613, Martinus Nijhoff, Dordrecht.

Krikorian, A.D., O'Connor, S.A. and Fitter, M.S., 1983, Chromosomal stability in cultured plant cells and tissues. In: *Handbook of Plant Cell Culture, Techniques for propagation and breeding* Vol. 1 (W.R. Sharp, D.A. Evans, P.V. Ammirato and Y. Yamada, eds.), pp. 541-581, Macmillan Publishing Company, New York.

Krikorian, A.D., Kann, R.P., O'Connor, S.A. and Fitter, M.S. 1986, Totipotent suspension as a means of multiplication. In: *Tissue Culture as a Plant Production System for Horticultural Crops,* (R. Zimmerman et al., eds.), pp. 61-72, Martinus Nijhoff/Dr.W. Junk, The Hague.

Kumar, P.P., Raju, C.R., Chandramohan, M. and Iyer, R.D., 1985, Induction and maintenance of friable callus from the cellular endosperm of *Cocos nucifera* L., *Plant Sci.,* 40: 203-207.

Lawrence, J.B., Villnave, C.A. and Singer, R.H., 1988, Sensitive, high-resolution chromatin and chromosome mapping *in situ:* Presence and orientation of two closely integrated copies of EBV in a lymphoma line, *Cell,* 52: 51-61.

Le Saint, J.P. and de Nuce de Lamothe, M., 1987, Dwarf coconut hybrids: Performance and value, *Oleagineux,* 42: 360-362.

Lioret, C. and Ollagnier, M., 1981, La culture *in vitro* de tissus chez le palmier a huile, *Oleagineux,* 36: 111-112.

Llyod, F.E., 1910, Development and nutrition of the embryo, seed and carpel in the date, *Phoenix dactylifera* L. *Missouri Bot. Garden Rep,* 21: 103-164.

Loomis, W.D., 1974, Overcoming problems of phenolics and quinones in the isolation of plant enzymes and organelles, *Methods in Enzymology,* 31A: 528-544.

Malaurie, B., 1987, L'embryogenese somatique en milieu liquide du palmier a huile: *Elaeis guineensis* Jacq. et *E. guineensis* x *E. melanococca,* premier resultats, *Oleagineux,* 42: 217-222.

Maramorosch, K., 1988, Non-chemical control of plant mycoplasma diseases. In: *Mycoplasma Diseases in Crops* (K. Maramorosch and S.P. Raychaudhuri), pp. 431-449, Springer-Verlag, New York.

Maramorosch, K. and Hunt, P., 1981, Lethal yellowing disease of coconut and other palms. In: *Mycoplasma Disease of Coconut and other Palms* (K. Maramorosch and S.P. Raychaudhuri, eds.), pp. 185-210, Academic Press, New York.

Martin, J.P. and Rabechault, H., 1978, Demande de brevet d'invention No. 7628361 Procede de multiplication vegetative et plantes ainsi obtenus, (See also Patent Specification 1584854, The Patent Office, London Feb. 18, 1981).

Mason, S.C., 1925, The minimum temperature for growth of the date palm and the absence of a resting period, *J. Agri. Res.,* 31: 415-453.

McClintock, B., 1978, Mechanisms that rapidly reorganize the genome, *Stadler Genetics Symposium,* Univ. of Missouri, Columbia, 10: 25-47.

Meunier, J., 1975, Le Palmier a huile americain *Elaeis melanococca, Oleagineux,* 30: 51-62.

Meunier, J., Vallejo, G. and Boutin, D., 1976 L'hybride *E. melanococca* x *E. guineensis* et son amelioration, *Oleagineux,* 31: 519-528.

Mohammed, S., 1985, Effects of gibberellin on fruit of date palm: A review, *Principes,* 29: 23-30.

Murashige, T. and Skoog, F., 1962, A revised medium for rapid growth and bioassays with tobacco tissue cultures, *Physiol. Plant.,* 15: 473-497.

Nevers, P., Shepherd, N.S. and Saedler, H., 1986, Plant transposable elements, *Adv. Bot. Res.,* 12: 103-203.

Ninan, C.A. and Raveendranath, T.G., 1965, A naturally occurring haploid embryo in the coconut palm (*Cocos nucifera* L.), *Caryologia,* 18: 619-623.

Nixon, R.W., 1966, Growing dates in the United States, *Agriculture Information Bulletin* No. 207, ARS, United States Department of Agriculture.

Nwankwo, B.A., 1981, Facilitated germination of *Elaeis guineensis* var. *pisifera* seeds, *Ann. Bot. (Lond.),* 48: 251-254.

Nwankwo, B.A. and Krikorian, A.D., 1982, Water as a storage medium for *Elaeis guineensis var. pisifera* seeds under aseptic conditions, *Ann. Bot. (Lond.),* 50: 793-798.

Nwankwo, B.A. and Krikorian A.D., 1983a, Aseptic storage of *Elaeis guineensis* var. *pisifera* seeds, *Principes,* 27: 34-37.

Nwankwo, B.A. and Krikorian, A.D., 1983b, Morphogenetic potential of embryo and seedling derived callus of *Elaeis guineensis* Jacq. var. *pisifera* Becc., *Ann. Bot. (Lond.),* 51: 65-76.

Obendorf, R.L. and Wettlaufer, S.H., 1984, Precocious germination during *in vitro* growth of soybean seeds, *Plant Physiol.,* 76: 1023-1028.

Oppenheimer, Ch. and Reuveni, O., 1972, *Development of a method for quick propagation of new and superior date varieties,* Agricultural Research Organization, The Volcani Center, Final Report of Research Conducted under grants authorized by U.S. Public Law 480, Bet Dagan, Israel, pp. 92.

Padmanabham, D., 1976, Rare axillary shoots in *Phoenix, Principes,* 20: 91-97.

Pannetier, C. and Buffard-Morel, J., 1982, Production of somatic embryos from leaf tissues of coconut (*Cocos nucifera* L.). In: *Plant Tissue Culture* 1982 (A. Fujiwara, ed.), pp. 755-756, The Jpn. Assoc. Plant Tissue Cult., Tokyo.

Pannetier, C. and Buffard-Morel, J., 1985, Coconut palm (*Cocos nucifera* L.). In: *Biotechnology in Agriculture and Forestry, Vol. I, Trees 1* (Y.P.S. Bajaj, ed.), pp. 430-450, Springer-Verlag, Berlin.

Paranjothy, K., 1982, A review of tissue culture of oil palm and other palms, Palm Oil Research Institute of Malaysia, (PORIM) Occasional Paper, No. 3.

Paranjothy, K., 1984, Oil palm. In: *Handbook of Plant Cell Culture, Vol. 3* (P.V. Ammirato, D.A. Evans, W.R. Sharp and Y. Yamada, eds.), pp. 591-605, Macmillan Publishing Company, New York.

Paranjothy, K., 1987, Recent developments in cell and tissue culture of oil palm. In: *Proceedings of a Symposium on Agricultural Applications of Biotechnology* (A.N. Rao and H.Y. Mohan Ram, eds.), pp. 83-95, COSTED and ANBS, Natn. Univ. Singapore.

Pesce, C., 1985, Oil Palms and other Oilseeds of the Amazon, Translated and edited from the original 1941 Portuguese, (D.V. Johnson, ed.), Reference Publications, Algonac, Michigan.

Popenoe, P., 1973, *The Date Palm* (H. Field, ed.), Field Research Projects, Coconut Grove, Miami, Florida.

Poulain, C., Rhiss, A. and Beauchesne, G., 1979, Multiplication vegetative en culture *in vitro* du palmier-dattier (*Phoenix dactylifera* L.), Academie d'Agriculture de France, *C. R. Seances,* 65: 1151-1154.

Purvis, C., 1956, The root system of the oil palm: Its distribution, morphology and anatomy, *West African Inst. Oil Palm Res. J.,* 1: 60-82.

Rabechault, H., 1962, Recherches sur la culture *in vitro* des embryos de palmier a huile (*Elaeis guineensis* Jacq.), 1. Effets de l'acide indolyl-acetique, *Oleagineux,* 17: 757-764.

Rabechault, H. and Martin, J.P., 1976, Multiplication vegetative du Palmier a huile (*Elaeis guineensis* Jacq.) a l'aide de cultures de tissus foliaires, *C. R. Acad. Sci. (Ser. D),* 283: 1735-1737.

Rabechault, H., Ahee, J. and Guenin, G., 1970, Colonies cellulaires et formes embryoides obtenues *in vitro* a partir de cultures de'embryons de palmier a huile (*E. guineensis* Jacq. var. *dura* Becc.), *C. R. Acad. Sci., (Ser. D),* 270: 3067-3070.

Rabechault, H., Martin, J.P. and Cas, S., 1972, Recherches sur la culture des tissus de palmier a huile (*Elaeis guineensis* Jacq.), *Oleagineux,* 27: 531-534.

Rajanaidu, 1980, Variation in the natural population of oil palm (*Elaeis guineensis* Jacq.) from Nigeria, Doctoral Thesis, University of Birmingham.

Raman, K., Walden, D.B. and Greyson, R.I., 1980, Propagation of *Zea mays* L. by shoot tip culture: A feasibility study, *Ann. Bot. (Lond.),* 45: 183-189.

Raveendranath, T.G. and Ninan, C.A., 1974, A study of somatic chromosome complements of tall and dwarf coconuts (*Cocos nucifera* L.) and its bearing on intervarietal variation and evolution in coconuts, *J. Plant. Crops,* 1: 17-22.

Rees, A.R., 1962, High temperature pre-treatment and the germination of seeds of the oil palm, *(Elaeis guineensis* Jacq.), *Ann. Bot. (Lond.),* 26: 569-581.

Reuveni, O., Adato, Y. and Lilien-Kipnis, H., 1972, A study of new and rapid methods for the vegetative propagation of date palms, pp. 17-24, In: *49th Ann. Rep. Date Growers' Institute,* Indio, California.

Reynolds, J.F., 1982, Vegetative propagation of palm trees, In: *Tissue Culture in Forestry* (J.M. Bonga and D.J. Durzan, eds.), pp. 102-107, Martinus Nijhoff/Dr. W. Junk, The Hague.

Reynolds, J.F. and Murashige, T., 1979, Asexual embryogenesis in callus cultures of palms, *In Vitro,* 15: 383-387.

Rhiss, M.A., 1980, Palmier dattier, Multiplication vegetative en culture *in vitro,* Doctoral Thesis, pp. 160, Universite de Paris-Sud Centre D'Orsay.

Ruer, P., 1968, Contribution a l'etude du systeme racinaire du palmier a huile, Universite de Paris.

Saaidi, M., Toutain, G., Bannerot, H. and Louvet, J., 1981, La selection du palmier-dattier (*Phoenix dactylifera* L.) pour la resistance au bayoud, *Fruits de'outre Mer* (Paris), 36: 241-249.

Sajise, J.U. and de Guzman, E.V., 1972, Formation of adventitious roots in coconut *maka puno* seedlings grown in medium supplemented with naphthaleneacetic acid, Kalikasan, *Philip. J. Biol.,* 1: 197-206.

Sarkar, S.K., 1957, Sex chromosomes in palms, *Genetica Iberica,* 9: 133-142.

Sato, D., 1946, Karyotype alteration and phylogeny, VI. Karyotype analysis in Palmae, *Cytologia,* 14: 174-186.

Schlichting, C.D., 1986., The evolution of phenotypic plasticity in plants, *Ann. Rev. Ecology Systematics,* 17: 667-693.

Schroeder, C.A., 1970, Tissue culture of date shoots and seedlings pp. 25-27, *Date Growers' Institute (Indio, California),* In: *47th Ann. Rep. Date Growers' Institute* Indio, Califonia .

Scorza, R., 1982, *In vitro* flowering, *Hortic. Rev.,* 4: 106-127.

Scowcroft, W.R., 1985, Somaclonal variation: The myth of clonal uniformity. In: *Genetic Flux in Plants* (B. Hohn and E.S. Dennis, eds.) pp. 217-245, Springer-Verlag, Vienna.

Selvaratnam, E.M., 1952, Embryo of the coconut, *Nature (Lond.),* 169: 714-715.

Sento, T., 1972, Studies on seed germination of palms, V. On *Chrysalidocarpus lutescens, Mascarena verschaffeltii,* and *Phoenix dactylifera, J. Jpn. Soc. Hortic. Sci.,* 41: 76-82.

Sento, T., 1974, Studies on seed germination of palms, VI. On *Cocos nucifera* L., *Phoenix humilis* Royle var. *hanceana* Becc. and *Phoenix sylvestris* Roxb., *J. Jpn. Soc. Hortic. Sci.,* 42: 350-358.

Sharma, D.R., Kumari, R. and Chowdhury, J.B., 1980, *In vitro* culture of female date palm (*Phoenix dactylifera* L.) tissues, *Euphytica,* 29: 169-174.

Smith, W.K. and Jones, L.H., 1970, Plant propagation through cell culture, *Chem. and Ind.* 44: 1399-1401.

Smith, W.K. and Thomas, J.A., 1973, The isolation and *in vitro* cultivation of cells of *Elaeis guineensis, Oleagineux,* 28: 123-127.

Srinivasan, C., Litz, R.E., Barker, J. and Norstog, K., 1985, Somatic embryogenesis and plantlet formation from Christmas palm callus, *Hortscience,* 20: 278-280.

Staritsky, G., 1970, Tissue culture of oil palm (*Elaeis guineensis* Jacq.) as a tool for its vegetative propagation, *Euphytica,* 19: 288-292.

Sudasrip, H., Kaat, H. and Davis, T.A., 1978, Clonal propagation of the coconut via the bulbils, *Philip. J. Coconut Stud.* 3: 5-14.

Tahardi, S. and Warda-Dalem, K., 1982, Kulture embrio kelapa kopyor *in vitro, Menara Perkebunan,* 50: 127-130.

Thanh-Tuyen, N.T. and de Guzman, E.V., 1983, Formation of pollen embryos in cultured anthers of coconut (*Cocos nucifera* L.), *Plant Sci. Lett.,* 29: 81-88.

Tisserat, B., 1979, Propagation of date palm (*Phoenix dactylifera* L.) *in vitro, J. Exp. Bot.,* 90: 1275-1283.

Tisserat, B., 1981, *Date palm tissue culture: Advances in Agricultural Technology,* ATT-W-17, United States Department of Agriculture Agricultural Research Service, pp.50.

Tisserat, B., 1983, Tissue culture of date palms - A new method to propagate an ancient crop - and a short discussion of the California date industry, *Principes,* 27: 105-117.

Tisserat, B., 1984a, Propagation of date palms by shoot tip cultures, *Hortscience,* 19: 230-231.

Tisserat, B., 1984b, Date palm, In: *Handbook of Plant Cell Culture, Vol. 2, Crop Species* (W.R. Sharp, D.A. Evans, P.V. Ammirato and Y. Yamada, eds.), pp. 505-545, Macmillan Publishing Company, New York.

Tisserat, B., 1984c, Clonal propagation: Palms. In: *Cell Culture* and *Somatic Cell Genetics, Vol. 1 Laboratory Procedures and their Applications* (I.K. Vasil, ed.), pp. 74-81, Academic Press, Orlando.

Tisserat, B., 1987, Palms. In: *Cell and Tissue Culture in Forestry Vol. 3, Case Histories: Gymnosperms, Angiosperms and Palms* (J.M. Bonga and D.J. Durzan, eds.), pp. 338-356, Martinus Nijhoff, Dordrecht.

Tisserat, B. and De Mason, D.A., 1985, Occurrence and histological structure of off-shoots and inflorescences produced from *Phoenix dactylifera* L. plantlets *in vitro, Bull. Torrey Bot. Club,* 112: 35-42.

Tomlinson, P.B., 1970, Monocotyledons - towards an understanding of their morphology and anatomy, *Adv. Bot. Res.,* 3: 207-292.

Torres, A.M. and Tisserat, B., 1980, Leaf isozymes as genetic markers in date palms, *Am. J. Bot.,* 67: 162-167.

Trabut, L., 1921, La multiplication du dattier par semis, *C. R. Acad. Agric. Sci.,* 33: 718-722.

Tsai, J.H., 1980, Lethal yellowing of coconut palms - Search for a vector. In: *Vectors of Plant Pathogens* (K.F. Harris and K. Maramorosch, eds.), pp. 177-200, Academic Press, New York.

Tsai, J.H. and Maramorosch, K., 1988, Lethal yellowing, *Rev. Trop. Plant Pathol.,* 4 (In Press).

Tsai, J.H. and Thomas, D.L., 1981, Transmission of lethal mycoplasma by *Mundus crudus.* In: *Mycoplasma Diseases of Trees and Shrubs* (K. Maramorosch and S.P. Raychaudhuri, eds.), pp. 211-229, Academic Press, New York.

Turner, P.D., 1977, Oil palm (*Elaeis guineensis* J.). In: *Plant Health and Quarantine in International Transfer of Genetic Resources* (W.B. Hewitt and L. Chiarappa, eds.), pp. 197-208, CRC Press, Boca Raton, Florida.

Turner, P.D., 1981, Oil palm diseases and disorders, Incorporated Society of Planters, Oxford University Press, Oxford.

Uhl, N.W. and Dransfield, J., 1987, Genera Palmarum. A classification of palms based on the work of Harold E. Moore Jr., The L.H. Bailey Hortorium and the International Palm Society, Allen Press, Lawrence, Kansas.

Vallade, J., 1966a, Aspect morphologique et cytologique de l'embryon quiescent d'*Elaeis guineensis* Jacq., *C. R. Acad. Sci., Ser. D.,* 262: 856-859.

Vallade, J., 1966b, L'evolution de l'embryon d'*Elaeis guineensis* Jacq. au cours de al germination, *C. R. Acad. Sci., Ser. D.,* 262: 989-992.

Vallade J., 1969, Organogenese embryonnaire chez *Elaeis guineensis* Jacq., *Rev. Cytol. Biol. Vegetale,* 32: 343-352.

Vouyouklis, G.V., 1981, Une methode de'isolement de protoplastes du Palmier a huile *Elaeis guineensis* Jacq., *Phyton* (*B. Aires*), 40: 169-178.

White, P.R., 1949, Growth hormones and tissue growth in plants. In: *Survey of Biological Progress, Vol. I* (G.S. Avery, Jr., ed.), pp. 267-280, Academic Press, New York.

Wicart, G., Mouras, A. and Lutz, A., 1984, Histological study of organogenesis and embryogenesis. in *Cyclamen persicum* Mill tissue cultures: Evidence for a single organogenetic pattern, *Protoplasma,* 119: 159-167.

Wooi, K.C., Wong, C.Y. and Corley, R.H.V., 1982, Tissue Culture of palms- A review. In: *Tissue Culture of Economically Important Plants* (A.N. Rao, ed.), pp. 138-144, COSTED and ANBS, Natn. Univ. Singapore.

Yuri, S.J.A., 1987, Propagation of Chilean wine palm *(Jubaea chilensis)* by means of *in vitro* embryo culture, *Principes* 31:183-186.

Zaid, A. and Tisserat, B., 1983, Morphogenetic responses obtained from a variety of somatic explant tissues of date palm, *Bot. Mag. Tokyo,* 96: 67-73.

Zaid, A. and Tisserat, B., 1984, Survey of the morphogenetic potential of excised palm embryos *in vitro, Hortic. Res.,* 24: 1-9.

Zuniga, L.C., 1959, Studies on *makapuno*-bearing trees, I. Segregation of the nut endosperm of artificially pollinated *Makapuno*-bearing and ordinary coconut trees, *Philip. J. Agric.,* 24: 51-67.

11

IN VITRO STRATEGIES FOR SANDALWOOD PROPAGATION

V.A. Bapat and
P.S. Rao

ABSTRACT

This paper comprises the details of investigations carried out on the regenerative potentialities of organ, callus, cell suspension and protoplast cultures of a tree species *Santalum album* (Sandalwood). Direct differentiation of multiple buds was observed from hypocotyl and nodal stem segments, whereas callus development and formation of somatic embryos was observed from stem internodal segments of mature 20-year-old trees. Callus tissues also yielded cell suspensions which turned out to be embryogenic. Endosperm isolated from mature fruits also callused and under appropriate conditions, differentiated somatic embryos. Somatic embryos obtained through such

V.A. Bapat and P.S. Rao * Plant Biotechnology Section, Bio-Organic Division, Bhabha Atomic Research Centre, Trombay, Bombay - 400 085, India.

diverse morphogenetic pathways developed into plantlets which were successfully established in the soil. Protoplasts were also successfully isolated from various explant sources. Stem callus protoplasts as well as protoplasts derived from cell suspensions underwent divisions, formed colonies and regenerated somatic embryos. Synthetic seeds were prepared by encapsulating somatic embryos in the alginate matrix. The potentiality of *in vitro* technology is discussed with respect to sandalwood multiplication and the preservation of germplasm.

INTRODUCTION

Numerous recent techniques of plant biotechnology are being widely used for rapid multiplication of elite plants either through the process of direct plant regeneration from cultured explants, or from callus and cell suspensions. Although several ornamental and herbaceous flowering plants are being widely propagated through tissue culture in several laboratories throughout the world (George and Sherrington, 1984), success in woody species is limited (Durzan, 1980; Mott, 1981; Bonga 1982; Bajaj, 1986). Afforestation projects launched by different agencies require a rapid system for multiplication of several forest trees. We have developed a model system for a tree species *(Santalum album)* which shows flexibility for the manipulation of several *in vitro* experiments. Cultures derived from various sources of this plant such as organ, callus, cell suspension and protoplasts, gave intense regeneration of somatic embryos and plantlets.

Santalum album (Sandalwood) is a prominent forest tree in India. Sandalwood occupies a pre-eminent position in Indian forestry and has been known from time immemorial for its sweet fragrance and medicinal properties and has been popularly called *fragrant gold*. India enjoys a near monopoly in the world sandalwood trade, which fetches millions of rupees annually in foreign exchange. However, the sandalwood tree suffers from a large number of diseases of which the spike disease is the most destructive. This disease has acquired a virulent epidemic status in major areas of sandalwood forest in India, greatly reducing the yield of heartwood. These factors call for concerted efforts in the preservation and protection of existing sandal plantations and the development of alternative techniques for rapid multiplication of resistant trees.

This paper describes the sustained investigations carried out over the past decade at the Plant Biotechnology Section of the Bhabha Atomic Research Centre (BARC) on developing micropropagation technology for sandalwood. The results include some of the published work (Rao and Bapat, 1978; Bapat and Rao, 1979, 1984; Rao and Raghava Ram, 1983; Rao et al., 1984; Bapat et al., 1985 and Rao and Ozias-Akins, 1985).

MATERIALS AND METHODS

Hypocotyl and Stem Segments

Hypocotyl segments of five mm length were obtained from four-week-old seedling grown *in vitro*. Surface sterilised stem segments from 20-year-old mature trees with desirable characteristics were also excised and placed aseptically on a basal nutrient medium. The composition of basal nutrient medium used in the initial experiments was as follows (mg l^{-1}): KNO_3 (1,900), NH_4NO_3 (1,650), $CaCl_2.2H_2O$ (440), $MgSO_4.7H_2O$ (370), KH_2PO_4 (170), $MnSO_4.4H_2O$ (25), H_3BO_3 (10), $ZnSO_4.7H_2O$ (10), Na_2EDTA (18.6), $FeSO_4.7H_2O$ (13.9), myo-inositol (100), nicotinic acid (5), folic acid (5), glycine (2), pyridoxine-HCl (0.5), thiamine-HCl (0.5), biotin (0.05), sucrose (2 percent) and agar (0.6 percent). The basal medium (BM) was supplemented with auxins, cytokinins and other growth adjuvants, as and when required, either singly or in combinations. In some experiments, MS (Murashige and Skoog, 1962) basal medium was used. The pH of the medium was adjusted to 5.8 before jellying the medium with agar. The cultures were grown under continuous fluorescent light (1,000 lux) at $25 \pm 2°C$ and an RH of 55-60 percent. The endosperm tissue was excised from mature seeds and was cultured on MS medium fortified with growth substances.

Cell Suspension Culture

Cell suspensions were initiated from friable stem callus growing on MS + 2,4-D (1 mg l^{-1}). 300 mg of callus was placed in 50 ml of liquid nutrient medium. Under agitation at 120 rpm, the cultures produced a fine suspension of single cells and few cell aggregates. Suspension cultures were sub-cultured every seventh day by removing half of the medium and replacing it with an equal volume of fresh medium. In order to ascertain the embryogenic potentiality of suspensions, cells were transferred to various liquid and agar media.

For the encapsulation of embryogenic cell suspensions, three-week-old actively growing cell suspensions were filtered on the nylon net and washed twice with the MS basal medium. Washed cells were mixed (1:1) with 10 ml of 2.5 percent Na - alginate solution. The mixture was pipetted dropwise into 50 ml of MS basal medium containing $CaCl_2.2H_2O$ (1.036 g/150 ml). Beads of 5-10 mm formed by this technique were shaken for 1 h and washed twice with the nutrient medium. One set of beads was transferred to 25 ml medium in 100 ml flasks (10 beads/flask) and kept on shaker at $25 \pm 2°C$ in continuous light. The other set of beads was stored at 4°C in a petri dish on a filter paper soaked in MS basal medium.

Protoplast Culture

Protoplasts were isolated from various sources, such as leaf mesophyll, stem and hypocotyl callus as well as cell suspensions. One gram callus tissue was transferred to 15 ml conical centrifuge tubes, gently macerated with a glass rod in 10 ml nutrient medium and centrifuged to obtain the pellet. The pellet was incubated in the enzyme mixture. For mesophyll cells, young leaves from mature trees were surface sterilised, macerated with 0.3 M mannitol and transferred to the enzyme mixture. Different enzyme mixtures containing combinations of cellulase (R-10, Onozuka), macerozyme (Kinki Yakult), hemi-cellulase (Sigma) and pectinase (Sigma) were employed in various concentrations along with 0.55 M mannitol as an osmoticum. The pH of the enzyme solution was adjusted to 5.5. Following the incubation of tissue in the enzyme, the suspension was passed through a nylon mesh (100μ) and centrifuged at 40 rpm for 3 min. The enzyme was decanted and the pellet was washed twice with the nutrient medium. To separate the protoplasts from the debris, the pellet was floated on 20 percent sucrose and following centrifugation, the top layer, which consisted only of the protoplasts, was pipetted out and washed twice with the nutrient medium. The protoplast suspensions were cultured in thin layers in 6 cm diameter plastic petridishes in various liquid nutrient media (2 ml). After the formation of microscopic colonies, osmoticum was reduced to 0.3 M level by transferring the cultures to a fresh medium. Subsequently, 2 ml of nutrient medium containing 0.4 percent soft agar was incorporated in petridishes containing 2 ml of protoplast suspension. The protoplast density was 4 to 5 x 10^5/ml.

Synthetic Seed Preparation

Somatic embryos, carefully isolated from suspension cultures, were blot dried on a filter paper and were mixed for a few seconds with a gel of sodium alginate (Sigma) prepared in the MS basal medium. The embryos, mixed well in the gel were picked up by means of a forcep and dropped into a solution of $CaCl_2.2H_2O$ (1.036g/ 150 ml). The drops (beads) containing a single embryo were allowed to remain in this solution for 40 min. on a gyratory shaker (80 rpm) in light (950 lux). After the incubation period, the beads were recovered by decanting off the $CaCl_2.2H_2O$ solution and the beads were then washed 3-4 times with the MS basal medium. These encapsulated embryos were subsequently cultured on the nutrient medium. One set of encapsulated embryos was stored at 4°C for 45 days on a filter paper soaked in the MS basal medium. After the storage period, the beads were again cultured in the nutrient medium.

RESULTS

Shoot Bud Differentiation in Hypocotyl Segments

Cultured hypocotyl segments differentiated shoot buds on BM and these appeared as small, green protuberances. The growth of buds did not continue on the BM. However, the incorporation of auxins such as IAA, IBA, NOA or NAA to BM proved effective for further differentiation of shoot buds. Among the cytokinins, Ad, BAP, Kn, and Zn, promoted bud initiation (Fig. 1.1). Bud formation was optimum on BAP and 15-20 buds originated on a single explant. The buds developed into young, green, leafy shoots which attained a height of 25 to 30 mm in 6-8 weeks. Approximately 90 percent of the hypocotyl explants excised from the basal region (with respect to germinated seedling) showed potential for producing buds. The potential decreased to 80 and 60 percent for those obtained from the middle and apical region, respectively. The position of the explant on the nutrient medium also influenced bud formation. Only when the root end of the hypocotyl segment was dipped in the nutrient medium, 100 percent bud formation was observed. On reversing the position, bud formation dropped to 10 percent. In another treatment, thin discs were excised from hypocotyl and cultured. In 70 percent of the cultures, shoot buds originated from all sides of the disc.

In an effort to develop complete plantlets with roots, regenerated shoot buds were excised and individually placed on a variety of root-inducing media. Rooting was observed in a few cultures on BM + NAA (0.5 mg l^{-1}) + IBA (0.5-5 mg l^{-1}).

Nodal stem segment consisting of two axillary buds, when placed on MS + BAP (1 mg l^{-1}), produced multiple shoot buds after three weeks of culture. Other cytokinins such as Kn and Zn did not induce such response from the axillary buds. To induce roots from isolated shoot buds, a variety of hormonal combinations were attempted, but the attempts were unsuccessful.

Somatic Embryogenesis in Hypocotyl and Internodal Stem Segments Callus

When the hypocotyl explants from which the regenerated shoot buds had been removed were transferred to a fresh medium of the same composition, new shoot buds arose and this process continued. However, in about 20 percent cultures, extensive proliferation took place in the explants, resulting in a large callus mass in which, four weeks later, numerous somatic embryos of pre-globular stages differentiated (Fig. 1.2). To maintain a continuous supply of embryos as well as to enhance the embryogenic potential of somatic callus, portions of embryogenic callus were grown in BM enriched with growth substances. On BM + IAA (1 mg l^{-1}), or NAA (1 mg l^{-1}), embryos rooted readily and shoot development was moderate, whereas on BM + 2,4-D (1 mg l^{-1}) all the embryos proliferated into a callus tissue.

The internodal stem segments proliferated into a rapidly growing callus on MS + 2,4-D (1 mg l^{-1}) + Kn (0.2 mg l^{-1}) within eight weeks of culture. Callus tissues, upon transfer to MS + IAA (1 mg l^{-1}) + BAP (1 mg l^{-1}), differentiated numerous somatic embryos and plants (Fig. 1.5).

In order to maintain a continuous supply of embryos, portions of embryogenic callus were excised and transferred to fresh media of the same composition. At the end of six weeks, the sub-cultured callus produced several mounds of fresh calli in which numerous embryos differentiated. In many cultures, young embryos, instead of developing into plantlets, differentiated again and produced a callus mass which again

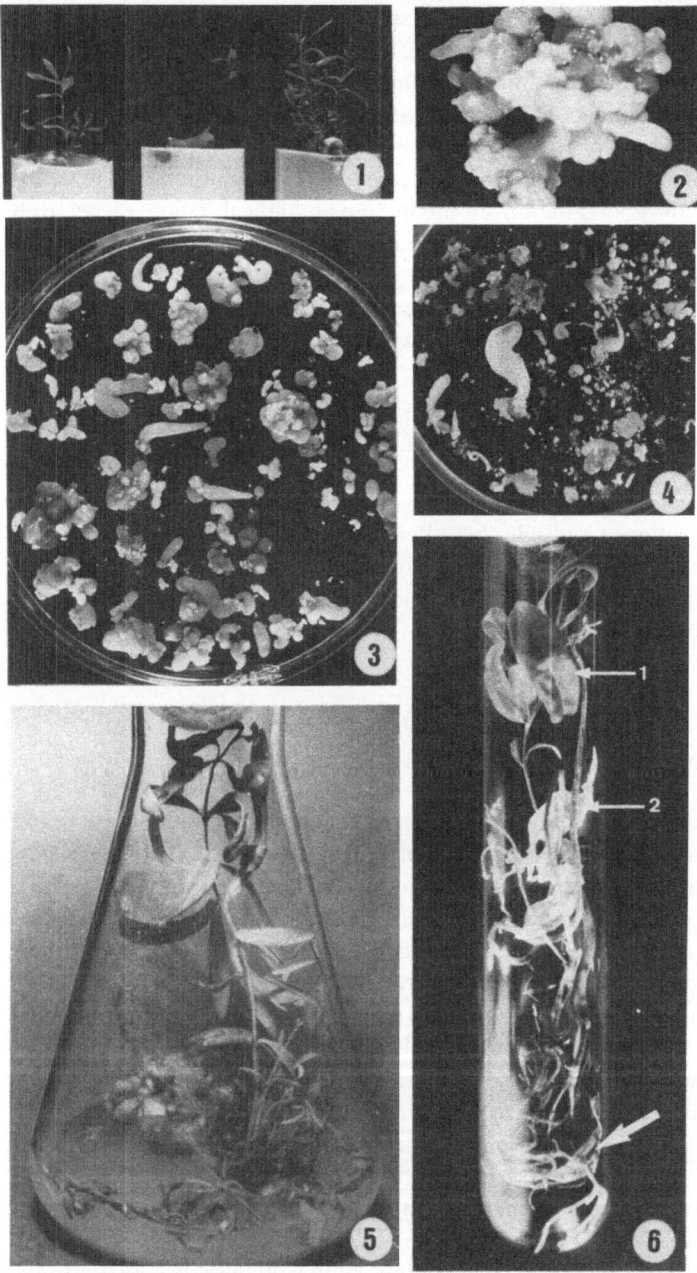

Fig. 1: Tissue Culture of Sandalwood. 1. Direct induction of multiple buds on hypocotyl explant, 2. Somatic embryos differentiated from callus. 3,4. Embryos from suspension cultures. 5. Plantlet development from somatic embryos. 6. Sandalwood and host plant. Note the attachement of *Santalum* with the roots of *Cajanus:* 1. *Santalum*, 2. *Cajanus*. Arrow indicates the attachment of roots.

showed intense regeneration of somatic embryos. The differentiation of embryos in the callus was a continuous and non-synchronous process, since embryos with just a few cells were found intimately mixed with mature embryos and young plantlets. Interestingly, there were many embryos arising by budding from other embryos. Periodic sub-culturing of the tissue was done at an interval of four to six weeks by transferring it to a fresh medium of the same composition. The tissue that had undergone several passages did not show any decline in the number of embryos being produced per culture.

Embryos at the dicotyledonous stage were isolated from the callus and were grown on MS + IAA (1 mg l^{-1}) + IBA (0.5 mg l^{-1}) + GA$_3$ (0.5 mg l^{-1}) + sucrose (5 percent) for plantlet development. The rooted plantlets were transferred to paper cups and were subsequently shifted to earthen pots. The plantlets grew vigorously and produced normal flowers within a span of 18 months.

Somatic Embryogenesis in Endosperm Callus

Endosperm tissue cultured on MS + 2,4-D (1 mg l^{-1}) alone or in combination with Kn (0.2 mg l^{-1}) proliferated into a callus. On transfer to a fresh medium, a good proliferating callus tissue was obtained which could be sub-cultured indefinitely. A large number of embryos differentiated on transfer of this tissue to MS + IAA (1 mg l^{-1}) + BAP (1 mg l^{-1}) + sucrose (5 percent).

One of the characteristic features of *Santalum* is that it is an obligatory root parasite. It has been observed that sandal draws its nutrients partly from the soil through the root ends and partly from host plants through a haustorial connection. In the conventional technique, a seedling of *Santalum* is placed near a host plant and such seedlings show vigorous growth as compared to those grown in the absence of the host. Hence, we investigated the possibility of growing the sandalwood plantlets with seedling of *Cajanus in vitro*. The sandalwood plantlet from somatic embryo produced the haustoria which got attached to the roots of *Cajanus* within eight weeks. After the connection was established (Fig. 1.6), the sandalwood plantlet grew rapidly and appeared very robust and healthy. This demonstration of *in vitro* host-parasite relationship would be of great significance for high percentage survival of tissue culture raised plants. In addition, many other studies could be conducted under aseptic conditions on the host-parasite relationship.

Suspension Cultures

Cell suspensions were established in MS liquid medium supplemented with 2,4-D (1 mg l^{-1}). The growth of the cells was so rapid that the medium was filled with small clump of single cells and cell aggregates. The nature of the suspension had an effect on the rapidity with which the somatic embryos were formed. Plating of cells from suspensions on MS + IAA (0.5 mg l^{-1}) + BAP (0.5 mg l^{-1}) resulted in the formation of a proembryogenic mass followed by the differentiation of somatic embryos. However, a large number of organised embryos were obtained in liquid medium when 2,4-D was replaced with IAA and BAP (0.5 mg l⁻¹ each; Fig. 1.3,4). Such embryos in liquid medium were independent of each other. Mature, organised embryos developed into plantlets on transfer from liquid to agar medium.

Cell suspension encapsulated in alginate as beads and stored at 4°C for 45 days started liberating cells in the surrounding liquid medium within 8-10 days after transfer to 25°C from 4°C. The free cells in the liquid medium [MS + IAA (0.5 mg l⁻¹) + BAP (0.5 mg l⁻¹) + GA₃ (0.1 mg l⁻¹)] grew rapidly after 25-30 days. Such suspensions mainly consisted of single cells, cell aggregates and chain of cells. On plating on agar medium, colonies were formed after three weeks which showed embryo differentiation.

Protoplasts

(i) Protoplasts from hypocotyl callus : In the case of hypocotyl, maximum release of protoplasts could be obtained by the enzyme mixture comprising cellulase (2 percent), pectinase (1 percent) and hemi-cellulase (1 percent) after 7 h of incubation. The first division occurred after five days of culture, followed by repeated divisions, resulting in clearly distinguishable colonies at the end of two weeks.

(ii) Protoplasts from stem callus : Maximum yield of protoplasts (8.73 x 10⁶) was obtained when the actively growing stem callus was incubated in cellulase (1 percent) + macerozyme (0.5 percent). The cultured protoplasts regenerated their cell wall within 36 to 48 h: this was detected by the asymmetrical shape assumed by viable protoplasts as well as by calcofluor staining. About 60 to 70 percent of the protoplasts divided 50 percent of such divided protoplasts showed sustained divisions leading to colonies (Fig. 2). The medium consisted of V-47 (Binding, 1974) major salts; MS minor salts and iron; LS vitamins, sucrose and glucose (2 percent each), powdered charcoal (0.1 percent) and hormones: 2,4-D, NAA and BAP (1 mg l⁻¹ each). On this medium, the colonies grew rapidly and within 8 to 10 weeks developed into microcalli. On MS + IAA (1 mg l⁻¹), such colonies grew into callus which differentiated into numerous somatic embryos. The embryos developed into plantlets.

(iii) Protoplasts from mesophyll cells : An enzyme combination of cellulase (2 percent), macerozyme (1 percent), hemi-cellulase (1 percent) and 0.8M mannitol gave a very satisfactory yield of protoplasts. Optimum release of protoplast was observed after 8 h. However, leaf mesophyll derived protoplasts did not show any divisions.

(iv) Protoplasts from suspension cultures : A satisfactory yield of protoplasts was obtained by using a combination of macerozyme (1 percent), driselase (1 percent) and cellulase (1 percent) for the isolation of protoplasts from suspension cultures. Various basal media with BA and 2,4-D in different concentrations and combinations were tested for their ability to induce divisions. Optimum protoplast division was noticed on V-47 medium supplemented with BA and 2,4-D. The first division in the regenerated protoplasts occurred on the third day followed by rapid subsequent divisions. Visible

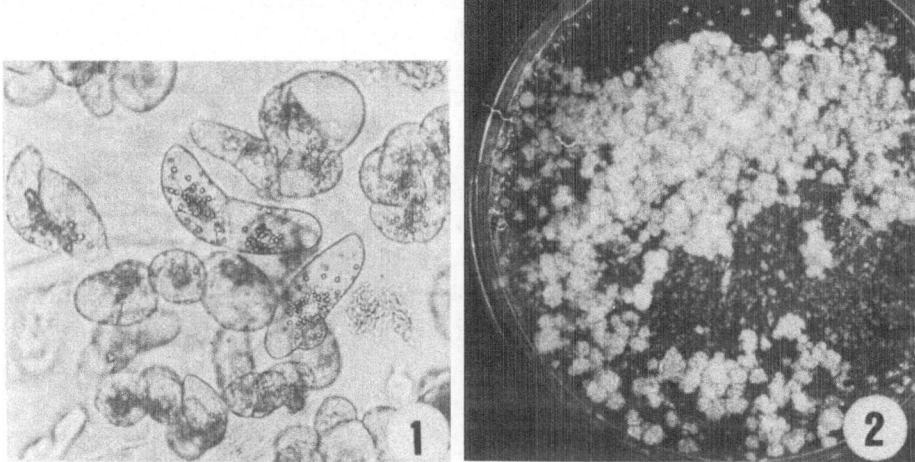

Fig. 2: Protoplast Culture of Sandalwood. 1. Dividing protoplasts. 2. Colonies from divided protoplasts

colonies had developed by the third week and these grew further when diluted with the fresh medium. On transfer of such colonies to MS + IAA (1 mg l⁻¹) + BAP (1 mg l⁻¹) + casamino acid (400 mg l⁻¹), proembryos developed. Many embryos were produced in scattered clusters on a large matrix of soft, embryogenic callus. Complete plantlets were obtained from somatic embryos upon transfer to an appropriate medium.

Synthetic Seeds

The cultured encapsulated somatic embryos (Fig. 3,1) on MS basal medium showed no signs of germination in the first six weeks of culture. At the end of eight weeks, in 10 percent of the prepared beads, a portion of the root from the encapsulated embryos pierced through the matrix and established a direct contact with the medium. The addition of IAA, NAA, IBA or BAP (1 mg l⁻¹) separately to MS did not influence the germination of embryos. Subsequently, the shoot portion also emerged from the beads. On further transfer to MS + IAA (1 mg l⁻¹) + IBA (0.5 mg l⁻¹) + GA₃ (0.5 mg l⁻¹) + sucrose (5 percent), the germinated portion elongated into shoot and the root (Fig. 3.2), eventually leading to a well developed plantlet at the end of 16 weeks. Among the various concentrations of alginate tested (2,3 and 4 percent), best results were obtained with 3 percent alginate. Depending on the size, embryos curved inside the alginate matrix or sometimes formed oblong shaped beads. The encapsulated embryos stored at 4°C for 45 days germinated on transfer to MS basal medium at

$25 \pm 2^\circ$C. However, the germination rate was much reduced (3 percent). Different nutrient sources were tested to observe the rate of germination of synthetic seeds; however, maximum germination was obtained on MS basal medium. On tap water without any nutrient, seeds initially showed signs of germination but subsequently necrosed. The synthetic seeds were also sown directly in the soil, but so far no germination has been observed. On MS + BAP (0.5 mg l^{-1}) + IAA (0.5 mg l^{-1}) the encapsulated embryos, instead of germinating, started producing secondary embryos which came out of the beads after six weeks of culture. Such secondary embryos, on separation from the original embryo, developed into plantlets. The set of encapsulated embryos stored at 4°C for 45 days also retained the capacity to produce secondary embryos when cultured on MS + BAP (0.5 mg l^{-1}) + IAA (0.5 mg l^{-1}) at $25 \pm 2^\circ$C.

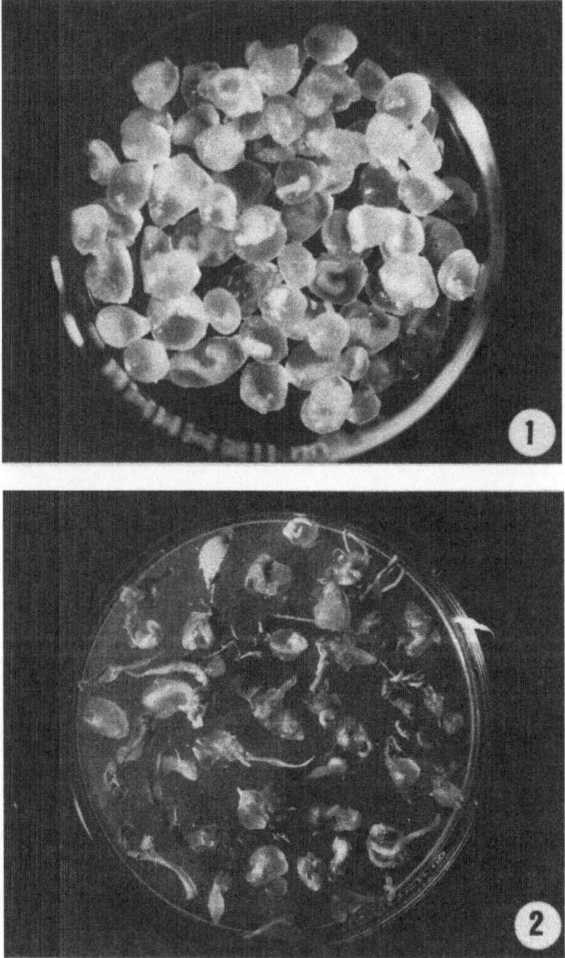

Fig. 3: "Synthetic Seeds" of Sandalwood . 1. Encapsualted embryos . 2. Germinated embryos

DISCUSSION

This paper has reviewed the different parameters studied during the investigations on sandalwood, relating to micropropagation, cell and protoplast culture and preparation of synthetic seeds. The results point to the standardisation of the technique for obtaining plant regeneration through organogenesis and embryogenesis in the cultured hypocotyl and stem explants of *Santalum album*. The embryogenic potentiality of *Santalum* cultures did not decline during serial sub-cultures, as is observed in many tissues. This is of considerable interest because it leads to the possibility of developing a large number of sandalwood plantlets from somatic embryos by harvesting them at regular intervals and on a continuous basis. The process would, therefore, offer a dependable system for efficient micropropagation. Significant additional benefit can accrue from the early flowering of *in vitro*-raised plants in the field. This could allow the breeding cycle to be shortened further.

The demonstrated feasibility of propagating plantlets from somatic embryos in *Santalum* is being followed up by several lines of research. Foremost is the success in protoplast culture and the regeneration of sandalwood protoplasts. This is significant as protoplast culture work has been attempted only in a restricted number of tree species. The results obtained in protoplast technology of sandalwood could form the basis of parasexual hybridisation and genetic modification in genetic engineering programs. It would be possible to expose protoplasts to mutagens or ionising radiations and then rear them to plants to introduce genetic variations. One could also envision the protoplast fusion between disease resistant and high oil-yielding but susceptible trees.

The encapsulation of somatic embryos offers a technology of practical application, both as a simple delivery system for distribution among different areas/zonal plantations as well as the preservation of germplasm. The main emphasis of our work was to demonstrate the retention of the viability of embryos after encapsulation in the alginate matrix and ensure the survival of encapsulated embryos from desiccation. At present, the germination rate of synthetic seeds is low; however, the refinement of the technique, the selection of a proper nutrient medium and the use of other matrices could possibly enhance the germination rate of encapsulated embryos. The revival of the embryogenic cell line after storage for 45 days at 4°C and the low but consistent survival of somatic embryos from such suspension, are of great advantage and would be a powerful tool to preserve the desirable elite genotypes. The results reported here can open up new strategies for the propagation approach, especially in trees.

REFERENCES

Bajaj, Y.P.S., 1986, Biotechnology of tree improvement for rapid propagation and biomass energy production. In: *Biotechnology in Agriculture and Forestry* (Y.P.S. Bajaj, ed.), pp. 1-23, Springer-Verlag, Berlin.

Bapat, V.A. and Rao, P.S., 1979, Somatic embryogenesis and plantlet formation in tissue cultures of sandalwood (*Santalum album* L.), *Ann. Bot.,(Lond.)* 44: 629-630.

Bapat, V.A. and Rao, P.S., 1984, Regulatory factors for *in vitro* multiplication of sandalwood tree (*Santalum album* L.). I. Shoot bud regeneration and somatic embryogenesis in hypocotyl cultures, *Proc. Indian Natn. Sci. Acad.,* 93: 19-27.

Bapat, V.A., Gill, R. and Rao, P.S., 1985, Regeneration of somatic embryos and plantlets from stem callus protoplasts of sandalwood tree *(Santalum album),* Curr. Sci., 54: 978-982.

Binding, H., 1974, Regneration von haploiden und diploiden Pflanzen aus protoplasten von *Petunia hybrida* L. *Z. Pflanzenphysiol.,* 74: 327-356.

Bonga, J.M., 1982, Vegetative propagation of mature trees by tissue culture. In: *Tissue Culture of Economically Important Plants,* (A.N. Rao, ed.), pp. 191-196, COSTED and ANBS, Natn. Univ. Singapore.

Durzan, D., 1980, Progress and promise in forest genetics, Paper Science and Technology, The cutting edge, Inst. Paper Chem. Milwaukee, WI, pp. 31-59.

George, E.F. and Sherrington, P.D., 1984, *Plant Propagation by Tissue Culture: Handbook and Directory of Commercial Laboratories Ltd.,* Exegetics Limited, Hants.

Mott, R.L., 1981, Trees. In: *Cloning Agriculture Plants via in Vitro Techniques* (B.V. Conger, ed.), pp. 217-254, CRC Press, Boca Raton, Florida.

Murashige, T. and Skoog, F., 1962, A revised medium for rapid growth and bioassays with tobacco tissue cultures, *Physiol. Plant.,* 15: 473-497.

Rao, P.S. and Bapat, V.A., 1978, Vegetative propagation of sandalwood plants through tissue culture, *Can. J. Bot.,* 56: 1153-1156.

Rao, P.S. and Ozias-Akins, P., 1985, Plant regeneration through somatic embryogenesis in protoplast cultures of sandalwood (*Santalum album* L.), *Protoplasma,* 124: 80-86.

Rao, P.S. and Raghava Ram, N.V., 1983, Propagation of sandalwood (*Santalum album* L.) using tissue and organ culture technique. In: *Plant Cell Cultures in Crop Improvement* (S.K. Sen and K.L. Giles, eds.), pp. 119-124, Plenum Press, New York.

Rao, P.S., Bapat, V.A. and Mhatre, M. 1984, Regulatory factors for *in vitro* multiplication of sandalwood tree *(Santalum album),* II. Plant regeneration in nodal and internodal stem explants and occurrence of somaclonal variations in tissue culture raised plants, *Proc. Indian Natn. Sci. Acad.,* 50: 196-202.

12

MICROPROPAGATION OF
FICUS AURICULATA LOUR *

Nirmala Amatya and
S.B. Rajbhandary

ABSTRACT

Cotyledonary nodes obtained from the aseptic culture of *Ficus auriculata* were cultured in the MS medium Murashige and Skoog (1962) supplemented with BAP (1 mg l⁻¹) and NAA (0.01 mg l⁻¹) to induce shoot formation. Multiple shoots thus formed were sub-cultured every eight weeks to produce more shoots. Rooting of microshoots occurred in a mixture of sand and dry leaf powder (2:1, v/v). Rooted plantlets were established in the field.

INTRODUCTION

Fodder trees play an important role in Nepal, because they constitute 40 percent of the livestock feed (Panday, 1982). *Ficus auriculata* is one of the popular

Nirmala Amatya * Tissue Culture Laboratory, Botanical Survey and Herbarium. Godawari, Kathmandu, Nepal. *S.B. Rajbhandary* * Department of Medicinal Plants. Thapathali. Nepal.

* Synonym *Ficus roxburghii* Wall.

fodder trees distributed over altitudes between 800-2000 m in Nepal. Despite their usefulness, seed propagation is rarely practised. Vegetative propagation by cuttings has also been reported (Singh, 1982). But from cuttings, only a small number of clonal plants can be produced in a year. Thus, for a large scale distribution of planting stock, rooted cuttings offer limited potential to meet the demand. An alternative approach to rooted cuttings for cloning is micropropagation which allows plant production on a scale of over one million plants from a single explant in a year.

This paper discusses *in vitro* plant regeneration of *Ficus auriculata* and its subsequent field establishment, using cotyledonary nodes as explants.

MATERIALS AND METHODS

Seeds of *Ficus auriculata* were washed for one hour in running tap water, followed by surface disinfection with a detergent (two drops of Teepol in 100 ml of water). The seeds were then washed three times with distilled water. Finally, they were sterilized in 0.1% mercuric chloride solution for 15 minutes and then washed three times with sterile distilled water to remove the sterilant. Fifty seeds were placed, 10 in each 100 ml flask containing MS basal medium. Seeds germinated within three weeks, and the germination percentage was 75. This experiment was repeated three times.

Cotyledonary nodes, of germinated plants measuring 2-4 mm, were cultured on MS medium with benzyl aminopurine (BAP) at 1 mg l^{-1}, naphthalene acetic acid (NAA) at 0.01 mg l^{-1} and casein hydrolysate (CH) at 1 gm l^{-1}. The medium was solidified with 0.7 percent agar and pH was adjusted to 5.8 before autoclaving. The medium was autoclaved under 15 lb sq. inch^{-1} for 15 min. The cultures were incubated at 16 h photoperiod at 25 $\pm$ 4°C. Light was provided by cool fluorescent tubes at 3000 lux.

Within four weeks, the explants gave rise to a number of shoot buds which subsequently developed into new shoots. Sub-culturing was carried out at intervals of 8-10 weeks. For rooting, microshoots measuring 2-4 cm were excised and treated with 100 ppm IAA solution for five minutes and were planted in boxes containing a non-sterile mixture of two parts of sand and one part of powdered dry leaves (v/v). To see the rooting behavior at different periods, 30 shoots were transferred to the boxes every week, throughout the year. Rooted plants that had grown to a height of 5-6 cm were transferred to the soil and maintained under greenhouse conditions before planting them out in the field.

RESULTS

After 10 weeks of culture, each cotyledonary node explant produced 10-12 shoots. The individual shoots, when sub-cultured every eight weeks on the identical medium, gave rise to 20-25 shoots in a 250 ml flask (Fig. 1). This rate of proliferation was maintained even after two years of 8-10-weekly sub-cultures. Treatment of the excised microshoots with IAA (100 ppm) apparently stimulated the induction of roots after four weeks of plantation in the non-sterile sand box.

Rooting of microshoots was not observed throughout the year (Table 1). Maximum rooting (80 percent) was observed during the months of May and June when the average day/night temperatures were 34°C/13°C and 34°C/17°C, respectively. There was no root development during the months of January and February when the average day/night temperatures were 24°C/4°C and 26°C/6°C, respectively. After three weeks of root development in the sand box, the rooted cuttings were transferred to soil in the pots for hardening in the greenhouse. After four weeks, the rooted cuttings were ready for transplanting in the field (Fig. 2). Once rooted, the rate of plant survival during soil transfer was 80 percent. To date, 200 plants of *F. auriculata* have been produced by the tissue culture technique described above and 75 plants have been distributed to the different altitudinal zones of Nepal for field trials. One such plant planted at Godawari in August 1987, grew to a height of 55 cm in six months (Fig. 3). The height of the plant was 12 cm at the time of planting.

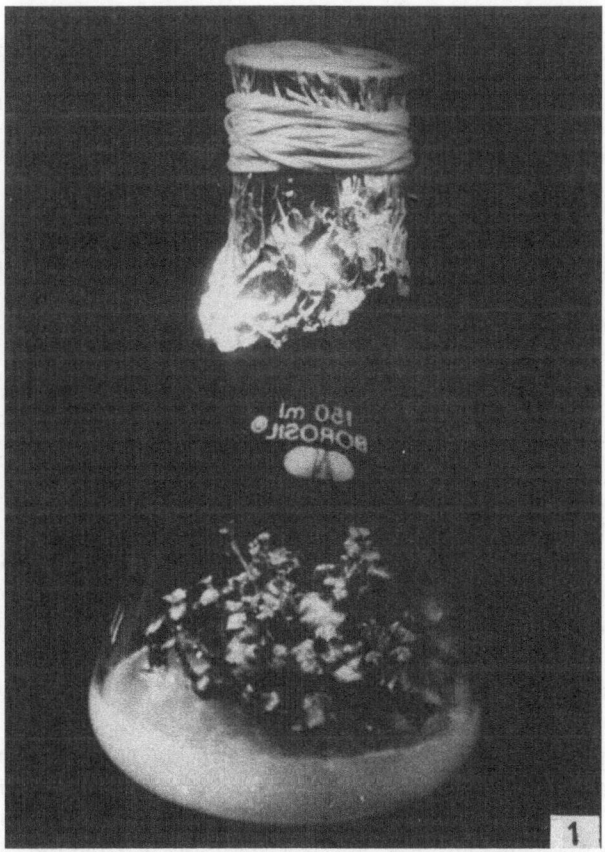

Fig. 1. Multiple shoot formation on MS with 1.0 mg l⁻¹ BAP and 0.01 mgl⁻¹ NAA after eight weeks in culture.

Table 1: Effect of monthly mean maximum and minimum day/night temperatures on rooting of microshoots of *Ficus auriculata*. Thirty shoots, each 2-4 cm long, were transferred each week.

Month	Average maximum temperature (°C)	Average minimum temperature (°C)	Rooting percentage
Nov. (1986)	29	11	31
Dec. (1986)	24	6	10
Jan. (1987)	24	4	0
Feb. (1987)	26	6	0
Mar. (1987)	26	8	10
Apr. (1987)	28	9	35
May (1987)	34	13	80
June (1987)	34	17	80
July (1987)	31	18	65
Aug. (1987)	27	18	50
Sept. (1987)	29	19	40
Oct. (1987)	29	15	38

DISCUSSION

There has recently been an increasing awareness of the importance of improvement programs for forest trees. As a result, various advances in the development of tissue culture techniques for forest trees species have been reported (Bonga, 1987).

A number of papers on tree tissue culture (Bonga, 1985) reflect the growing interest of tissue culturists in exploiting the technology to produce clean and healthy planting stock. Emphasis has also been laid on the development of cost-effective micropropagation methods for forest trees (Brown and Sommer, 1982; McCown, 1985).

In tree breeding, a rapid gain in genetic improvement can be achieved through cloning of selected individuals. Tissue culture provides the means to produce a large number of clonal propagules for commercial forestry plantations. In order to maintain genetic stability in regenerants, it is, however, desirable to avoid the callus phase because it is known that plant regeneration via the callus phase generally results in chromosome changes in its progeny (D'Amato, 1977).

Fig. 2: Rooted plantlet seven weeks after the microshoot was transferred to sand bed box.

Since plant regeneration in the present study is through multiple shoot formation from the cotyledonary node (pre-formed meristem), it may be assumed that the established plants are all true-to-type. Further, the method can possibly be applied to clone the selected mature trees using explants with meristems such as shoot tips and nodal segments.

The method of avoiding *in vitro* rooting by directly transferring the multiple shoots into a non-sterile sand box, as employed in the present study, and our earlier communications (Karki and Rajbhandary, 1984; Manandhar and Rajbhandary, 1986; and Suwal et al., 1987) indicate that the production cost of tissue culture plants can be drastically reduced. Hartney (1982) and Ahuja (1986) have also advocated that the production cost of tissue culture plants can be reduced if rooting is done under *in vivo* conditions.

Since the rate of shoot bud proliferation remained unchanged over two years of 10-weekly sub-cultures, it may be reasonable to assume that the method described above offers a means to obtain over a million plants of *F. auriculata* in a year from a single cotyledonary node.

Fig. 3: Six-month-old field established plant.

ACKNOWLEDGEMENTS

The authors are grateful to Dr. S.B. Malla, Director General, Department of Medicinal Plants, Ministry of Forest and Soil Conservation, Nepal, for the facilities provided.

REFERENCES

Ahuja, M.R., 1986, Perspectives in plant biotechnology, *Curr. Sci.,* 55: 217-224.

Bonga, J.M., 1977, Applications of tissue culture in forestry. In: *Applied and Fundamental Aspects of Plant Cell Tissue and Organ Culture* (J. Reinert and Y.P.S. Bajaj, eds.), pp. 93-107, Springer-Verlag, Berlin.

Bonga, J.M., 1987, Clonal propagation of mature trees: Problems and possible solutions. In: *Cell and Tissue Culture in Forestry, Vol. 1* (J.M. Bonga and D.J. Durzan, eds.), pp. 249-271, Martinus Nijhoff *Publishers,* Dordrecht.

Brown, C.L. and Sommer, H.E., 1982, Vegetative propagation of dicotyledonous trees. In: *Tissue Culture in Forestry* (J.M. Bonga and D.J. Durzan, eds.), pp. 109-149, Martinus Nijhoff/Dr. W. Junk, The Hague.

D'Amato, F., 1977. Cytogenetics of differentiation in tissue and cell cultures. In: *Applied and Fundamental Aspects of Plant Cell, Tissue and Organ Culture* (Y.P.S. Bajaj, ed.) pp. 343-357, Springer-Verlag, Berlin.

Hartney, V.J., 1982, Tissue culture of *Eucalyptus, Proc. Int. Plant Prop. Soc.,* 32: 98-109.

Karki, A. and Rajbhandary, S.B., 1984, Clonal propagation of *Chrysanthemum cinerariaefolium* VIS (Pyrethrum) through tissue culture, *Pyrethrum Post,* 15: 118-121.

Manandhar, A. and Rajbhandary, S.B., 1986, Rooting in non-sterile potting mix of in vitro potato and its field establishment, *Indian. J. Hort.,* 43: 235-238.

McCown, B.M., 1986, Woody ornamentals, shade trees and conifers. In: *Tissue Culture as a Plant Production System for Horticultural Crops,* (R. H. Zimmerman, R. J. Griesbach, F. A. Hannerschlag and R. H. Lawson, eds.) pp. 333-342, Martinus Nijhoff,Dordrecht.

Murashige, T. and Skoog, F., 1962, A revised medium for rapid growth and bioassays with tobacco tissue cultures, *Physiol. Plant,* 15: 475-497.

Panday, K.K., 1982, Importance of fodder trees and tree fodder. In: *Fodder Trees and Tree Fodders in Nepal,* pp. 48-53, Birmensdorf, Switzerland, Swiss Federal Institute of Forestry Research, Birmensdorf.

Singh, R.V., 1982, *Ficus roxburghii* Wall., In: *Fodder Trees of India,* pp. 157-160, Oxford & IBH Publishing Co., New Delhi,

Suwal, B., Karki, A. and Rajbhandary, S.B., 1987, The *in vitro* proliferation of forest tree, *Dalbergia sissoo* Roxb., Ex. D.C., *Silvae Genet.* 37: 26-28.

Thorpe, T.A. and Biondi, S., 1983, Conifers. In: *Handbook of Plant Cell Culture,* Vol.1 (D.A. Evans, W.R. Sharp, P.V. Ammirato and Y. Yamada, eds.), pp. 449-451, Macmillan Publishing Company, New York.

13

GENETIC VARIATION IN TISSUE CULTURE AS A CONSEQUENCE OF THE MORPHOGENIC PROCESS

Robert D. Locy

ABSTRACT

Propagule costs currently limit the utility of plant tissue culture technology. Most commercially useful systems for propagation rely on axillary bud enhancement as the multiplication process. The ability to utilise more efficient multiplication techniques such as shoot organogenesis and somatic embryogenesis could reduce propagation costs, but these techniques give rise to more variation among the regenerated plants than does axillary bud enhancement.

By studying the process of adventitious shoot formation in tomato and by studying a variant cell line of tomato that is able to grow on ribose as the sole carbon source, we have determined that *tissue culture induced variation* results from processes occurring during shoot formation. Cells competent to form shoots arise during the growth of a callus in a spontaneous and independent manner, much like a mutation. The mechanism for the induction of competent cells may also be the same which gives rise to

---

Robert D. Locy * Native Plants Inc., 417 Wakara Way, Salt Lake City, Utah 84108, U.S.A.

genetically altered cells. This implies that a propagation system which allows competent cells to be propagated *in vitro,* rather than a system which requires the *de novo* induction of competent cells, would give less variation.

INTRODUCTION

Plants can be regenerated *in vitro* by one of the three general procedures, viz. axillary bud enhancement, shoot organogenesis, and somatic embryogenesis. Shoot micropropagation via axillary bud enhancement differs from both shoot organogenesis and somatic embryogenesis, since both the latter procedures involve the production of shoots or plants adventitiously from an unorganised callus tissue. Axillary bud enhancement allows plant regeneration from organised meristems, and multiplication occurs by the growth of organised shoot tips to produce additional axillary meristems *in vitro.* This critical biological difference has made axillary bud enhancement the choicest method for plant micropropagation for over a decade (Murashige, 1974; Holdgate, 1977; Hughes, 1981).

The major advantage gained by maintaining such highly organised cultures is that clonal fidelity is assured (Murashige, 1974; Lutz et al., 1985). In comparison, shoot organogenesis is known to produce a great deal of variation among regenerated plants (Shepard et al., 1980; Evans and Sharp, 1983; Dulieu, 1986), and somatic embryogenesis may (Benzion et al., 1986; Fukui, 1986; Larkin, 1986 and Orton, 1986) or may not (Swedlung and Vasil, 1985) produce substantial variation in regenerated plants.

The major disadvantage of axillary bud enhancement procedures is that the cost of a propagule produced by such a procedure is high. The utility of tissue/organ culture techniques for mass propagation of many species is limited by the cost of propagules obtained through axillary enhancement procedures. This is particularly true for forestry species, since planting material costs are a major part of the cost of reforestation. It has been estimated that 80 percent of the cost of propagule production via axillary bud enhancement is the cost of labor. Thus, if we are to lower propagule costs which would allow the tissue/organ culture propagation technology to be more widely adapted in the forest replanting industry, it is critical to examine other ways of developing more labor-efficient systems for propagule production.

One way to obtain such efficiency is through the utilisation of more labor-and space-efficient procedures of shoot organogenesis and somatic embryogenesis, especially the latter. Growing masses of unorganised callus tissue that give rise to great numbers of shoots is clearly more efficient than propagating highly organised shoot tip cultures.

Two major problems have historically kept adventitious shoot formation and somatic embryogenesis procedures from wide acceptance. Firstly, appropriate protocols using these techniques of propagation have been lacking for many important crop species. Secondly, the degree of clonal fidelity provided by shoot organogenesis or somatic embryogenesis is either far too low or uncertain, making such techniques unreliable for commercial propagation. With the recent advent of shoot organogenesis or somatic embryogenesis protocols for many important forest trees and woody plant species (Durzan, 1985; Litz, 1985), only the problem of genetic variation, arising in such systems, remains to be addressed to allow these techniques to be used with impunity.

We have been studying a variant tomato cell line selected for the ability to grow on ribose as the sole carbon source, and the relationship of this variant to the process of shoot organogenesis. Our observations may shed some light on how variation arises during the process of shoot organogenesis, and how best to design both shoot organogenesis systems and somatic embryogenesis systems which at least minimise such genetic variation and maintain a high degree of clonal fidelity.

MATERIALS AND METHODS

Media

All experiments were conducted using an agar-solidified basal medium containing Murashige and Skoog salts and vitamins (Murashige and Skoog, 1962), 100 mg l^{-1} m-inositol, 0.8 percent Difco Bacto agar, 1 mg l^{-1} Zn, and 0.1 mg l^{-1} NAA. This basal medium containing 3 percent sucrose is referred to as CFM (callus-forming medium) and was used for the initiation and culture of callus tissue. The same basal medium containing no sucrose but 1.5 percent ribose-referred to as RCM (ribose-containing, medium), was used for the selection and growth of ribose-adapted cultures. SFM (shoot-forming medium) contained the basal medium listed above, with 3 percent sucrose but the Zn and NAA were deleted and 3 mg l^{-1} BAP and 0.3 mg l^{-1} IAA were added to the medium. The pH of the medium was adjusted such that a pH of 5.8 was obtained after autoclaving. All media were autoclaved at 1.06 kg cm^{-2} (121°C) for 15 min, cooled to 45-50°C and poured into 15 x 100 mm pre-sterilised disposable plastic petri plates.

Initiation and Growth of Tomato Callus Cultures

Tomato callus cultures were initiated from sterile tomato hypocotyl segments. Tomato seeds of the cultivar UC82-B or other varieties, as specified in the text and tables, were surface sterilised in 2.5 percent sodium hypochlorite solution (50 percent diluted laundry bleach) for 30 min. The seeds were rinsed 3 times in sterile distilled water and transferred to sterile petri dishes containing three thicknesses of filter paper moistened with sterile distilled water. The seedlings were grown in a growth room at 22-24°C with lighting provided by cool white fluorescent lamps. The photoperiod was 16 hours of light and 8 hours of darkness. From 10-14 days old aseptically raised seedlings, hypocotyl segments 1-2 mm in diameter were excised and placed on CFM. The culture conditions for seedling explants were as described above. Hypocotyl segments formed prolific, dark green callus on this medium. The callus was sub-cultured on fresh CFM every 28-32 days and grown as described above.

Mechanical Maceration and Plating of Callus Cultures

For certain experiments, the callus tissue was mechanically macerated and plated on various media. This was accomplished by forcing a clump of callus tissue through a stainless steel mesh screen. The macerated tissue that passed through an 810 micron mesh screen but did not pass through a 520 micron mesh screen was collected. The clumps of cells were transferred to sterile centrifuge tubes and the packed cell volume of the clumps was measured by suspending the clumps in liquid CFM or RCM medium. The suspended cells were centrifuged at 200 x g for 10 min and the packed cell volume of the cells was measured. The washed cells were suspended in 20 times their packed cell volume of liquid CFM or RCM containing 0.1 percent Difco Bacto-agar. This concentration of agar kept the medium liquid at room temperature but was sufficient to keep the cell clumps suspended such that the suspension of clumps could be uniformly pipetted to give a reproducible titer. Seven ml of the suspension was pipetted onto a sterile 90 mm diameter disc of miracloth placed on sterile paper towelling. The liquid medium was slowly absorbed into the towelling, leaving a uniformly distributed population of cell clumps resting on the miracloth disc. The disc was then transferred to an appropriate medium, and maintained under the growth conditions cited above.

RESULTS

Selection of Cell Lines Capable of Growing on Ribose as the Sole Carbon Source

Hypocotyl tissues of the cultivars UC 82-B, Bonnie Best, Castlong, VFNT Cherry and Marglobe cultures were initiated on CFM medium. After 30 days, the callus arising from the hypocotyl explants was sub-cultured on CFM. Following the second sub-culture, the callus was mechanically macerated and plated on RCM. Table 1 shows that between 1.5 and 2.4 percent of the callus clumps plated on RCM grew actively, forming organised bright green tissue. The remaining clumps turned dark brown and ceased to grow on RCM.

The bright green tissue clumps which grew on RCM could be sub-cultured on fresh RCM and subsequent growth showed a high degree of organisation. The ability to form organised structures resembling shoot primordia and to subsequently form shoots was particularly surprising, since the hormone composition of the medium was precisely the same as that normally used to culture callus tissue.

Table 1 : The frequency of formation of ribose-adapted cultures by callus derived from tomato hypocotyl explants of various tomato cultivars

Cultivar	Percent of plated clumps growing on ribose after 30 days
Bonnie Best	2.2
Castlong	2.4
Marglobe	1.9
UC 82-B	2.4
VFNT Cherry	1.5

Stability of Cell Lines Able to Grow on Ribose

Cultures which were selected to grow on ribose were transferred to CFM containing sucrose. Such cultures resumed growth as an unorganised callus rather than maintaining the organised growth habit characteristic of growth on ribose. When cultures selected on RCM were transferred to CFM for 4-12 weeks and were again placed on RCM, the response was virtually identical to that obtained when cultures that had never been on RCM were placed on RCM, i.e., the tissue turned brown and a selection process was required to obtain cultures that would be able to grow on ribose. The repetition of this transfer cycle from RCM to CFM and back to RCM several times, always gave the same result. Thus, at least *in vitro,* the ability to grow on ribose does not appear to be stable in the absence of selection.

It should be noted that the selected callus material was not single-cell-cloned at any time. In fact, all attempts to clone single cells failed because the ability to grow on ribose appears to be lost if single cells or even cell aggregates containing a few cells are plated on RCM. Thus, the cultures selected for growth on ribose could be a mixture of both selected and unselected cells. Since the selected cell types appear to grow much more slowly than the unselected cell types, it is possible that the apparent instability of the ribose-adapted phenotype was due to normal callus cells mixed with the selected cell population.

Regeneration of Plants from Cultures Growing on RCM

From the results presented above, it seems unlikely that the phenotype adapted to ribose arises from a genetic mutation. First, the rate of occurrence of the variant cell type is too high (1.5-2.5 percent) to suggest a mutational event. Second, the phenotype selected may not be stable *in vitro* in the absence of selection. To further confirm if the phenotype adapted to ribose was a genetically stable mutation, plants were regenerated from selected cultures that had been growing on RCM for over six months. Ten explants of leaf, petiole and stem from 30 regenerated plants obtained by this procedure were cultured on RCM. No culture capable of growing on ribose was obtained (data not shown). However, if the cultures were first initiated on CFM, then ribose-adapted cultures could be selected from the callus tissue regenerated on all the explants by transferring the callus tissue formed on CFM to RCM. This behavior was identical to the behavior of seedling-derived control plants or plants derived by regeneration from explants without selection for the ability to grow on ribose.

The 10 plants regenerated from cultures growing on ribose were allowed to flower. All 10 plants were found to be sterile. No pollen was shed by the flowers, and no seed was set when freshly opened flowers were pollinated by pollen from tomato plants raised from seed. When root tips derived from the selected plants were squashed and chromosome counts were performed, in all cases the plants showed a normal complement of 24 chromosomes (B. Swedlund, personal communication). Thus, the sterility did not appear to be related in any way to chromosomal aberration or aneuploidy.

Despite the fact that the ribose-adapted phenotype was unstable in culture, and tissue pieces from regenerated plants showed no improved ability to grow on ribose, the argument could be made that the plant phenotype adapted to ribose in culture might be sterile and that this phenotype was the result of mutation. To further support this hypothesis, a cellular-based means of assessing the adaptive versus the mutational nature of the ability to grow on ribose, was sought.

Fluctuation Test for Ability to Grow on Ribose

A cellular method was developed for studying the process of mutation in bacteria over four decades ago by Luria and Delbruck (1943). This method was used to infer whether a selected trait arose spontaneously and independently in the absence of selection or whether mutations are induced physiological adaptations to selection factors in the medium. This procedure, now classic, is referred to as a fluctuation test.

The essence of the fluctuation test is that the number of cells per clone in a population of clones that acquired a trait, will be distributed according to Poisson's Law, i.e., a sampling distribution, if they are induced after being plated on selective media. Whereas, the distribution will be non-Poisson and shall have a much higher variance than mean if the cells arise by mutation. The accuracy of this assertion can be tested by comparing the variance of several samples taken from one large clone (these should be a sampling distribution) with the variance of independent samples taken from a series of clones. The validity of a fluctuation test of plant cell cultures consisting of multicellular aggregates has been previously established by Murphy (1982).

To perform a fluctuation test for ribose adaptation, 60 tomato hypocotyl explants were placed on CFM for four weeks. The explant and callus formed were then mechanically macerated, as described in *Materials and Methods*. The callus clumps derived from each individual hypocotyl explant were separately plated as callus clones, and the number of shoot-forming clumps per clone was determined. For the 90 clones examined, the mean number of ribose-adapted clumps per clone was 5.21, and the variance between the clones was 43.56. The goodness of fit for the observed values to the expected values for a Poisson distribution was calculated using a Chi-square test. The probability that the observed values would vary from the expected values to the extent they did, even if the two distributions were the same, was less than 0.001 (see Table 2).

To establish that the non-Poisson nature of the distribution of shoot-forming clumps did not arise as a result of the process of mechanical maceration or plating and thus represented some type of artifact that might cause misinterpretation, a control experiment was performed. The control consisted of mechanically macerated callus tissue as before, but the clumps from each clone were then mixed together and uniform aliquots were removed from the mixture of clumps and plated. This mixing control verified that the fluctuation test is valid since the mixing control mean was 5.48 ribose-adapted clumps per sample and the variance was 7.92. By a Chi-square test, these data were not significantly different from a Poisson distribution with a mean of 5.48 (see Table 2).

Table 2 : Fluctuation test for the ability of 60 callus clones to grow on ribose as the carbon source. The probability of obtaining a variance as high as observed if the distribution were Poisson is shown

	Number of clones	Mean	Variance	Goodness of fit to Poisson
Experiment	90	5.21	43.569	$P < 0.001$
Mixing	55	5.48	7.922	$P < 0.10$

The above observations are consistent with the notion that the trait of ribose adaptation arises spontaneously and independently during the growth of callus in the absence of selection.

The Relationship Between Ribose Adaptation and Shoot Formation

The data presented above, when taken as a whole, do not conclusively show the mutational nature of ribose adaptation, nor do they conclusively allow the rejection of the hypothesis that ribose adaptation is a mutation. Genetic transmission of the trait cannot be established because the plants regenerated from ribose-adapted cultures are sterile. Therefore, let us turn our attention to an observation mentioned earlier that ribose-adapted cultures prolifically form shoot primordia and shoots on media containing hormone levels normally used for callus formation.

Tomato callus loses its totipotency with each sub-culture. By approximately the fourth or fifth monthly sub-culture, it is nearly impossible to regenerate plants from tomato callus, and the ease with which plants are obtained decreases with each sub-culture (Table 3). Tomato hypocotyl explants were cultured on either CFM, RCM or SFM. The callus generated on CFM was further sub-cultured on CFM. At each sub-culture, some callus was mechanically macerated and plated on SFM and RCM. For the original explants cultured on SFM, 100 percent explants grew and formed shoots. On RCM, none of the hypocotyl explants grew or formed shoots. Approximately 20 percent of the callus clumps derived from first sub-culture callus demonstrated the ability to form shoots on SFM or to grow and form shoots on RCM. For second sub-culture callus, the percentage of clumps able to form shoots on SFM and the percentage of clumps able to grow and form shoots on RCM fell to about five percent (Table 3). This parallel loss in the ability to form shoots on SFM and loss in the ability to grow and form shoots on RCM continued for the third, fourth and fifth sub-cultures as well (Table 3).

Table 3. : The loss of ability to form shoots and the ability to grow on ribose as a function of callus sub-culture interval is shown for callus derived from tomato hypocotyl segments

Callus sub-culture	Percent of pieces forming shoots	Percent of pieces growing on ribose
Explant	100.0	0.0
1st sub-culture	21.0	19.0
2nd sub-culture	5.5	4.9
3rd sub-culture	1.8	1.4
4th sub-culture	0.2	0.3
5th sub-culture	0.0	0.0

This observation suggests that ribose adaptation may be induced as a property of some stage of the developmental process of shoot formation. However, this hypothesis does not easily explain why 100 percent hypocotyl explants are able to form shoots on SFM, while none of the hypocotyl explants directly plated on RCM would grow or form shoots. Clearly, events which occur during the first callus sub-culture interval must somehow be related to ribose adaptation, since after the first callus sub-culture interval, the loss in shoot-forming ability and the loss in the ability to grow on ribose are highly correlated.

To further investigate the timing of these early occurring events, the following experiment was executed. Tomato hypocotyl explants were placed in culture on CFM for 0,2,4,7,9 or 11 days. Following this, cultures on CFM explants from each treatment were then transferred to SFM for 0,3,5,7,10,12,14 or 17 days. Subsequently, the explants were transferred to a hormone-free basal medium containing three percent sucrose. On this medium, shoot elongation occurs but not shoot formation. After 30 days on the hormone-free medium, the number of shoots and shoot primordia that had formed and elongated so as to be visible at 20X magnification were counted.

Table 4. : Tomato hypocotyl explants were initiated into culture on CFM. After the number of days designated in the table, the explants plus associated callus were transferred to SFM for the number of days designated in the table. The explants and callus were then transferred to a hormone-free medium, and the number of shoots formed per explant was determined for each treatment after 30 days in culture.

Days CFM	Shoots formed per explant Days on SFM							
	0	3	5	7	10	12	14	17
0	0.03	0.00	0.03	0.12	0.38	0.73	1.20	0.47
2	0.06	0.08	0.13	0.37	0.63	1.05	0.37	0.47
4	0.12	0.17	0.30	0.73	1.08	1.47	1.07	1.27
7	0.23	0.38	0.67	0.35	0.36	0.43	1.03	0.68
9	0.23	0.45	0.67	0.13	0.60	0.48	0.70	1.17
11	0.30	0.42	0.70	0.53	0.38	0.47	0.28	0.60

When hypocotyl explants were placed directly on SFM (0 days on CFM) and allowed to remain there for up to 10 days prior to their transfer to a hormone-free medium, very few shoots were formed per explant (Table 4).

However, if the explants remained on SFM for 12 days or more, a statistically significant increase was observed in the number of shoots formed per explant. This can be interpreted to show that more than 10 but less than 12 days are required for the induction of cells competent to form a shoot and the subsequent determination of these cells to form a shoot primordium. A determined shoot primordium is then capable of growing in a hormone-free medium to manifest itself as a visible shoot. Until 12 days of culture on SFM, the process of determination is not completed; and thus, few or no shoots are formed when explants are transferred from SFM.

The culture of explants on CFM for two, four or seven days prior to the transfer of those explants to SFM can reduce the number of days required on SFM for shoot determination to occur. If explants are left on CFM for more than nine days, then further culture on CFM does not reduce the time required on SFM below the minimal three required days. The interpretation of this phenomenon is that an event occurs on CFM or SFM which has been referred to as the induction of competence to form a shoot (Christianson and Warnick, 1983; Ammirato, 1985). More than seven but less than nine days are required to induce the competent state for tomato hypocotyl segments.

Fig. 1 summarises these events graphically. The induction of the competent state can occur on either CFM or SFM and requires five to nine days. Following the induction of the competent state, the process of determination requires approximately three to five additional days to occur. Determination occurs only on SFM and thus, a total of 12 days are required for both competence to be induced and for the competent state to be determined to form a shoot. The critical question that arises is, when is the ability to grow on ribose acquired? Table 5 shows the results of a media transfer experiment similar to that discussed above, except that instead of the transfer from CFM to SFM to a hormone-free medium, the transfer was made from CFM to RCM to hormone-free medium.

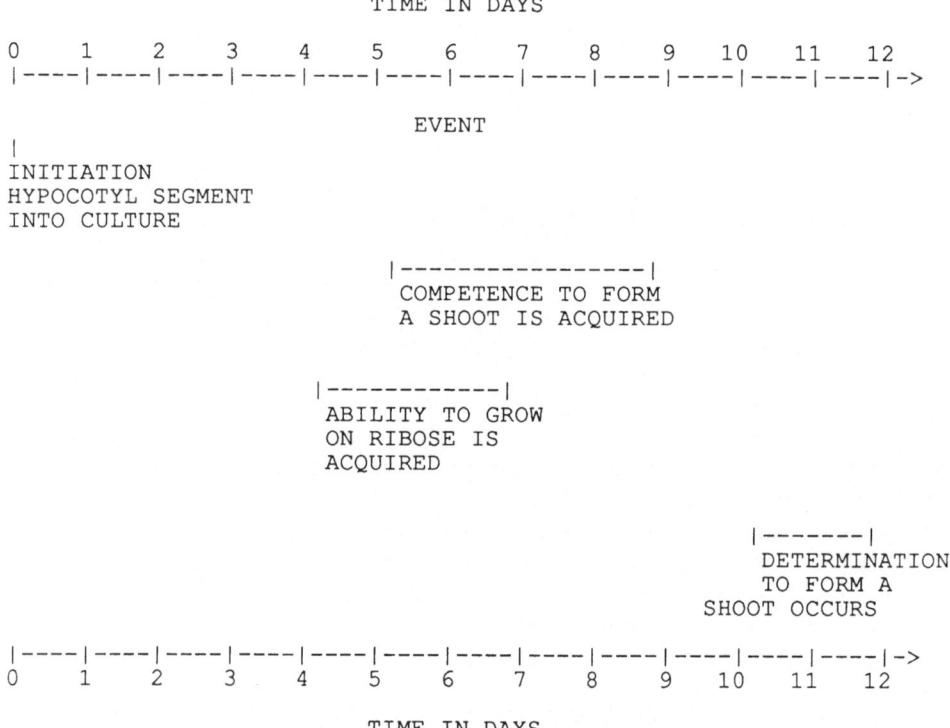

```
                          TIME IN DAYS

0     1     2     3     4     5     6     7     8     9    10    11    12
|-----|-----|-----|-----|-----|-----|-----|-----|-----|-----|-----|-----|->
                              EVENT

|
INITIATION
HYPOCOTYL SEGMENT
INTO CULTURE

                    |-----------------|
                    COMPETENCE TO FORM
                    A SHOOT IS ACQUIRED


               |------------|
               ABILITY TO GROW
               ON RIBOSE IS
               ACQUIRED


                                               |-------|
                                               DETERMINATION
                                               TO FORM A
                                         SHOOT OCCURS

|-----|-----|-----|-----|-----|-----|-----|-----|-----|-----|-----|-----|->
0     1     2     3     4     5     6     7     8     9    10    11    12

                          TIME IN DAYS
```

Fig. 1: A schematic summary of the conclusions from the data presented in Tables 5 and 6.

When explants are directly plated onto RCM (0 days on CFM), shoot formation and shoot growth is inhibited (Table 5). Similarly, if explants are plated on CFM for two or four days, limited growth and few shoots are obtained when the explants are transferred to RCM and then to hormone-free medium for 30 days. However, after seven or more days of culture on CFM, the ability to grow on the ribose-containing medium is acquired, and more than three but less than five days of growth on RCM will allow shoot determination to occur (Table 5).

Table 5: Tomato hypocotyl explants were cultured on CFM. After the number of days designated in the table, the explants plus associated calli were transferred to RCM for the number of days designated in the table. The explants and calli were then transferred to a hormone-free medium, and the number of shoots formed per explant was determined for each treatment after 30 days in culture

Days CFM	Shoots formed per explant days on RCM							
	0	3	5	7	10	12	14	
0	0.02	0.02	0.08	0.02	0.12	0.00	0.06	0.02
2	0.04	0.16	0.24	0.12	0.18	0.23	0.17	0.16
4	0.08	0.12	0.40	0.34	0.34	0.34	0.30	0.40
7	0.23	0.40	0.58	0.84	0.80	0.62	0.60	0.74
9	0.46	0.50	0.84	1.02	1.20	0.72	0.38	1.12
11	0.34	0.42	0.76	0.74	0.44	0.50	0.70	0.40

These observations suggest that at about the time competence to form a shoot is induced during the growth of tomato callus, i.e., four to seven days of culture on RCM, the ability to grow on ribose is acquired. Furthermore, the substitution of ribose for sucrose alters the hormonal requirements for the determination to occur, so that determination can occur on RCM (same hormones as CFM). However, cells which have not acquired the competence cannot grow on RCM.

This leads to the conclusion that the ability to grow on ribose is acquired as a part of the normal process of shoot formation at about the time of competence induction and definitely prior to the time of determination. It is also obvious from the fluctuation analysis performed above for the ability to grow on ribose that this developmental state must arise spontaneously and independently during the growth of a callus, and it must be stable enough to allow the fluctuation analysis. If this hypothesis is correct, then it could be argued that if a fluctuation analysis was performed for clumps of callus capable of forming shoots, the distribution should be non-Poisson, with a much greater variance than mean, just as it was in the analysis of the ribose-adapted phenotype.

Fluctuation Test for Shoot Formation

Tomato hypocotyl explants were placed on CFM and grown for 30 days. Each explant and associated callus was mechanically macerated, as described above, and the callus clumps were plated on SFM. After 30 days of culture, the number of shoot-forming clumps in each callus clone was determined. The mean number of shoot-forming clumps per clone was 15.35, and the variance of the population of 60 clones was 97.55. As shown in Table 6, this distribution does not follow a Poisson distribution. A mixing control similar to that described above was also performed, and the mixing control did fit a Poisson distribution.

These observations are consistent with the notion that the acquisition of competence to form a shoot occurs spontaneously and independently during callus growth and involves at least semi-stable changes in gene expression, not unlike mutational events. The ability to grow on ribose is acquired at about the same time as competence is acquired.

Table 6: Fluctuation test for the ability of 60 callus clones to form shoots. The probability of obtaining a variance as high as observed if the distribution were Poisson is shown

	Number of clones	Mean	Variance	Goodness of fit to Poisson
Exp	60	15.35	97.553	$P < 0.001$
Mixing control	35	14.40	14.211	$P < 0.25$

DISCUSSION

The growth of plant cells on ribose and other pentose sugars is not common. In studies where the ability of plant cells to grow on a wide range of carbon sources has been examined, it has been seen that ribose is not a suitable carbon source (Maretzki et al., 1974). The behavior of typical tomato callus cells does not appear to contradict this observation, since over 95 percent of the cell clumps derived from a tomato callus are unable to grow on ribose.

A relatively facile process of selection is involved in the acquisition of cell lines able to grow on ribose as the sole carbon source. Whether such cells occur as a result of a mutational event is more difficult to infer. The frequency with which such cultures are obtained (Table 1), and the fact that cultures do not appear to retain the ability to grow on ribose in the absence of selection, support the hypothesis that the ribose adapted phenotype is not arrived at by a stable genetic change, i.e., a mutation, but rather the ability to grow on ribose is an *epigenetic adaptation* occurring in culture (Meins, 1983).

Not surprisingly, this hypothesis is further supported by the observation that the ability to grow on ribose is not transmitted to regenerated plants. However, the fact that all plants obtained from cultures that are able to grow on ribose are sterile, suggests that some stable change leading to sterility has been selected at the cellular level, and this change is not related to gross chromosomal aberration. Furthermore, fluctuation analysis performed at the cellular level implies that the ribose-adapted phenotype arises in a spontaneous and independent manner during the growth of a callus as a mutation would.

Whether the ribose adapted phenotype is a mutant or is *epigenetic,* it does not seem to be present in the original hypocotyl explant (Tables 4,5 and 6), but during the process of callus formation and growth, a new cell type is acquired in a relatively high frequency (approximately one to three percent of the cells). Furthermore, the process by which cells acquire the ability to grow on ribose is a random, spontaneous process occurring in an independent probabilistic manner during each callus cell generation. That ribose adapted cells must arise in such a manner and are not simply induced once the callus is plated on ribose-containing medium, is clearly shown by the results of the fluctuation analysis for the ability to grow on ribose (Table 2).

The nature of the cell type which arises by such a process cannot be ascertained with any certainty from the data presented here. However, it appears that this cell type is, in some way, related to the process of shoot organogenesis since selected cultures become highly organised, and retain morphogenic potential for at least two years. Typical tomato callus cultures lose all morphogenic potential in less than five months (Table 3).

The results presented here indicate that a differentiated or differentiable cell type arises in an independent and spontaneous manner in the absence of the inducing agent, i.e., in a manner analogous to a mutational event, which subsequently leads to the ability to grow on ribose as the sole carbon source. The timing of the appearance of this ribose-adapted cell type corresponds, within the precision of the experiments presented here, with an early stage of the regeneration process and once selected, may or may not be stable at the cellular and/or whole plant level. The process of plant cell differentiation involves not only changes in the level of gene expression regulated by hormonal action, but also the changes in the expression of some genes arising spontaneously in a probabilistic matter, during the growth of the callus tissue. The hormone-regulated changes leading to visible shoots occur only during the latter stages of shoot determination, since only after competence has been induced, changes in the hormones are required to determine a shoot. Apparently, the spontaneous and independent changes arise as part of the normal process of competence-induction, and are an essential part of the mechanism by which the competent state arises.

If such phenotypic changes arising by a mutation-like process are required to induce competence as part of the normal process of shoot organogenesis, then there must be a mechanism by which such changes are produced. It is possible to explain the genetic variation arising during shoot organogenesis is derived from the same process and mechanism. Such variation, thus, seems to be the inevitable consequence of competence-acquisition.

The above infers that what we must avoid is the process of competence-induction *in vitro* if we are to obtain an adventitious system for *in vitro* propagation to give clonal fidelity. While this may at first seem like a contradiction, competent cells must be obtained from the intact plant, and such competent cells must be propagated *in vitro*. This is precisely what we do when axillary bud enhancement procedures are employed. By maintaining a high degree of organisation in shoot tip cultures, we take competent cells contained in meristems and multiply them *in vitro*.

To make axillary enhancement systems more efficient, the growth of a less organised or unorganised tissue was also required. To obtain the highest degree of clonal fidelity, the unorganised and undifferentiated tissue must be grown together with meristematic, competent cells.

To date, at Native Plants Incorporated, several regeneration systems for important crop species have been established. The challenge now is to understand these systems well enough to regulate what type of cells to grow *in vitro* and how to grow them so as to expand the utility of micropropagation to species which are recalcitrant to tissue culture techniques.

ACKNOWLEDGEMENTS

The author gratefully acknowledges the assistance of Brad Swedlund in the preparation of the manuscript.

REFERENCES

Ammirato, P.V., 1985, Patterns of development in culture. In: *Tissue Culture in Forestry and Agriculture* (R.R. Henke, K.W. Hughes, M.J. Constantin and A. Hollaender, eds.), pp. 9-31, Plenum Press, New York.

Benzion, G., Phillips, R.L. and Rines, H.W., 1986, Case histories of genetic variability *in vitro*: Oats and Maize. In: *Cell Culture and Somatic Cell Genetics of Plants, Vol. 3, Plant Regeneration and Genetic Variability* (I.K. Vasil, ed.), pp. 435-448, Academic Press, New York.

Christianson, M.L. and Warnick, D.A., 1983, Competence and determination in the process of *in vitro* shoot organogenesis, *Dev. Biol.,* 95: 288-293.

Dulieu, H., 1986, Case histories of genetic variability *in vitro*: Tobacco. In: *Cell Culture and Somatic Cell Genetics of Plants, Vol. 3, Plant Regeneration and Genetic Variability* (I.K. Vasil, ed.), pp. 399-418, Academic Press, New York.

Durzan, D.J., 1985, Tissue culture and improvement of woody perennials: An overview. In: *Tissue Culture in Forestry and Agriculture* (R.R. Henke, K.W. Hughes, M.J. Constantin and A. Hollaender, eds.), pp. 233-256, Plenum Press, New York.

Evans, D.A. and Sharp, W.R., 1983, Single gene mutations in tomato plants regenerated from tissue culture, *Science,* 221: 949-951.

Fukui, K., 1986, Case histories of genetic variability *in vitro*: Rice. In: *Cell Culture and Somatic Cell Genetics of Plants, Vol. 3, Plant Regeneration and Genetic Variability* (I.K. Vasil, ed.), pp. 385-398, Academic Press, New York.

Holdgate, D.P., 1977, Propagation of ornamentals by tissue culture. In: *Applied and Fundamental Aspects of Plant Cell, Tissue and Organ Culture* (J. Reinert and Y.P.S. Bajaj, eds.), pp. 18-44, Springer-Verlag, Berlin.

Hughes, K.W., 1981, Ornamental species. In: *Cloning Agricultural Plants via In Vitro Techniques* (B.V. Conger, ed.), pp. 5-51, CRC Press, Boca Raton, Florida.

Larkin, P.J., 1986, Case histories of genetic variability *in vitro:* Wheat and Triticale. In: *Cell Culture and Somatic Cell Genetics of Plants, Vol. 3, Plant Regeneration and Genetic Variability* (I.K. Vasil, ed.), pp. 367-383, Academic Press, New York.

Litz, R.E., 1985, Somatic embryogenesis in tropical fruit trees. In: *Tissue Culture in Forestry and Agriculture* (R.R. Henke, K.W. Hughes, M.J. Constantin and A. Hollaender, eds.), pp. 179-194, Plenum Press, New York.

Luria, S.E. and Delbruck, M., 1943, Mutations of bacteria from virus sensitivity to virus resistance, *Genetics,* 28: 491-511.

Lutz, J.D., Wong, J.R., Rowe, J., Tricoli, D.M. and Lawrence, R.H., 1985, Somatic embryogenesis for mass cloning of crop plants. In: *Tissue Culture in Forestry and Agriculture* (R.R. Henke, K.W. Hughes, M.J. Constantin, and A. Hollaender, eds.), pp. 105-117, Plenum Press, New York.

Maretzki, A., Thom, M. and Nickell, L.G., 1974, Utilization and metabolism of carbohydrates in cell and callus cultures. In: *Tissue Culture and Plant Science* (H.E. Street, ed.), pp. 329-361, Academic Press, London.

Meins, F., 1983, Heritable variation in plant cell culture, *Ann. Rev. Plant Physiol.,* 34: 327-346.

Murashige, T., 1974, Plant propagation through tissue cultures, *Ann. Rev. Plant Physiol.,* 25: 135-166.

Murashige, T. and Skoog, F., 1962, A revised medium for rapid growth and bioassays with tobacco tissue cultures, *Physiol. Plant.,* 15: 473-497.

Murphy, T.M., 1982, Analysis of distributions of mutants in clones of plant cell aggregates, *Theor. Appl. Genet.,* 61: 367-372.

Orton, T.J., 1986, Case histories of genetic variability *in vitro:* Celery. In: *Cell Culture and Somatic Cell Genetics of Plants, Vol. 3, Plant Regeneration and Genetic Variability* (I.K. Vasil, ed.), pp. 345-366, Academic Press, New York.

Shepard, J.F., Bidney, D. and Shahin, E., 1980, Potato protoplasts in crop improvement, *Science,* 208: 17-24.

Swedlung, B. and Vasil, I.K., 1985, Cytogenetic characterization of embryogenic callus and regenerated plants of *Pennisetum americanum, Theor. Appl. Genet.,* 69: 575-581.

14

PERFORMANCE CRITERIA IN RESPONSE SURFACES FOR METABOLIC PHENOTYPES OF CLONALLY PROPAGATED WOODY PERENNIALS

Don J. Durzan

ABSTRACT

Response surfaces for metabolic phenotypes describing intermediary nitrogen metabolism are portrayed by 3-dimensional computer assisted graphic methods. Response surfaces describe:

* biosynthetic hierarchies of families or *networks* of free amino acids

* temporal events in development, involving discrete physiological states, arbitrarily defined by current and past compositional behavior, and

Don J. Durzan * Department of Environmental Horticulture, University of California, Davis, CA 95616. U.S.A.

*   levels of metabolites as a percentage distribution or as a flux in a state network map.

      Surfaces are correlated to the protein complement of cells through a series of 2-dimensional gel electrophoretograms and to the activity of synthetic plant growth regulators. Performance criteria associated with response surfaces are identified for clonally propagated woody perennials.

INTRODUCTION

      Analysis of the dynamic changes in metabolites can be used to quantify true-to-type development in clonal propagation for tree improvement programs. This review highlights how response surfaces for metabolic phenotypes are applied to:

*   zygotic embryo development in the pistachio
*   rescued somatic embryos of several *Prunus* sp., and
*   the development of carpellary tissues in selected cultivars.

      One important outcome of tree improvement is the mass propagation of elite, true-to-type clonal products (Libby, 1983). A corollary to mass propagation is the need to assess the expression of genetic gains of trees under field conditions.
      In this review, the author has encouraged new approaches to the diagnostic study of metabolic phenotypes of clonally propagated trees. The review starts by providing some background information, interpretations of phenotypic variance, construction of response surfaces, statistical considerations, results with the reference model systems (pistachio kernel and peach mesocarp), and a discussion of performance criteria associated with response surfaces. Computer-assisted graphic systems (Jurs, 1986; Durzan, 1987a,c) are employed to construct response surfaces depicting metabolic phenotypes. Concepts for this work were derived from :

*   phenetics (Yablokov, 1986)
*   Sewall Wright's adaptive landscapes (Wright, 1968)
*   earlier notions of nitrogen metabolism that portray physiological states and metabolic networks as 3-dimensional maps in relation to the activity of endogenous and exogenously applied plant growth regulators (Steward, 1968; Firn, 1986; Durzan, 1987a,b,c);
*   more recently, from notions of the metabolic control theory (Kacser and Burns, 1973, 1981; Ochs et al., 1985; Sauro et al., 1987).

      While the ultimate fate or application of such an approach is uncertain, the author anticipates that it will lead to better and more useful performance models of the real world that can be extended to forest trees and forest biotechnology.

Background Concepts

Population genetics provides a theoretical basis for tree improvement programs using clonal forestry (Libby, 1983; Timmis et al., 1987). Sewall Wright (1968) made the point that large populations tend to be heteroallelic at all loci, and strongly so at many loci. Furthermore, a genetically improved tree may be considered as located as a point in a gene frequency space with several dimensions. The gene frequency space describes the number of alleles at each locus and provides a summation over all loci. A 2-dimensional portrayal of this space is often called the *Sewall Wright landscape.* According to Agur and Slobodkin (1986), Sewall Wright's concept has become the single most suggestive metaphor in speculative evolutionary theorisation. For our purposes, it also provides a genetic data set for correlation with metabolic phenotypes based on varied genetic backgrounds in the landscape.

We must also consider the field performance of clonally propagated trees with elite attributes and long-term adaptive properties of the trees to their local environment. This *environmental fix* or performance interface can be represented as an adaptive metabolic phenotype. Many difficulties remain in constructing realistic theories of genotypic and phenotypic expression under field conditions.

Response surfaces are useful for metabolic phenotypes, diagnostic purposes, qualitative and quantitative reasoning (artificial intelligence) and comparatively large-scale integration of considerable numerical data over a long time. The latter is a requirement that derives from the analysis of many variables (alleles) associated with the long, complex life cycles of genetically improved and locally adapted trees.

The properties of elite genes in clonal products can be studied on at least three levels (Brink, 1962; Holliday, 1987):

* the mechanism of their transmission from generation to generation
* the mode of action of genes during the development of the tree from the fertilised egg to the adult. The changes in gene activity during development are generally referred to as *epigenetic,* and
* the nature of gene expression and epigenetic changes as influenced by external or environmental factors that produce either transient changes in gene activity or a permanent (irreversible) pattern of physiological activities.

Each level involves the characterisation of the clone in terms of its life support system units (reproductive structures, bioreactors, etc.) and of the clone's ecological potential (Fig. 1; Feedback relations exist among these levels cf. construction of response surfaces).

The author focuses on :

* changes in the shape of the epigenetic response surface of metabolic phenotypes, rather than on genetic gain or gene frequency change
* on temporal properties that involve the physiological state, and
* on the flux of nitrogen in biosynthetic families of free amino acids that are responsive to environmental disturbances, given a fixed clonal genotype.

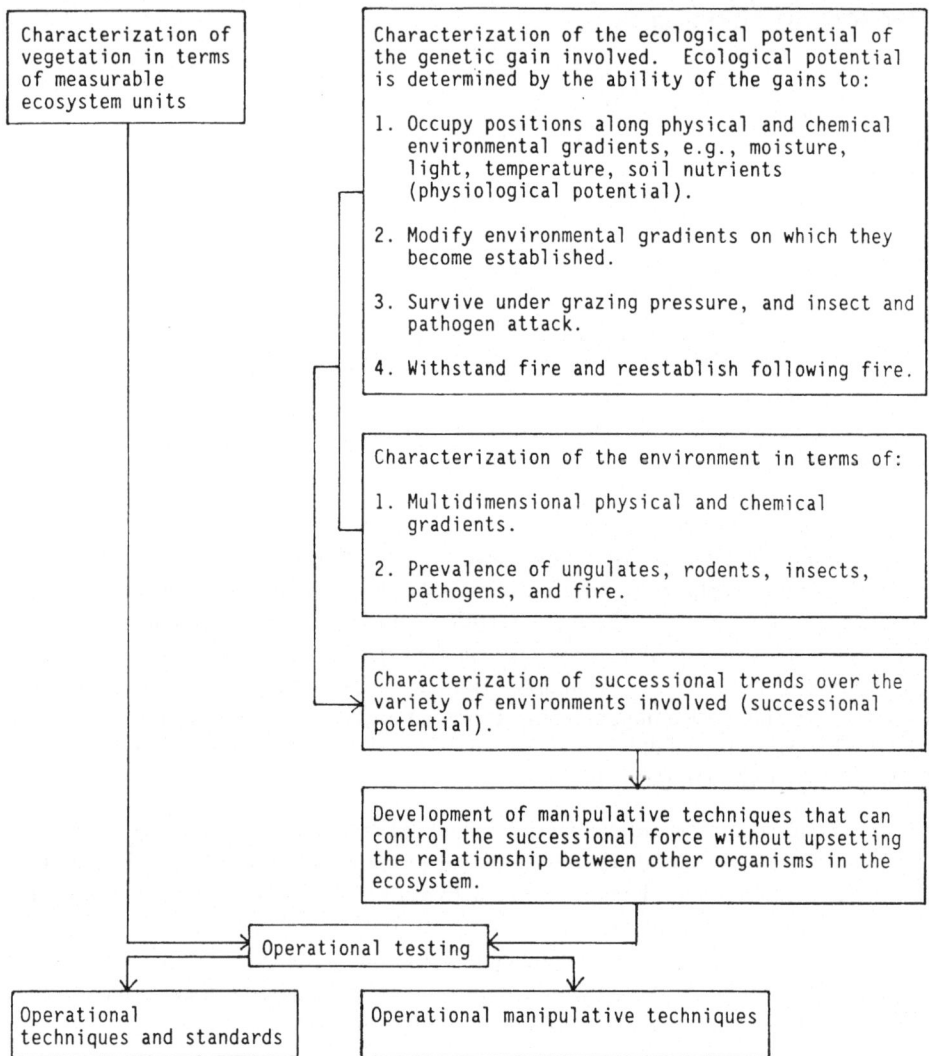

Fig. 1: Information required to develop manipulative techniques for vegetation preservation that can be applied to the gains sought for woody perennials. After Stone (1965) (cf. Shugart 1984).

Using selected examples, based on fruit and nut trees, performance criteria for growth and development should become evident from comparisons of response surfaces from a wide range of tissues, organs and cultural conditions. Epigenetic response surfaces, which represent subsets of the metabolic phenotype, may enable better quality and process control (Durzan, 1988a,b). Furthermore, should blocks

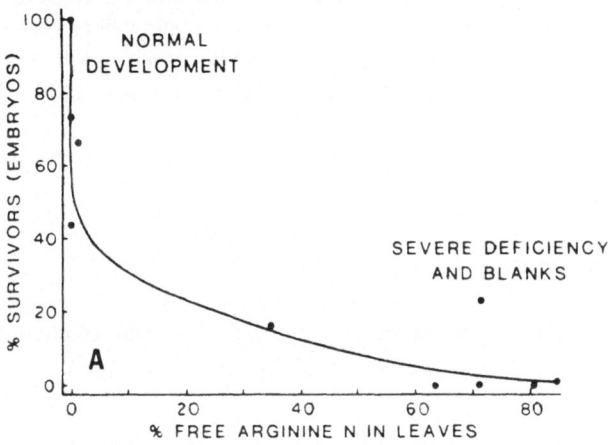

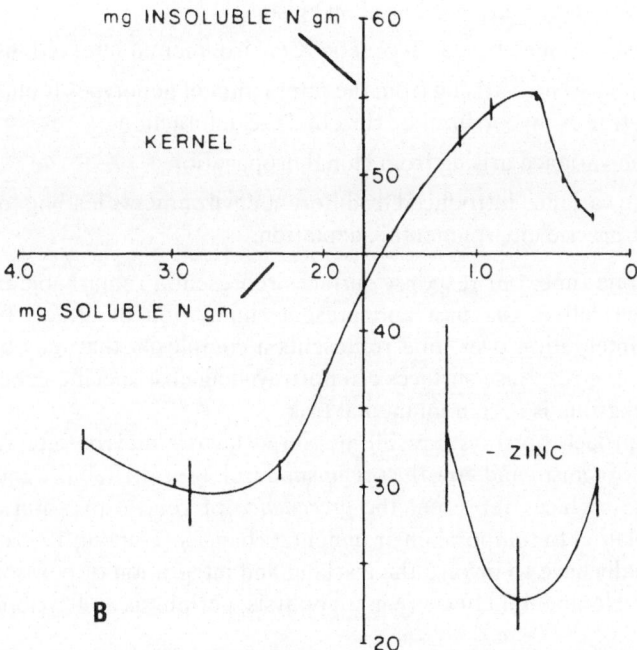

Fig. 2: Compositional changes in developing pistachio kernels on branches of trees over a range of zinc deficient soils. A. Relation between the free arginine nitrogen in leaves as a percentage of the total soluble nitrogen and the recovery of fully developed and normal kernels in 20-years-old trees over a range of zinc deficiency. B. The composition of developing kernels in terms of changes in total soluble nitrogen (mg nitrogen per g fresh wt) and insoluble (protein) nitrogen (mg nitrogen per g fresh wt) over the course of a season. Lower inset shows composition of kernels from severely zinc deficient trees. Vertical bars on curves represent standard deviations.

occur in genetic mechanisms, biochemical reactions and developmental processes (Fig. 2), the response surfaces should reveal how, when, and where such events express themselves. From these data sets, simulations could be attempted in areas that develop the subject theoretically and where further experimentation would be impossible or prohibitive.

METHODS

Phenotypic Variance

Metabolic response surfaces may be considered in terms of phenotypic variance (V_p) as follows:

$$V_p = V_g + V_d + V_{ep} + V_{gxe} + V_{gxg} + V_c + V_e$$

where V_g is the average effect of genes

V_d is the genetic variance due to dominance

V_{ep} is the genetic variance based on epistasis

V_{gxe} is the variance based on genetic X environmental interactions

V_{gxg} is the variance arising from the interaction of genotypes found in rootstock x scion, tree X mycorrhizal or rhizobial etc. interactions

V_c is the variance arising from clonal propagation

V_e is the variance introduced by different environments leading to the requirement of precise environmental adaptation.

At any one time, our response surfaces represent a remarkable integration of variances which reflect the past and present summation of effects from discrete sources. This integration over time represents a complexity that may be difficult to analyse, unless the response surfaces are portrayed against specific genetic and environmental backgrounds - a monumental task.

Our approach portrays how all measured factors may be integrated and expressed by the organism and how the organism develops its irritability and responsiveness. Response surfaces represent the *precedence* of gene expression of metabolic networks in relation to temporal environmental changes (Durzan, 1988a). Our analyses will eventually have to involve the tracking and integration of positional, environmental and developmental effects (e.g. topophysis, periphysis and cyclophysis).

Construction of Response Surfaces for Metabolic Phenotypes

It is often problematic to describe the response surface of a metabolic phenotype for a clonal system using experimental data without a model reference. In an attempt to solve this problem, state network model references have been constructed, based on the organisation of cell metabolism (Durzan, 1984; Ochs et al., 1985; Welch and Clegg, 1987) and metabolic control theory (Kacser and Burns, 1973, 1981; Sauro et al., 1987), with reference to metabolites in the soluble nitrogen pool (Durzan, 1982, 1986). These models portray the step-wise sequential biosynthesis of amino acid metabolites in families or networks, starting from a single precursor (source), leading to the final products (sinks). The precursor is the first compound in the pathway accepting nitrogen. The formation of the amino acid, first contributing to the hierarchy of biosynthesis, is usually accomplished by transamination reactions (Durzan and Steward, 1983).

Metabolite families are portrayed on one axis of a 3-dimensional graph. The behavior of metabolites in all families is depicted over time (Fig. 3).

The final point on the time axis represents a defined or discrete developmental outcome. At any given time, the line describing concentrations of metabolites across all networks is arbitrarily taken as representing the *physiological state* of the sampled cells or tissues at that time. This state may be further characterised by separation of proteins by 2-dimensional gel electrophoresis of total proteins (Young, 1984), and more specifically, by the non-histone chromosomal proteins (Pitel and Durzan, 1987a,b,c, 1988). In reality, the physiological state in question is the outcome of earlier concentrations and physiological states. The remaining vertical axis arbitrarily describes the nitrogen (N) flux in all measured compounds.

Taken together, the axes portray a 3-dimensional map of events or an epigenetic response surface with maxima, minima, transient states, etc. Experimental data in this model are readily handled as a numerical matrix.

As a significant extension of this approach, metabolic control theory and its analysis provides a formal matrix algebra procedure for determining the *flux control coefficients* in metabolic pathways (Sauro et al., 1987). While limitations still exist in the application of the theory to cyclic pathways, a single matrix equation permits calculations of the control coefficients for flux and concentration, and distribution of flux at branch points in pathways. This means that the response of a pathway to the alterations of the enzyme content or to the modulations by an effector or forcing factor can be determined.

We are assuming a simplified, linear and connected sequence of families of metabolites that are acted upon enzymatically from an initial compound containing nitrogen to the final product. Network portrayals of flux should contain branches, substrate cycles or conserved cycles. The modification of the design of the response surface to meet these needs is not yet fully addressed. For an indication of how such networks could be presented, and how data sets are related to larger systems, the reader is referred to the paper by Sauro et al. (1987). The present goal is to provide a simple format for qualitative and quantitative reasoning, rather than for actual or real system network representation. Our analytical skills and insights are not yet sufficiently strong or penetrating to consider the latter.

Stated in another way, the response surface is a pragmatic attempt to explore crucial questions involving epigenetic process control, system predictability and model simulation and assessment. The level of resolution in the response surface remains relatively crude because we cannot yet measure all intermediary metabolites or

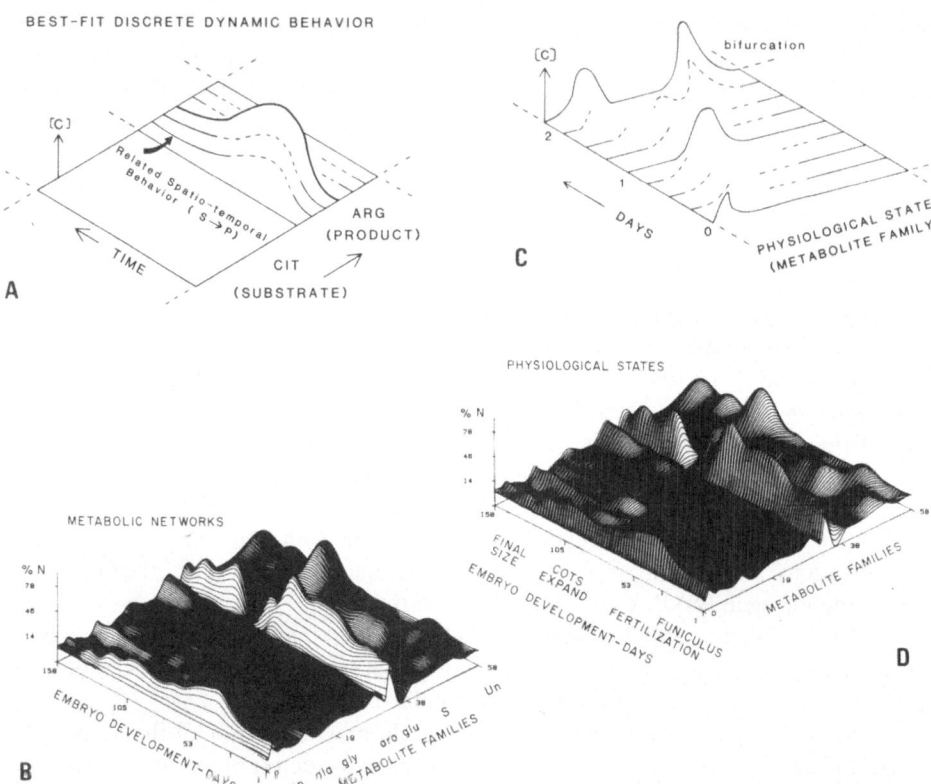

Fig. 3: Construction of a response surface for amino acid families in the alcohol-soluble nitrogen pool being translocated to a developing pistachio kernel over 158 days when final size is reached. For the estimations of biosynthetic networks for a family of amino acid, lines are mathematically fitted to concentrations over time to reveal the behaviour of metabolites in that family. Metabolites are arranged in a linear sequence or hierarchy of biosynthesis, e.g., citrulline is a substrate for the formation of arginine. When lines are fitted to experimental data in this way, we obtain the partial response surface for metabolic networks shown in B. This surface shows the percent distribution of nitrogen in each amino acid in all families observed over time. Metabolite families are illustrated for asp, aspartate; ala, alanine; gly, glycine; aro, aromatic amino acids; glu, glutamate; S, sulfur amino acids; Un, unidentified compounds; β , beta amino acids. Part of the smoothness of the network map derives from the arrangement of compounds in a precursor-product sequence. C. When lines are fitted to the same data at any given time and across all networks, we can estimate the physiological state of the pistachio embryo by the shape of the line. A partial map is illustrated for a 2-day period for a given metabolite family. At day 2, a branch-point in the distribution of nitrogen is indicated by the two peaks. D. When the approach in C is applied to experimental data, we obtain the partial response surface describing changes in physiological states over 158 days. Each state is separated by an arbitrary interval of time. A state-network matrix (e.g. 158 days x 56 metabolites) is constructed by combining or overlaying surface B and dimensional (e.g. Figure 4). Each unit of response space (rectangle) within the surface, is bounded by two metabolites (substrate and product) and by two physiological states (current and recent past). The resolution of such maps is based on the number of samples and sampling intervals in the experimental design.

coefficients in biosynthetic sequences, nor do we always have all the information and statistical reproducibility for this step. However, we can simplify and follow the connectivity of most metabolites of the glutamate family of amino acids, e.g., γ - aminobutyric acid, glutamine, ornithine, citrulline, urea, proline and in some cases, glutamic - γ - semialdehyde, Δ^1-pyrroline-5-carboxylic acid, argininosuccinate, etc. (Durzan, 1973).

Metabolic families of amino acids circumscribe the current model and define parameters for the interaction matrix. As pointed out by Sauro et al. (1987), this approach always isolates some part of a larger system and a test will always be needed for the validity of this isolation.

The portrayal of time is limited by the experimental design and cost of analytical procedures. The time axis can be hourly, daily, or weekly. The choice is usually between a simple, critical model, or a comprehensive, general model. In practice, a simple model is not always realistic when we look at tree behavior under field conditions or cells in bioreactors. Such complex systems tend to have non-linear components.

In Figs. 3 to 6, no fewer than duplicate and as many as four samples are taken at any one time. More sampling and replication, though prohibitive in cost, are required for thorough statistical testing of the model. Currently, we are primarily interested in the numerical (digital) tracking of nitrogen through carbon skeletons of metabolites identified in the model. We are also interested in selecting variables for the characterisation of physiological states over several intervals.

Mathematical curve fitting and statistical analyses (see next section) are performed through SAS or ISML programs, using a desktop personal computer (SAS, 1985). Representations of data can be enhanced using the graph theory, metabolic organisation and control, diffusion theory, catastrophe theory, and games and probability theory (Durzan, 1987c). These representations enable simulations that are currently impossible in field experiments. However, simulations are often fraught with spurious assumptions and leaps of faith based on inadequate experimental data.

Given the pros and cons, we still have the opportunity to mimic the operation and dynamics of the experimentally determined sets of differential equations. We can attempt to uncover logical patterns of behavior associated with experimentally measured discrete variables. For instance, can we find a developmental algorithm for embryogenesis or fruit ripening, or decipher a metabolic control theory based on the overall stability and fluctuations of the data set?

In reference and experimental state network models, we can identify material (e.g. nitrogen, carbon, tritium, metabolite), and logic flows (e.g. signals, thresholds, stability, resistance) throughout the response surface. We have not yet placed values on the variables, coefficients or constants in equations for sensitivity analyses. Sauro et al. (1987) used elasticity coefficients to describe the elasticity of enzymes with respect to metabolites, external effectors, system parameters, etc. In this review, we are not immediately concerned with the elasticity of enzymes, but with epigenetic events at the gene X environment level for a discrete developmental outcome.

Given the initial attempts at constructing response surfaces (Figs. 4,5 and 6), current theories related to the accuracy, precision and behavior of data set at the epigenetic level are described in Figs. 8,9 and 10. Over a range of cultivars, we can see for epicarp and mesocarp development, a dominant strategy common to the *Prunus* species in allocating carbon and nitrogen, through the aspartate family. The *Prunus* strategy contrasts with conifer or pistachio strategies where the glutamate family is dominant in the processing of nitrogen through the free amino acid pool. The goodness of fit of carbon 4 (aspartate) or carbon 5 (glutamate) strategies among cultivars is achieved by comparing or matching outputs with model response surfaces. The outcome is that we have useful data sets that can employ some of the parameters of metabolic control theory.

Statistical Description of Response Surfaces

Metabolic phenotypes concern the study of relationships between hierarchies of biochemical networks and the external environment. In woody perennials, response surfaces often show lagges, seasonal and diurnal changes. The evaluation of relationships among parameters requires suitable descriptive statistics and a suitable methodology appropriate for this type of problem.

There are at least two distinct aspects to the temporal analysis of response surfaces. First, there is the variation with the two sharply tuned natural environmental cycles, viz. the diurnal and seasonal cycles, both of which are *forced* oscillations. Second, there may be irregular cyclic variability, which is not strictly periodic and which extends over a continuous range of time scales, with a tendency to favor some time scales to others. Such time scales relate to positional receptors in the tree, developmental time, and local micro environmental and climatic change. When explants are taken for clonal propagation, this variability translates into the recovery of topoclones, periclones and cycloclones (Durzan, 1984).

Metabolic fluxes *tuned* to the forced oscillations in the environment can be mentioned in terms of their relative phases and amplitudes. We are also interested in perturbing fluxes by forcing changes in environmental and cultural conditions. One might wish to examine, for example, whether or not a flux lags an environmental parameter in some simple manner, as in the case of a partially insulated thermometer. A flux may also lag because of a series of inherent oscillations involving precursors and products (Higgins, 1967).

Statistical techniques involving single and cross spectral analysis are very powerful in examining all the aspects mentioned above (Shumway, 1988). In some cases, these offer considerable advantages over simple harmonic analysis, the superposed epoch method, etc. For single time-series records, reasonably satisfactory significance testing is available. For the comparisons of two time records (cross-spectral analyses), the frequency dependent relationship (coherence) may be tested with assumptions. Moreover, the individual spectra of change in environment and metabolism would have to be suitably smoothened, as depicted in Figs. 3 to 6.

For cross spectral analyses by computers (Salas et al., 1980; Shumway, 1988), the main requirement becomes the collection of meteorological and metabolic data in suitable sets of experimental design. This would involve the construction of time flux response surfaces which are dependent in part upon physiological processes, cultivational practices and clonal fidelity among different genetic backgrounds. In such a study, one aim would be to estimate the combined effect of a number of contributing processes.

RESULTS

Two Preliminary Applications

The state network reference models are being used as a research tool to answer questions about metabolic process control in clonal tissues having discrete developmental outcomes. Responses, imposed by various sets of experimental designs, are also assessed, using trees under field conditions. The general presentation of the response surface of a metabolic phenotype is shown in Fig. 2. This format is used throughout the studies described below.

1. Zygotic embryo development in *Pistacia vera* cv. Kerman

Embryo development was studied in clones (12 trees) over 158 days under field conditions in 20 year-old trees with and without zinc deficiency (Uriu, 1983; Durzan and Uriu, 1986). In a range of zinc deficient trees, stages of kernel development ranged from early abortion (blank formation) to complete filling of the fruit by cotyledons. The developing fruit contained over 100 free, naturally occurring nitrogenous compounds. The subset of free amino acids is shown in Fig. 3. The composition of the soluble nitrogen pool was determined by a computer-assisted analytical system with an automated amino acid analyser employing specific post-column reactions for amino acids, polyamines and monosubstituted guanidines (Dilley and Rocek, 1982; Ventimiglia and Durzan, 1987).

In the developing fruit (one to 90 days after fertilisation), the total alcohol-soluble nitrogen accumulated and declined as the kernel began rapid growth and development (90 to 158 days; Fig. 2). The decline in total soluble nitrogen was related non-linearly to an increase in bound nitrogen (protein; Durzan and Uriu, 1986).

The nitrogen metabolism of the kernel was dominated by the glutamate family, especially by free arginine nitrogen. The soluble nitrogen in the funiculus attaching the embryo in the fruit was mainly asparagine. During rapid embryo growth and development, the arginine nitrogen level gave way to glutamate nitrogen, as the total soluble nitrogen declined and seed proteins were synthesised, primarily in cotyledons.

In zinc deficient trees, both total soluble nitrogen and protein nitrogen always remained low. The overall linear correlation coefficient between blank production and zinc content of leaves at harvest date (Sept. 9) was -0.94. With increasing zinc deficiency, blank production increased, and arginine nitrogen accumulated in not only the aborting embryo, but also in leaves of branches bearing the rachis with its fruit (Fig. 4).

Fully developed embryos were characterised by a high and positive correlation coefficient between zinc content and percent free glutamate nitrogen (0.82). The metabolic phenotype for the soluble nitrogen of healthy kernels is shown in Fig. 3. This response surface becomes severely disrupted in zinc deficiency with the accumulation of arginine nitrogen (Fig. 2). The temporal analyses of these surfaces have not yet been evaluated.

2. Development of Early and Late Ripening Peaches

The marketability of fruit is a function of many factors, one of which is seasonal availability in terms of ripening date. The free amino acids, including those normally indispensable in the diet, are components of the soluble solids of Clingstone (Carson, Everts) and Freestone (Rio Oso Gem) varieties of *Prunus persica*.

The percentage distributions of nitrogen in free amino acids in soluble solids during the growth and development of the epicarp and mesocarp of early and late ripening cultivars are shown in Fig. 5.

All *Prunus* response surfaces obtained to date have a remarkably similar topology, in spite of known gradients in maturity and composition of fruits within peach trees (Dann and Jerie, 1988). All peach mesocarps were dominated by asparagine (major peak) in the aspartate family of amino acids, i.e. the peach mesocarp emphasises 4-carbon metabolites rather than 5- and 6 carbon metabolites, as seen in the pistachio hull (mesocarp) (Fig. 6). Generalisation can be made about strategies relating to the distribution or allocation of nitrogen in these species through transaminase activities. Differences characterising each cultivar also exist, but in reality, the significance of these remains unknown.

Metabolic control theory suggests that given statistically sound data, four levels are examined:

*   flux control coefficients
*   the connectivity relationships among flux control coefficients and enzyme elasticities
*   concentration control coefficients and connectivity with enzyme elasticities, and
*   temporal relationships to external environmental variables.

DISCUSSION

Performance Criteria

In judging tree performance by response surfaces in relation to metabolic control

STATE-NETWORK

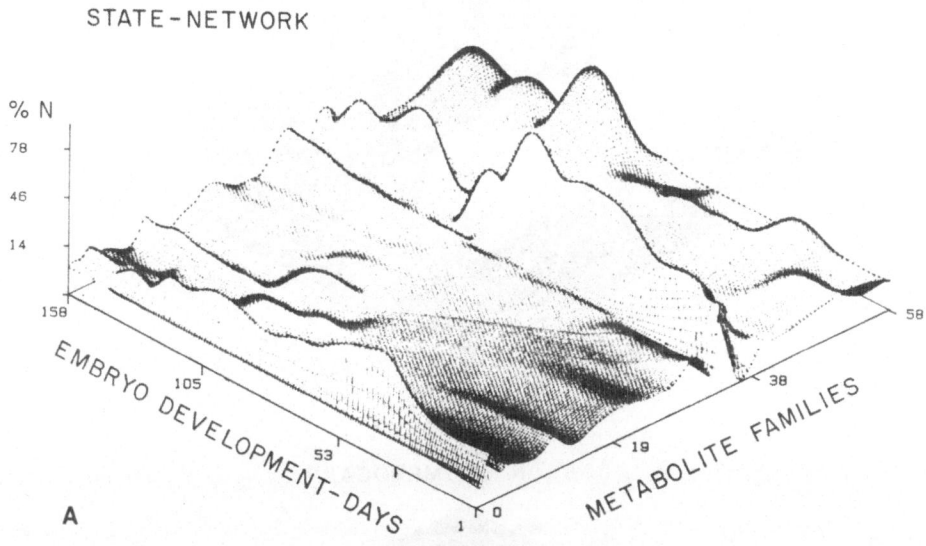

A

HULLS

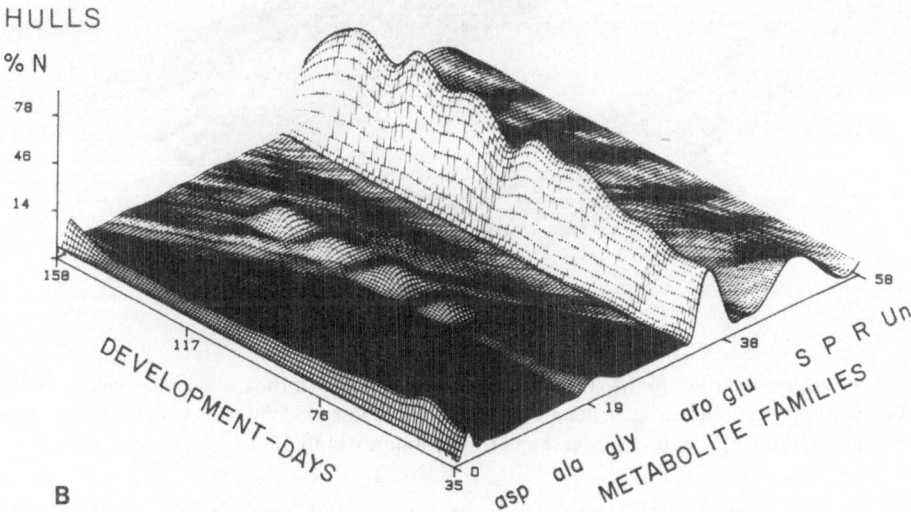

B

Fig. 4: The flux of nitrogen in each free amino acid arranged in a hierarchial network of metabolic families over 112 days of development of the epi/mesocarp of Clingstone peach cv Regina. A. Response surface showing positive maxima of nitrogen flux in soluble solids. B. Negative response surface for the same data obtained by inverting Fig. A (note change in sign of Z axis).

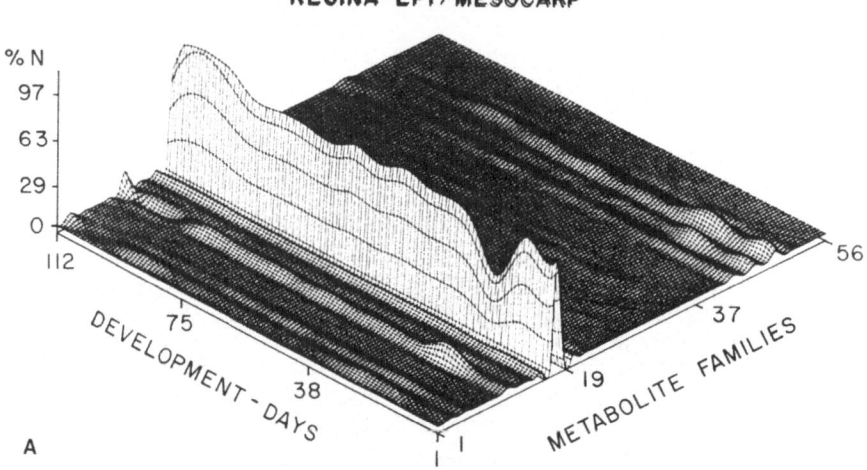

REGINA EPI/MESOCARP

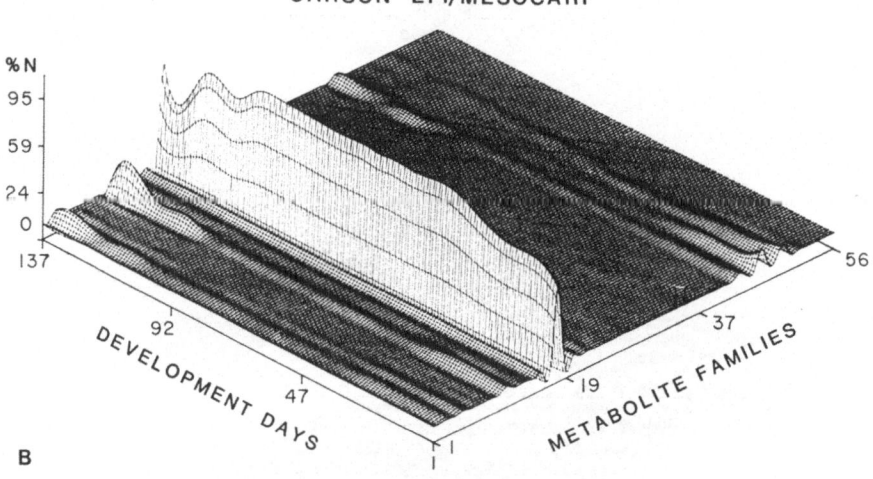

CARSON EPI/MESOCARP

Fig. 5: Response surface for (A) the pistachio kernel and (B) for the pistachio mesocarp to show distribution of soluble nitrogen, mainly to the glutamate family of amino acids. Note the preponderance of aromatic amino acids in the hull (metabolites 34 to 37).

theory (Sauro et al., 1987), we need to briefly examine perturbations to the control system of the tree. If an external modifier perturbation acts on an enzyme to change the enzyme's activity, the response coefficient for the flux is given by the flux control coefficient of the enzyme, multiplied by its elasticity with respect to the external modifier. The response coefficients for the metabolite concentrations are the concentration control coefficients with respect to the enzyme, multiplied by the elasticity. All of these coefficients are available from the solution of matrix equations and are based on standardised terminology in this field (Sauro et al., 1987).

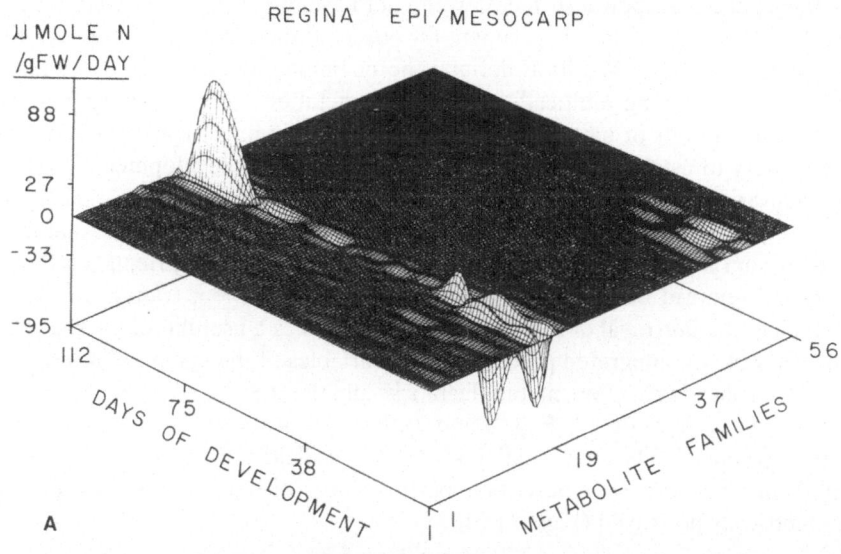

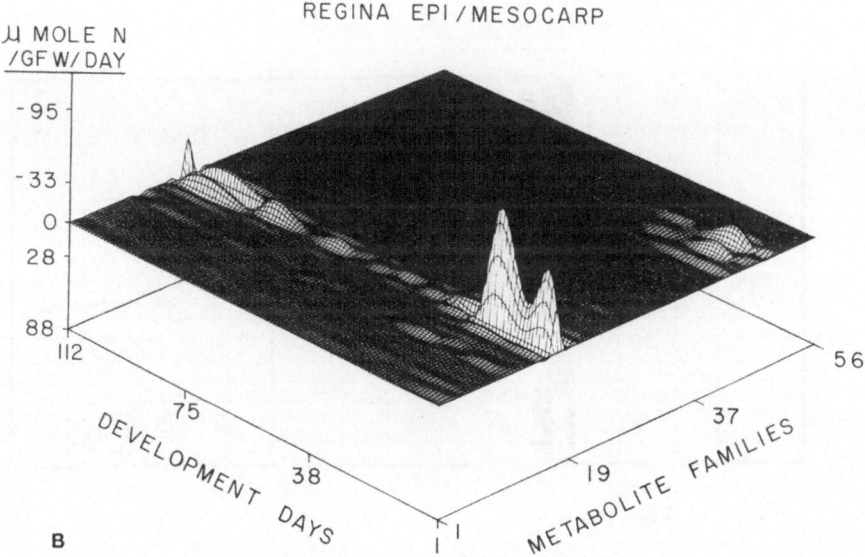

Fig. 6: Response surface for the free amino acids in the soluble solids of the epi/mesocarp from two Clingstone peach cultivars. A. Regina ripens early by 112 days after full bloom and the composition of soluble solids is dominated by high levels of asparagine nitrogen in the aspartate family (major peak just before metabolite 19). B. Carson ripens by 137 days and also reveals high level of free asparagine nitrogen. This same pattern is found in freestone varieties such as Rio Oso Gem and Everts. Taken together, the response surface for peach epi/mesocarp development have a common strategy involving the allocation of nitrogen to soluble solids as asparagine. This pattern is distinctly different from the pistachio kernel (Fig. 4) and its carpellary tissues (Fig. 5). The key to metabolite families is given in Fig. 3B.

In epigenetic response surfaces, we do not have direct data on enzymatic elasticities. Nevertheless, we have experimentally observed the rescaling of flux over time and physiological states. We have defined the beginning and end of our metabolic pathways in a developing multicellular organ. In metabolic control theory, however, the inputs and outputs of pathways (sources and sinks) remain invariant on the time scale necessary to establish the steady state. In true-to-type developmental systems, the same invariance is required, but the steady state can be, and often is, perturbed. Process control is characterised by internal correlations among plant parts that maintain the appropriate rate of protein turnover, provision of input nutrients, co-factors, etc., and the removal of enzymatic products by polymerisation, transport, etc. For these reasons, the portrayal of response surfaces becomes a useful tool for characterising the overall and integrated process control variables of the system of interest.

In the following section, metabolic blocks and the stability of response surfaces are considered in more detail (Figs. 4 and 7). Notions of *adaptive surfaces* and *resilience* are introduced. These are analogous to the elasticities used in the metabolites control theory. Terms used in describing performance attributes of response surfaces are geometrically portrayed (Figs. 7 to 10).

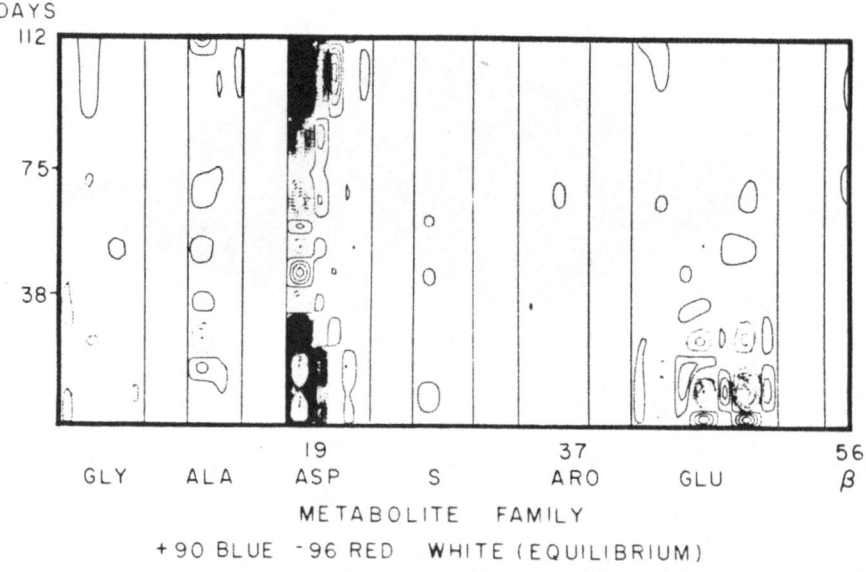

Regina μ mole/GFW/DAY EPI/Mesocarp

Fig. 7: Compression of a pulsating positive and negative response surfaces for selected amino acid families into a 2-dimensional surface as viewed directly overhead. Normally, fluxes are color-coded (blue: positive, red: negative) so that alternative signal fluxed can be seen over 112 days, each representing a discrete physiological state. Maximal signal or pulse strength for a positive flux was 90 μM gf wt per day. For a negative flux, the maximum signal was -96 μM gf wt per day. The key to metabolite families is given in Fig. 3B.

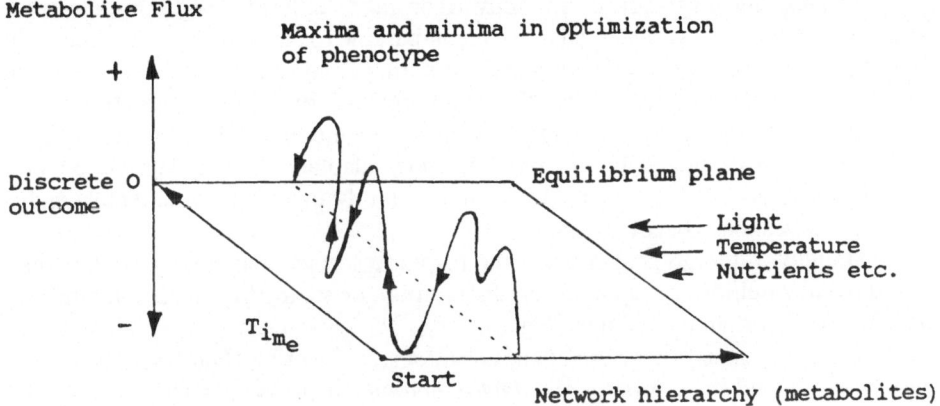

Fig. 8: Interpretation of flux stability for a pulsating response surface for one metabolite over time. Departures from an equilibrium plane are positive and negative and perturbed by factors leading to a discrete outcome at the final sampling time.

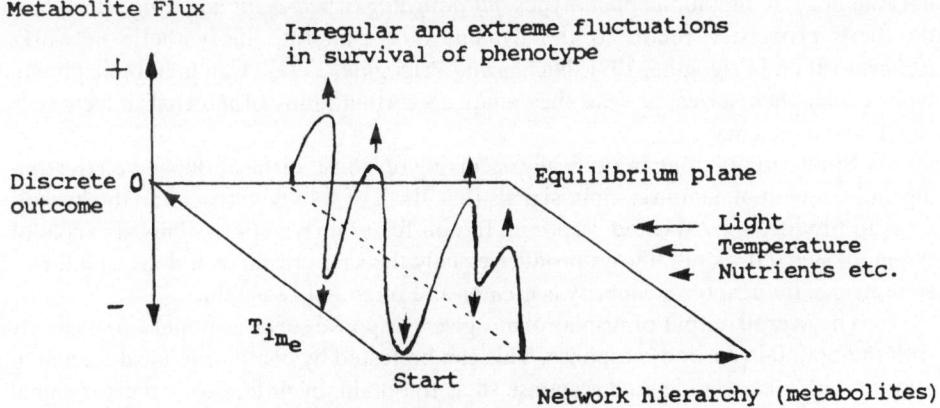

Fig. 9: Interpretation of resilience on pulsating response surface showing the flux of one metabolite over time. Departures from an equilibrium plane are outward (positive and negative) and not necessarily unstable. The phenomena in Figs. 8 and 9 help to describe the capacity and sensitivity of the response surface to deal with forcing factors of several types, e.g. changes in light, temperature, nutrients, plant growth regulators, etc.

To adapt means to adjust or modify according to different conditions, environments, etc. In response surfaces, adaptive is comparable to elasticity of an enzyme with respect to a metabolite or other forcing factors. In biology, adaptive usually refers to any alteration in the structure or function of an organism or any of its parts that result from natural selection and by which the organism becomes better fitted to survive and multiply in its environment. In *physiological terms,* adaptation is translated into a *decrease* in a response surface of a plant part to constantly applied environmental conditions, i.e. the response becomes stable.

Stability refers to properties of the response surface that causes it, when disturbed from conditions of equilibrium, distribution, or steady flux, to develop forces that restore the original condition (Fig. 8).

Using the signal theory in communications engineering (Gardner, 1986), the author has applied the notion of an *adaptive automaton* to the behavior of response surfaces. This approach postulates that the structure of a plant's metabolic phenotype is alterable or adjustable in such a way that its behavior or performance, according to some desired criterion, improves through contact with its environment and even possibly to the earth's climate precession cycle (Short and Menzel, 1986).

From an examination of the response surfaces, we can attempt to find evidence for self-adjustment (self-optimisation) in the face of changing diurnal and seasonal environments around the mother tree. Can the environmental fluctuations program recognisable, adaptive events into the organ of a tree ? How does genetics (different cultivars) affect the response of hierarchical metabolic phenotypes to a specific external change? Are metabolic phenotypes *self-designing* or *self-demultiplexing?* If so, how do these properties relate to the dynamics of elaborate biosynthetic networks (Glansdorff and Prigogine, 1971; Nicolis and Prigogine, 1977)? Can metabolic phenotypes repair themselves, i.e. can they adapt to certain kinds of internal defects (e.g. slight zinc deficiency)?

Similarities among periodically occurring response surfaces depend on the timing and strength of common input signals (e.g. flush of soluble nitrogen to the kernel) so as to produce the expected response. In non-linear enzymatic systems, a repeat of the input signal does not always produce exactly the same result as it does with linear systems, i.e. the response capacity is greater and often unpredictable.

The overall output or display of any given response surface should agree closely with the optimal (desired) response. This can be tested by comparing local area map outputs with a known, desired response so as to obtain, by difference, an error signal (Durzan, 1987a,c; 1988a,b). Factors that minimise the *error signal* can be identified so as to measure the extent of process control in the system.

This leads us to consider the terminology associated with variations in the response surfaces. *Measurement error* is the unintentional deviation from accuracy. It represents the difference between observed or calculated value and true value. *Chaos* refers to situations where chance or confusion is supreme. By contrast, *homeostasis* represents a relatively stable state of equilibrium or a tendency towards such a state. We can also distinguish *resilience* (as distinct from instability and enzyme elasticity) as the capacity of a stressed system to recover its size, shape, and behavior after

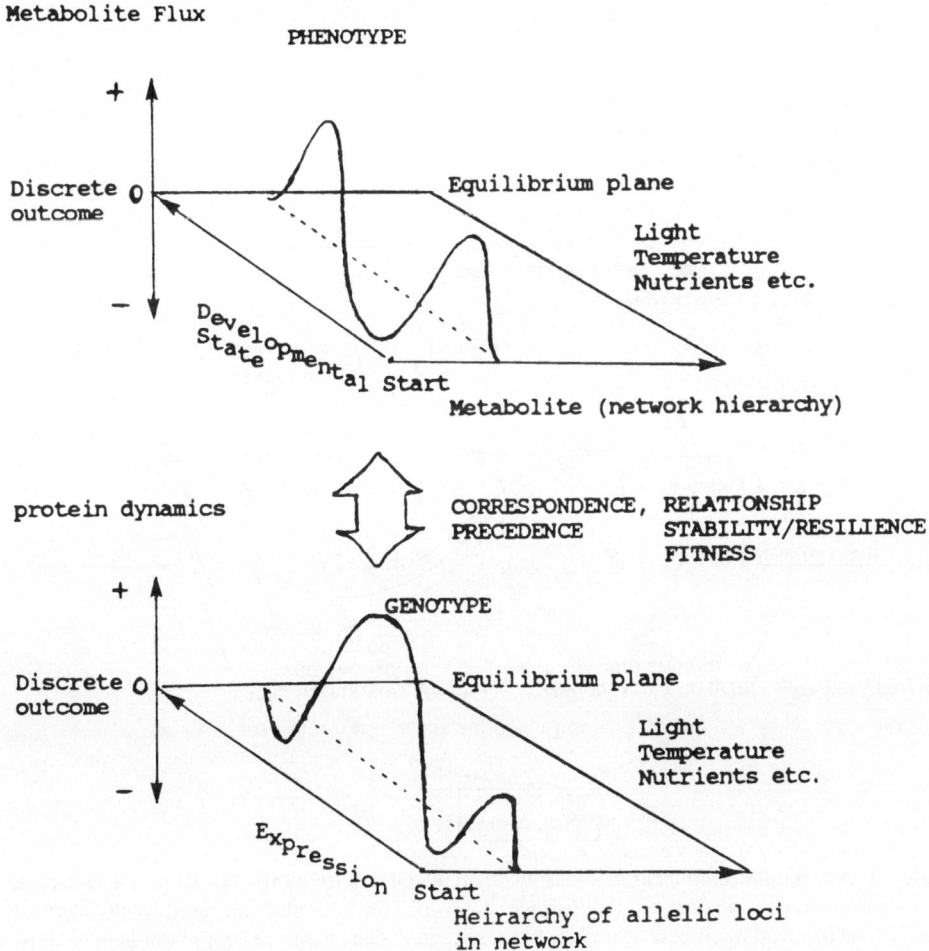

Metabolite Flux

PHENOTYPE

protein dynamics

Fig. 10 : Relationships are sought between a flux response surface for a single metabolite (top) and a single protein (bottom) from a 2-dimensional gel (not shown), as determined by the genotype or Sewall Wright's adaptive landscape (1968, not shown). In the precedence of events, the correspondences or matches between surfaces may be of diagnostic value.

deformations that are caused by forces, misfortune, or change (Fig. 9). On response surfaces, resilience describes the persistence of relationships (patterns) and measures the ability or capacity of the plant or tree to absorb changes after perturbation and deformation, and still survive or remain *fit*.

A resilient phenotype is capable of withstanding shock without permanent deformation, as in mild zinc deficiency, or adjust easily to disease, misfortune or change without becoming significantly unstable (outward arrows in Fig. 9). Durzan (1988a,c) has introduced the notion of *costs* in threshold optimisation.

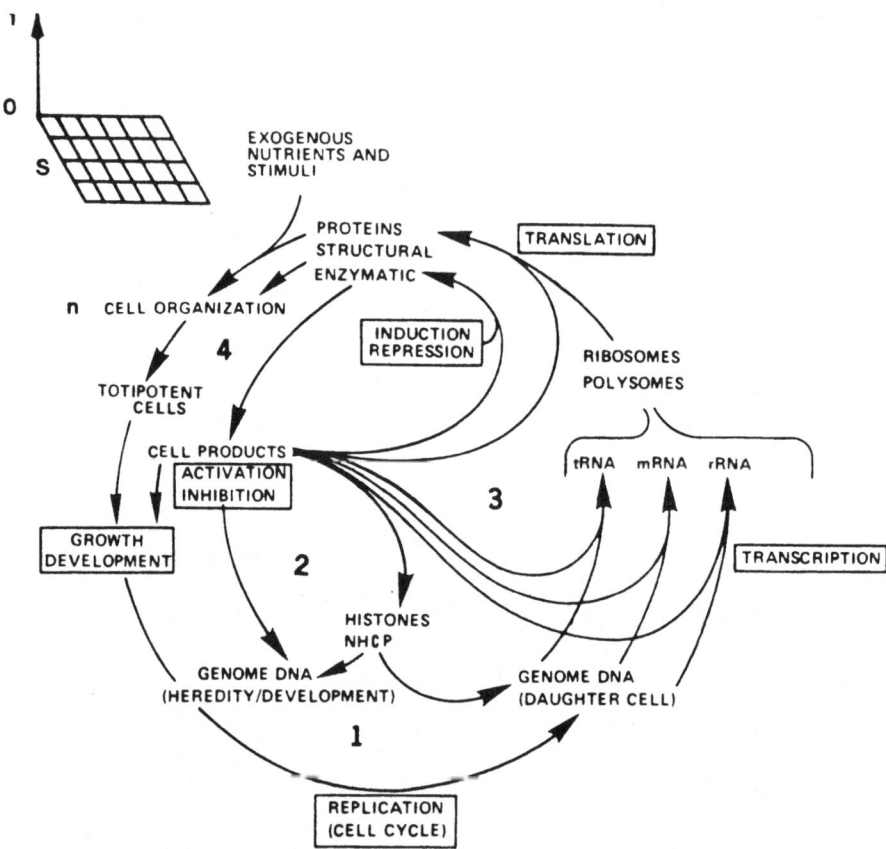

Fig. 11: A hypothetical nuclear-cytoplasmic cycle of cell determination (cf. Brink 1962) leading to the characterization of a response surface as shown in the upper left corner. Internal feedback cycles are involved in process central at 4 levels (1. gene-gene action, 2. cytoplasm-gene, 3. cytoplasm-cytoplasm protein, 4. cytoplasm-cytoplasm phenotype).

Over a period of time, it might be possible to distinguish among inherited traits that exploit environmental changes that are favorable, tolerable, inimicable (Figs. 10 and 11). There are three useful performance criteria for response surfaces in such interpretations: parsimony, probability and compatibility. Criteria of this type should be imposed after the biologist is thoroughly familiar with the assumptions and arguments of the model. *Parsimony* represents patterns that require the fewest character changes in relation to a forcing factor. Parsimony minimises the need to involve convergence or reversals (homoplasy) of variables or allelic traits in explaining the response. The use of *probability* method is based on initial assumptions regarding the rate of change in response surfaces. So far, probabilistic methods have been based on notions of Signal Theory (Durzan, 1987a,c).

Probabilisitic theories have not yet been experimentally applied to our model. *Compatibility* criteria attempt to find the largest sets of characters that show correspondence between parts of responses acquired as the result of parallel evolution or covergence. Here, we are dealing with pattern matching and the tracking of specific characters as in phylogenetic analysis (Fink, 1986). Forcing factors that contribute to genes X environment interactions should be manifested experimentally by response surfaces. Forcing factors should reflect important attributes of the organism, e.g. fertility, development, etc. The responsiveness or precise local environmental adaptation of trees must also consider ecological processes that affect the abundance, distribution and succession of species (Shugart, 1984; Robertson, 1987).

REFERENCES

Agur, Z. and Slobodkin, L.B., 1986, Environmental fluctuations: How do they affect the topography of the adaptive landscape?, *J. Genet.,* 65: 45-54.

Brink, R.A., 1962, Phase change in higher plants and somatic cell heredity, *Q. Rev. Biol.,* 37: 1-22.

Dann, I.R. and Jerie, P.H., 1988, Gradients in maturity and sugar levels of fruit within peach trees, *J. Am. Soc. Hortic. Sci.,* 113: 27-31.

Dilley, K.J. and Rocek, P.H., 1982, *Protein chemistry notes* PCN-11 LKB Biochrome Ltd., Cambridge, England.

Durzan, D.J., 1973, Nitrogen metabolism of *Picea glauca,* V. Metabolism of uniformly labelled ^{14}C-L-proline and ^{14}C-L-glutamine by dormant buds in late fall, *Can. J. Bot.,* 51: 359-369.

Durzan, D.J., 1982, Nitrogen metabolism and vegetative propagation of forest trees. In: *Tissue Culture in Forestry* (J.M. Bonga and D.J. Durzan, eds.), pp. 256-324, Martinus Nijhoff/Dr. W. Junk, The Hague.

Durzan, D.J., 1984, Special problems: Adult vs. juvenile explants. In: *Handbook of Plant Culture: Vol. 2. Crop Species.* (W.R. Sharp, D.A. Evans, P.V. Ammirato and Y. Yamada, eds.), pp. 472-503, Macmillan Publishing Company, New York.

Durzan, D.J., 1986, Ammonia: Its analogues, metabolic products and site of action in somatic embryogenesis. In: *Cell and Tissue Culture in Forestry. Vol. 2. Specific Principles and Methods: Growth and Developments* (J.M. Bonga and D.J. Durzan, eds.), pp. 92-136, Martinus Nijhoff, Dordrecht.

Durzan, D.J., 1987a, Plant growth regulators in cell and tissue culture of woody perennials, *Plant Growth Regulation,* 6: 95-112.

Durzan, D.J., 1987b, Plant growth regulator-directed phase specificity in cell and tissue culture for tree improvement, *Proc. Ann. Conf. Soc. Am. Foresters,* pp. 218-222, Birmingham Atlanta.

Durzan, D.J., 1987c, Physiological states and metabolic phenotypes in embryonic development. In: *Cell and Tissue Culture in Forestry. Vol. 2. Specific Principles and Methods: Growth and Developments* (J.M. Bonga and D.J. Durzan, eds.), pp. 405-439, Martinus Nijhoff, Dordrecht.

Durzan, D.J., 1988a, Metabolic phenotypes in somatic embryogenesis and polyembryogenesis. In: *Genetic Manipulation of Woody Plants* (J. Hanover, ed.), pp. 293-312, Plenum Press, New York.

Durzan, D.J., 1988b, Somatic polyembryogenesis and plantlet regeneration in selected tree crops, *Biotech. Gen. Eng. Rev.,* 6: 339-376.

Durzan, D.J. and Steward, F.C., 1983, Nitrogen metabolism. In: *Plant Physiology, A Treatise,* Vol. 8 (F.C. Steward, ed.), pp. 55-265, Academic Press, New York.

Durzan, D.J. and Uriu, K., 1986, Metabolic networks in developing pistachio embryos (*Pistacia vera* cv. Kerman). In: *Progress in Developmental Biology,* Part A (H.C. Slavkin, ed.), pp. 199-202, Alan R. Liss, Inc.

Fink, W.L., 1986, Microcomputers and phylogenetic analysis, *Science,* 234: 1135-1139.

Firn, R.D., 1986, Growth substance sensitivity: The need for clearer ideas, precise terms and purposeful experiments, *Physiol. Plant.,* 67: 267-272.

Gardner, W.R., 1986, *Introduction to Random Processes with Applications to Signals and Systems,* Macmillan Publishing Company, New York.

Glansdorff, P. and Prigogine, I., 1971, *Thermodynamic Theory of Structure Stability, and Fluctuations,* John Wiley, New York.

Higgins, J., 1967, The theory of oscillating reactions, *Ind. Eng. Chem.,* 59: 19-62.

Holliday, R., 1987, The inheritance of epigenetic defects, *Science,* 238: 163-170.

Jurs, P.C., 1986, Pattern recognition used to investigate multivariate data in analytical chemistry, *Science,* 232: 1219-1224.

Kacser, H. and Burns, J.A. 1973, The control of flux, *Symp. Soc. Exp. Biol.,* 27: 65-104.

Kacser, H. and Burns, J.A., 1981, The molecular basis of dominance, *Genetics,* 97: 639-666.

Libby, W.J., 1983, Potential of clonal forestry. In: *Clonal Forestry: Its Impact on Tree Improvement and Our Future Forests* (L. Zuffa, R.M. Rauter and C.W. Yeatman, eds.), Proc. 19th Can. Tree Improv. Assoc., Toronto.

Nicolis, G. and Prigogine, I., 1977, *Self Organization in Nonequilibrium Systems: From Dissipative Structures to Order through Fluctuations,* Wiley-Interscience, New York.

Ochs, R.S., Hanson, R.W. and Hall, J., 1985, *Metabolic Regulation,* Elsevier, New York.

Pitel, J.A. and Durzan, D.J., 1987a, Changes in the composition of the soluble and chromosomal proteins of jack pine (*Pinus banksiana* Lamb.) seedlings induced by fenitrothion, *Environ. Exp. Bot.,* 18: 153-162.

Pitel, J.A. and Durzan, D.J., 1987b, Chromosomal proteins of conifers, 1. Comparison of histones and nonhistone chromosomal proteins from dry seeds of conifers, *Can. J. Bot.,* 56: 1915-1927.

Pitel, J.A. and Durzan, D.J., 1987c, Chromosomal proteins of conifers, 2. Tissue-specificity of the chromosomal proteins of jack pine *(Pinus banksiana), Can. J. Bot.,* 56: 1928-1931.

Pitel, J.A. and Durzan, D.J., 1988, Chromosomal proteins of conifers, 3. Metabolism of histones and nonhistone chromosomal proteins in jack pine *(Pinus banksiana)* during germination, *Physiol. Plant.,* 50: 137-194.

Robertson, A., 1987, The centroid of tree crowns as an indicator of abiotic processes in a balsam fir wave forest, *Can. J. For. Res.,* 17: 746-755.

Salas, J.D., Delleur, J.W., Yevjevich, V. and Lane, W.L., 1980, *Applied Modelling of Hydrologic Time Series,* Water Resources Publ., Littleton, Colorado.

SAS, 1985, SAS/GRAPM User's Guide, Version 5, SAS Institute Inc. Cary NC.

Sauro, R.M., Small, J.R. and Fell, D.A., 1987, Metabolic control and its analysis, extensions to the theory and matrix method, *Eur. J. Biochem.,* 165: 215-221.

Short, D.A. and Mengel, J.G., 1986, Tropical climate phase lags and earth's precession cycle, *Nature (Lond.),* 323: 48-50.

Shugart, H.H., 1984, *A Theory of Forest Dynamics: The Ecological Implications of Forest Succession Models,* Springer-Verlag, New York.

Shumway, R.H., 1988, *Applied Statistical Time Series Analysis,* Prentice Hall, Engelwood Cliffs, New Jersey.

Steward, F.C., 1968, *Growth and Organization in Plants,* Addison-Wesley Publ. Co., Reading, Mass.

Stone, E.C., 1965, Preserving vegetation in parks and wilderness, *Science,* 150: 1261-1265.

Timmis, R., Abo El-Nil, M.M. and Stonecypher, R.W., 1987, Potential genetic gain through tissue culture. In: *Cell and Tissue Culture in Forestry, Vol. 2. Specific Principles and Methods: Growth and Development* (J.M. Bonga and D.J. Durzan, eds.), pp. 198-215, Martinus Nijhoff, Dordrecht.

Uriu, K., 1983, Diagnosis and correction of nutritional problems including the *crinkle-leaf* disorder, *Ann. Rep., Calif. Pistachio Industry,* Fresno, California pp. 47.

Ventimiglia, F.M. and Durzan, D.J., 1987, The determination of monosubstituted guanidines using a dedicated amino acid analyzer, *Liquid Chromatography/Gas Chromatography,* 4: 1121-1124.

Welch, R.G. and Clegg, J.S., (eds.), 1987, *The Organisation of Cell Metabolism,* Plenum Press, New York.

Wright, S., 1968, *Evolution and the Genetics of Populations,* Univ. Chicago Press, Chicago.

Yablokov, A.V., 1986, *Phenetics, Evolution, Population Traits,* Columbia Univ. Press, New York.

Young, D.A., 1984, Advantages of separations on giant two-dimensional gels for detection of physiologically relevant changes in the expression of protein gene products, *Clin Chem.,* 30: 2104-2108.

15

ALTERATION OF GROWTH AND MORPHOGENESIS BY ENDOGENOUS ETHYLENE AND CARBON DIOXIDE IN CONIFER TISSUE CULTURES

Prakash P. Kumar and
Trevor A. Thorpe

ABSTRACT

We studied the production of ethylene and carbon dioxide in tissue cultures of two coniferous tree species, namely, *Pinus radiata* (radiata pine) and *Picea glauca* (white spruce) to understand their roles in growth and morphogenesis *in vitro*. When cultured under shoot-forming conditions, cotyledons of radiata pine produced more ethylene

Prakash P. Kumar and *Trevor A. Thorpe* * Plant Physiology Research Group, Department of Biological Sciences, University of Calgary, Calgary, Alberta T2N 1N4, Canada.

and carbon dioxide than under non-shoot-forming conditions. Growth (fresh- and dry-weights) and shoot formation were enhanced by the accumulation of these gases during the initial stages of shoot bud differentiation. Reducing their concentrations by allowing the gases to diffuse out or by absorbing them into chemical traps, resulted in reduced growth and shoot formation. In contrast, when these gases were allowed to accumulate in flasks containing embryogenic suspension cultures of white spruce, the growth of the callus and the number of embryos produced were reduced by about 50 percent over the control, where the gases were allowed to diffuse out. The pattern of accumulation of these gases in the two conifer tissue cultures in relation to morphogenesis is discussed.

INTRODUCTION

Plant cells and tissues cultured *in vitro* produce significant amounts of ethylene and carbon dioxide (De Proft et al., 1985; Thomas and Murashige, 1979; Kumar et al., 1987). Depending on the closures used for sealing the culture vessels, these gases may or may not accumulate in the headspace. It has been pointed out earlier that the volatile emissions from cultured tissues can play an important role in modifying the pattern of morphogenesis *in vitro* (Thomas and Murashige, 1979, Kumar et al., 1987). It has also been reported that when cotyledons of radiata pine, a coniferous tree species, were cultured under shoot-forming conditions, the level of ethylene and carbon dioxide in the headspace affected shoot bud induction (Kumar et al., 1987). We initiated a series of experiments to determine if organised development in other coniferous tree species was similarly affected by the gaseous environment. We chose embryogenic cell suspension cultures of white spruce for this purpose. The selection of this system gave us an additional advantage, in that the response of an embryogenic culture system to ethylene and carbon dioxide could be compared to that of the organogenic cultures of radiata pine.

MATERIALS AND METHODS

Culture conditions for *Pinus radiata* D.Don cotyledons are described in Kumar et al. (1987). Cotyledons were excised aseptically from five-day-old seedlings and cultured on a modified Schenk and Hildebrandt medium supplemented with 25 μMN[6]-benzyl adenine (Aitken et al., 1981) in 25 ml Erlenmeyer flasks. The flasks were sealed with either gastight serum caps or foam bungs that are permeable to gases, and incubated in light. Embryogenic callus of *Picea glauca* (Moench) Voss. initiated in 1986 (Lu and Thorpe, 1987) was used to start cell suspension cultures. Modified DCR medium (Gupta and Durzan, 1985) supplemented with 500 mg l[-1] myo-inositol, 250 mg l[-1] glutamine and 500 mg l[-1] casein hydrolysate was used for the liquid suspension cultures. The cell suspension cultures (25 ml in 50 ml Erlenmeyer flasks sealed with gastight serum caps or foam bungs) that were maintained in the dark at 26°C on rotary shaker set at 150 rpm remained embryogenic. Ethylene was estimated by gas chromatography and carbon dioxide measurement was done using an infra-red gas analyser, as described earlier (Kumar et al., 1987).

The packed cell volume of the suspension culture was obtained after centrifuging the cultures at 300 X g for 5 min. Dry weight was estimated after drying the cells in an oven at about 80°C for 48 hours.

RESULTS AND DISCUSSION

The appearence of the cultures of both radiata pine and white spruce at the end of the culture periods (21 and 10 days, respectively) under the two culture-conditions used can be seen in Figs. 1 to 5. Distinct stages of shoot formation in the cotyledons of radiata pine have been described elsewhere (Villalobos et al., 1985). Briefly, the first organised structures observed in the cotyledons, called the promeristemoids, appeared in culture by day five.

These become meristematic nodules by day 10 and by day 15, they began to differentiate into shoot buds. The process of bud differentiation reached an optimum level by day 21 (Fig. 2), when the cotyledons could be transferred to cytokinin-free medium to allow elongation of the shoots. These shoot-forming cotyledons produced significant amounts of ethylene and carbon dioxide. The concentrations of these two gases increased continuously when the flasks were sealed with gastight serum caps until the end of the 21-day culture period (Fig. 6). Based on studies where ethylene and carbon dioxide were absorbed by traps, mercuric perchlorate and potassium hydroxide, respectively, we concluded that the accumulation of the gases up to day 15 is essential for bud differentiation and subsequent development of shoots (Kumar et al., 1987). During this time, meristematic nodules, which ultimately differentiate into shoot buds, were being formed (Villalobos et al., 1985). However, the excess ethylene (about 20 µl l^{-1}) and carbon dioxide (about 200 ml l^{-1}), accumulated at the end of the 21-day culture period, caused some degree of de-differentiation, and as a result, the majority of the shoot buds failed to elongate on transfer to a cytokinin-free medium. The fresh- and dry-weights, taken as a measure of growth, were found to be about 50 percent lower in the foam bung-closed flasks when compared to those in flasks sealed with serum caps (Table 1 and Figs. 1 and 2). The number of shoot buds produced was nearly 2.5 times higher in the cotyledons grown in serum-capped flasks (Table 1). Under non-shoot-forming conditions, ethylene and carbon dioxide accumulation was lowered by about ten and 100 fold, respectively (Kumar et al., 1987).

Embryogenic cell suspension cultures of white spruce also produced high concentrations of ethylene and carbon dioxide (Fig. 7). This is consistent with earlier reports that high levels of ethylene are produced during the active cell division and growth phase of suspension cultures of various plants (MacKenzie and Street, 1970; La Rue and Gamborg, 1971). In the cultures we used, the somatic embryos were at the

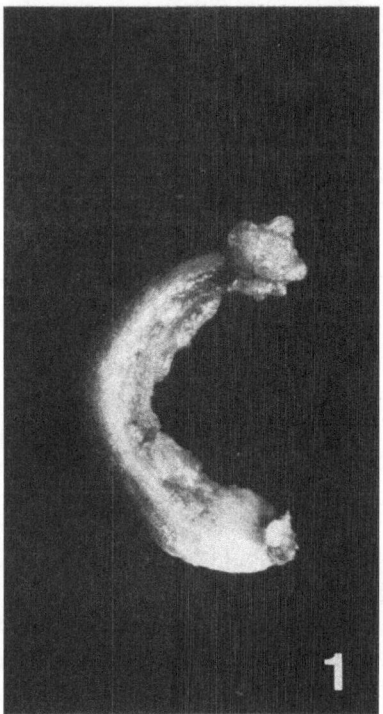

Fig. 1: A cotyledon of radiata pine *(Pinus radiata)* after 21 days of culturing on shoot-forming medium in flasks closed with foam bungs. The cotyledon is thin and has very few shoot buds.

proembryo stage (Fig. 5) with elongated suspension cells and a compact head which eventually developed into the embryo axis (Lu and Thorpe, 1987). We measured the gases during the first 10 days after sub-culturing the embryogenic cells into fresh medium that supports the induction of more somatic embryos. During this period, there was a continuous accumulation of carbon dioxide, until it reached the level of about 280 ml l^{-1} (Fig. 7). The ethylene level increased sharply during the first two days and from then on, until day 6, a slight increase was observed (Fig. 7). In contrast to the observation in radiata pine cultures, the growth of cells in the suspension culture of white spruce, as measured by their fresh-and dry-weights, was about 50 percent higher when the flasks were closed with foam bungs, which allow free gas exchange, instead of gastight serum caps (Table 1).

Also, the packed cell volume and the number of embryos per ml of culture were about twice as high: two fold higher when the gases were allowed to diffuse out of the flasks as they were produced. This is consistent with the observation of Dalton and Street (1976) that the accumulation of ethylene lowered the dry-weight of the cells and the cell number in *Spinacea oleracea* suspension cultures. Also, Horner et al. (1977) had observed that androgenesis in several species of *Nicotiana* remains unaffected by the removal of ethylene from the culture vessels or continuous flushing with air.

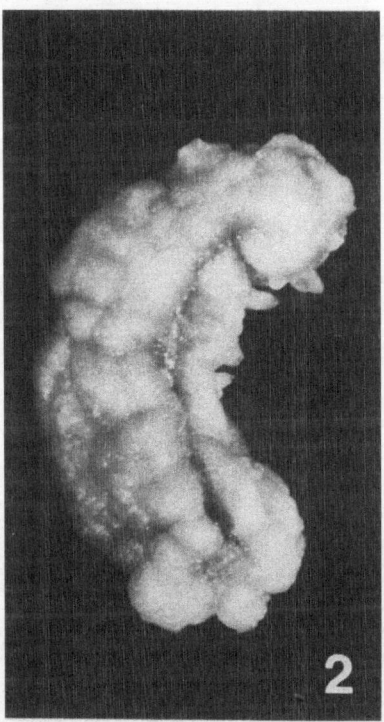

Fig. 2: A cotyledon of radiata pine, also after 21 days of culture. Culture conditions were as for cotyledon in Fig. 1, except that the flask was sealed with a gastight serum cap. The cotyledon has differentiated shoot buds all along its length. This cotyledon was from the same embryo as that in Fig. 1.

Table 1: Fresh- (FW) and dry- (DW) weights (mg per cotyledon) of *Pinus radiata* (Pr) cotyledons cultured for 21 days under shoot-forming conditions with the number of shoot buds per cotyledon, compared to the FW and DW (mg per 25 ml culture), packed cell volume (PCV, ml per 25 ml culture) and the number of embryos (per ml culture) of embryogenic cell suspension cultures of *Picea glauca* (Pg,) 10 days after sub-culturing. Data from flasks closed with foam bungs (FB) and gastight serum caps (SC) are given for each species ($\pm$SD).

FW		DW		PCV		No. of shoots or embryos	
FB	SC	FB	SC	FB	SC	FB	SC
Pr 12.7$\pm$2.6	22.5$\pm$2.8	2.0$\pm$0.5	3.9$\pm$0.5	--	--	14	34
Pg 619	295	51	22	5.25	2.75	1162$\pm$107	508$\pm$47

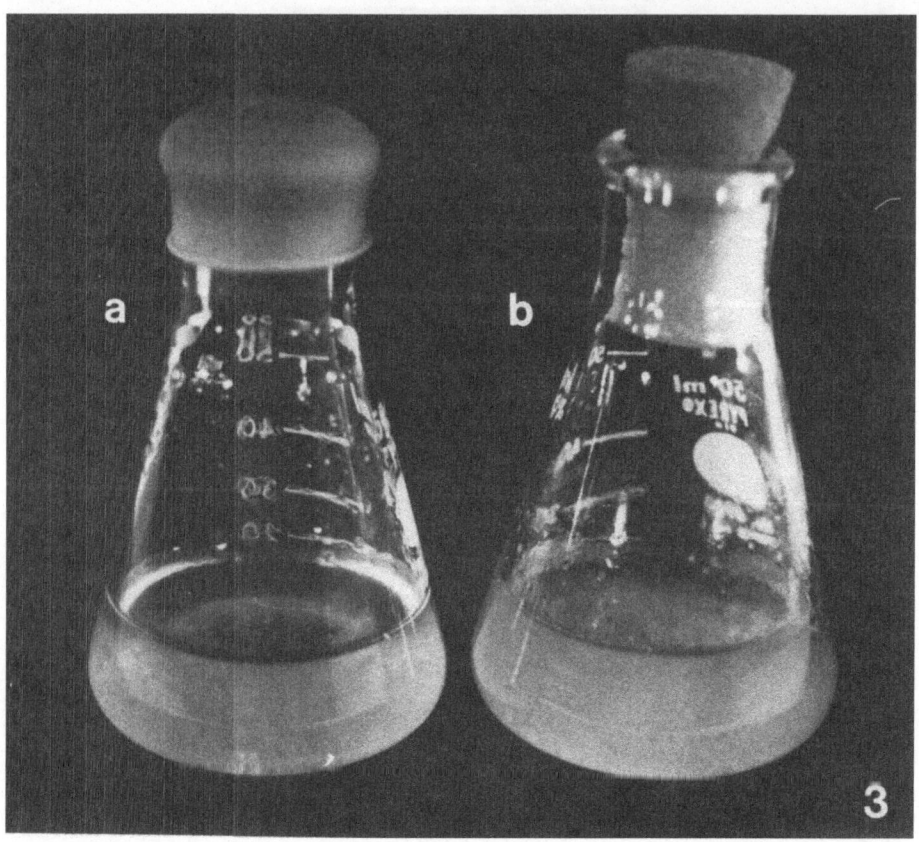

Fig. 3: Embryogenic cell suspension culture of white spruce *(Picea glauca)*. Flask (a) is sealed with a gastight serum cap and flask (b) is closed with a foam bung.

It has been suggested that while much of the ethylene produced in tissue cultures may be the result of enhanced metabolism, at least some of the ethylene formed early during the culture may have a growth regulatory role (Huxter et al., 1979; Kumar et al., 1987). Our results give further credence to this hypothesis, attributing a dual role (cause and effect) to ethylene in influencing growth and morphogenesis *in vitro*. We have evidence suggesting that non-autotrophically fixed carbon dioxide as a result of increased phosphoenolpyruvate carboxylase activity may be important during organogenesis in radiata pine cotyledon cultures (data not shown), as has been shown to be important for exponentially growing cell cultures (Bender, 1987; Rogers et al., 1987). The contrasting observations regarding the effect of allowing ethylene and carbondioxide to accumulate in the culture vessels on morphogenesis, allow us to speculate that the unique processes of organogenesis and somatic embryogenesis are probably regulated differently.

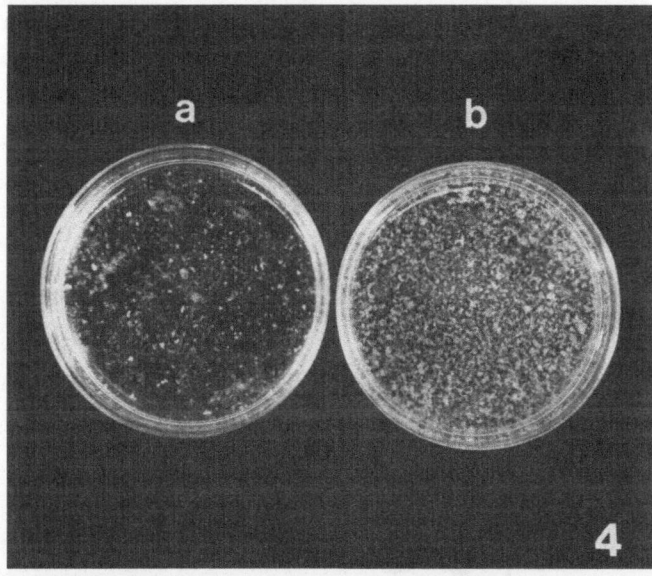

Fig. 4: A close-up view of embryogenic suspension cultures of white spruce. (a) Culture from a flask sealed with a serum cap. (b) Culture from a flask closed with a foam bung. Note that the number of cells per unit volume is about two-fold higher in (b) than in (a).

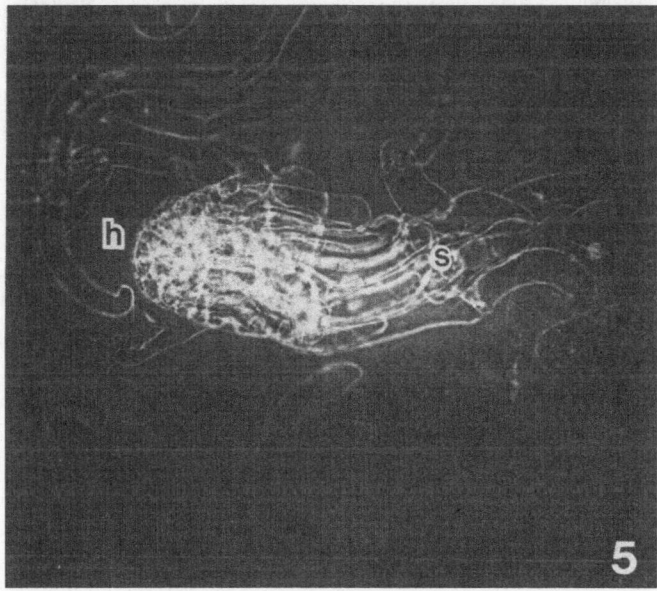

Fig. 5: The stage of development of somatic embryos in the suspension cultures of white spruce used in the experiments. Note the compact head(h) and elongated cells of the suspensor (s).

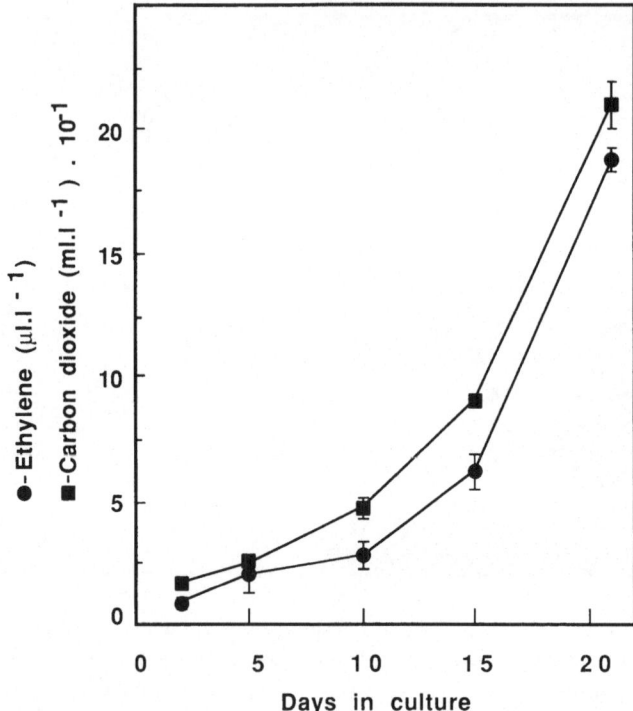

Fig. 6: The pattern of accumulation of ethylene and carbon dioxide in the flasks containing shoot-forming cotyledons (16 per flask) of *Pinus radiata*. The flasks were sealed with serum caps on day 0.

Currently, a wide variety of closures are being used for the different types of culture vessels employed in plant tissue culture. These include plastic caps, aluminium foil, plastic films, parafilm, plugs made from foam and cotton wool, etc. All these have different levels of permeability to gases. In general, the researcher uses the closures that are readily available. In the light of our observations, we suggest the selection of appropriate closures for the culture of each plant species after a simple preliminary experiment using closures of a wide range of permeability. Our results further support the idea that *in vitro* morphogenesis can be optimised (e.g., in commercial tissue culture laboratories) by appropriately manipulating the gaseous environment within the culture vessels.

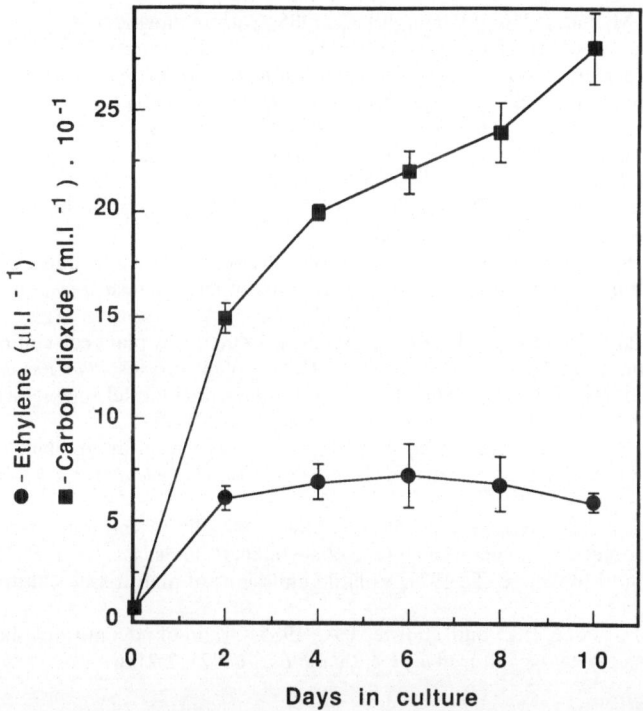

Fig. 7: The pattern of accumulation of ethylene and carbon dioxide in the flasks containing 25 ml of embryogenic suspension cultures of *Picea glauca*. The flasks were sealed with serum caps on day 0.

ACKNOWLEDGEMENT

Prakash P. Kumar acknowledges a travel award as a part of an Izaal Walton Killam Memorial Scholarship.

REFERENCES

Aitken, J., Horgan, K.J. and Thorpe, T.A., 1981, Influence of explant selection on the shoot-forming capacity of juvenile tissue of *Pinus radiata.*, *Can. J. For. Res.*, 11: 112-117.

Bender, L., 1987, Fixation and metabolism of carbon dioxide by photosynthetically active carrot callus cultures, *Can. J. Bot.*, 65: 1768-1770.

Dalton, C.C. and Street, H.E., 1976, The role of the gas phase in the greening and growth of illuminated cell suspension cultures of spinach (*Spinacea oleracea* L.), *In Vitro*, 12: 485-494.

De Proft, M.P., Maene, L.J. and Debergh P.C., 1985, Carbon dioxide and ethylene evolution in the culture atmosphere of *Magnolia* cultured *in vitro, Physiol. Plant.,* 65: 375-379.

Gupta, P.K. and Durzan, D.J., 1985, Shoot multiplication from mature trees of Douglas-fir *(Psedotsuga menziesii)* and sugar pine *(Pinus lambertiana), Plant Cell 'Rep.,* 4: 177-179.

Horner, M., McComb, J.A., McComb, A.J. and Street, H.E., 1977, Ethylene production and plantlet formation by *Nicotiana* anthers cultured in the presence and absence of charcoal., *J. Exp. Bot.,* 28: 1365-1372.

Huxter, T.J., Reid, D.M. and Thorpe, T.A., 1979, Ethylene production by tobacco *(Nicotiana tabacum)* callus, *Physiol. Plant.* 46: 374-380.

Kumar, P.P., Reid, D.M. and Thorpe, T.A., 1987, The role of ethylene and carbon dioxide in differentiation of shoot buds in excised cotyledons of *Pinus radiata in vitro, Physiol. Plant.,* 69: 244-252.

La Rue, T.A.G. and Gamborg, O.L., 1971, Ethylene production by plant cell cultures: Variations in production during growing cycle and in different plant species. *Plant Physiol.,* 48: 394-398.

Lu, C.-Y. and Thorpe, T.A., 1987, Somatic embryogenesis and plantlet regeneration in cultured immature embryos of *Picea glauca, J. Plant Physiol.,* 128: 297-302.

MacKenzie, I.A. and Street, H.E., 1970, Studies on the growth in culture of plant cells., VIII. The production of ethylene by suspension cultures of *Acer pseudoplatanus* L., *J. Exp. Bot.,* 21: 824-834.

Rogers, S.M.D., Ogren, W.L. and Widholm, J.M., 1987, Photosynthetic characteristics of a photoautotrophic cell suspension culture of soybean, *Plant Physiol.,* 84: 1451-1456.

Thomas, D.S. and Murashige, T., 1979, Volatile emissions of plant tissue cultures, *In Vitro,* 15: 654-658.

Villalobos, V.M., Yeung, E.C. and Thorpe, T.A., 1985, Origin of adventitious shoots in excised radiata pine cotyledons cultured *in vitro, Can. J. Bot.,* 63: 2172-2176.

16

STORAGE OF FOREST TREE GERMPLASM AT SUB-ZERO TEMPERATURES

M.R. Ahuja

ABSTRACT

The cold storage potential of seeds and dormant buds for the preservation of forest tree germplasm was investigated. Seeds from European beech (*Fagus sylvatica*), hybrid aspen (*Populus tremula* x *P. tremuloides*), larch (*Larix decidua*), Norway spruce *(Picea abies)*, Scots pine *(Pinus sylvestris)*, and silver fir (*Abies alba*) were stored for up to six days in liquid nitrogen (-196°C). Seed germination data indicated that in five of the six forest tree species, there was practically no loss of viability in the seeds stored in liquid nitrogen, as compared to controls. Beech seeds did not germinate following storage at -196°C. We are employing tissue culture technology to rescue beech embryo/cotyledon from seed stored in liquid nitrogen.

M.R. Ahuja * Federal Research Centre for Forestry and Forest Products, Institute of Forest Genetics and Forest Tree Breeding, Sieker Landstrasse 2, D-2070 Grosshansdorf, Federal Republic of Germany.

Dormant buds, along with twigs from 14 aspen (*Populus tremula, P. tremuloides*) and hybrid aspen clones, and only dormant buds from four aspen and hybrid aspen clones, were stored at zero and four different sub-zero temperatures (0°C, -5°C, -8°C, -18°C, -80°C) for up to two years. Before storage, one lot of dormant buds was coated with wax and stored under vacuum at low temperatures. After different periods of storage, bud explants were cultured on a modified Woody Plant Medium to monitor the growth and differentiation of microshoots. Results of these investigations indicate that the cold storage potential of an aspen clone is largely determined by the storage conditions especially temperature and the genotype. Of the five temperatures investigated, -80°C seemed to be optimal for long-term storage of excised dormant buds and dormant buds on the twigs of those aspen clones that are "high shooters". The results of the present study are discussed in relation to the cold storage of forest tree germplasm in gene banks.

INTRODUCTION

Forest decline is a world-wide phenomenon. Forest trees are cut down every day for wood, energy and industrial use. After the loss of forests, the land is readily used for agriculture, industry, human settlements and roadways. Abiotic factors (for example, air pollution) and biotic agents (for example, microbes, pests) are also causing forest decline in the form of *Waldsterben* (dying forests). *Waldsterben* is probably due to a multiplicity of causes, although air pollution seemingly plays a significant role in this multistep process. In the Federal Republic of Germany, more than 50 percent of the forest (both conifers and hardwoods) are dying or are already dead (Breloh and Dieterle, 1985). In countries such as India, the forest decline may be caused more by the irresponsible cutting down of forest trees than by environmental factors. In view of the depletion of genetic resources, far-sighted approaches need to be developed for the preservation of the vast pool of genetic variability in the forest tree species for future exploitation.   In recognition of this goal, the preservation of forest tree germplasm has become an important aspect of forest tree genetics, breeding and improvement programs. Broadly speaking, the preservation of gene resources (Melchior et al., 1986) may be carried out by :

* *In situ* preservation (for example, stands conserved by natural regeneration); and
* *Ex situ* preservation: In this category, the preservation of germplasm involves the evacuation of plantations, the establishment of seed or clonal orchards, and the storage of seed, pollen, and other plant parts such as twigs, buds, meristems, tissues etc.

In this article, the storage potential of seeds and dormant buds of forest tree species at sub-zero temperatures will be presented and discussed in relation to the long-term preservation of forest tree germplasm.

MATERIALS AND METHODS

Species and Storage Temperatures

1. Seeds: Seeds from six forest tree species were employed to test their storage potential in liquid nitrogen (-196°C). These included European beech (*Fagus sylvatica* L.), hybrid aspen, *Larix decidua* Mill., Norway spruce (*Picea abies* (L.) Karst.), scots pine (*Pinus sylvestris* L.), and silver fir (*Abies alba* Mill.). At least 50 to 100 seeds per species were cooled to 0°C and then directly immersed in liquid nitrogen. The same number of seeds were kept at 0°C, for control germinations. Procedures for rapid freezing in liquid nitrogen and thawing have already been described (Ahuja, 1986a).

2. Storage of twigs with dormant buds: Twigs from six *Populus tremula* clones [W(Wedesbuttel) 3, W 7, W 14, W 25, W 90, and W 97], two *P. tremuloides* clones [Turr (Turresson) 141 and Schreiner], and six hybrid aspen, *P. tremula* x *P. tremuloides*, clones [Se (Seedorf) 1, Se 9, Se 11, Esch (Escherode) 3, Esch 9, and Esch 11] were wrapped in plastic bags and stored at zero and four sub-zero temperatures (0°C, -5°C, -8°C, -18°C and -80°C). Twigs were collected during December-January, 1985.

3. Dormant buds: Dormant buds from two *P. tremula* clones (W 29 and W 80) and two hybrid aspen clones (Esch 2 and Esch 8) were stored at the above temperatures with or without wax-vacuum treatments, as described earlier (Ahuja, 1988).

Survival and Differentiation Tests

1. Seeds: The seeds were stored for up to six days in liquid nitrogen. Germination tests were performed to establish the viability of the seeds following storage in liquid nitrogen.

2. Dormant buds: After different periods of storage, bud explants were cultured on a modified Woody Plant Medium (Ahuja, 1983) to monitor the growth and differentiation of microshoots. At least eight to 10 bud explants per treatment per clone were cultured. Results are expressed as the mean number of microshoots per primary bud explant per clone, after eight weeks of growth, along with standard errors. The procedures for the preparation and culture of bud explants were essentially the same as previously described (Ahuja, 1986b).

RESULTS

Cold Storage of Seed

There were significant differences in the germination percentages between seed lots of the forest tree species stored at 0°C (controls); silver fir, *Abies alba*, showed the lowest seed germination (26 percent) and hybrid aspen, *P. tremula* x *P. tremuloides*, showed 100 percent seed germination (Fig. 1). In these experiments, control germination values for each species were compared to germination percentage,

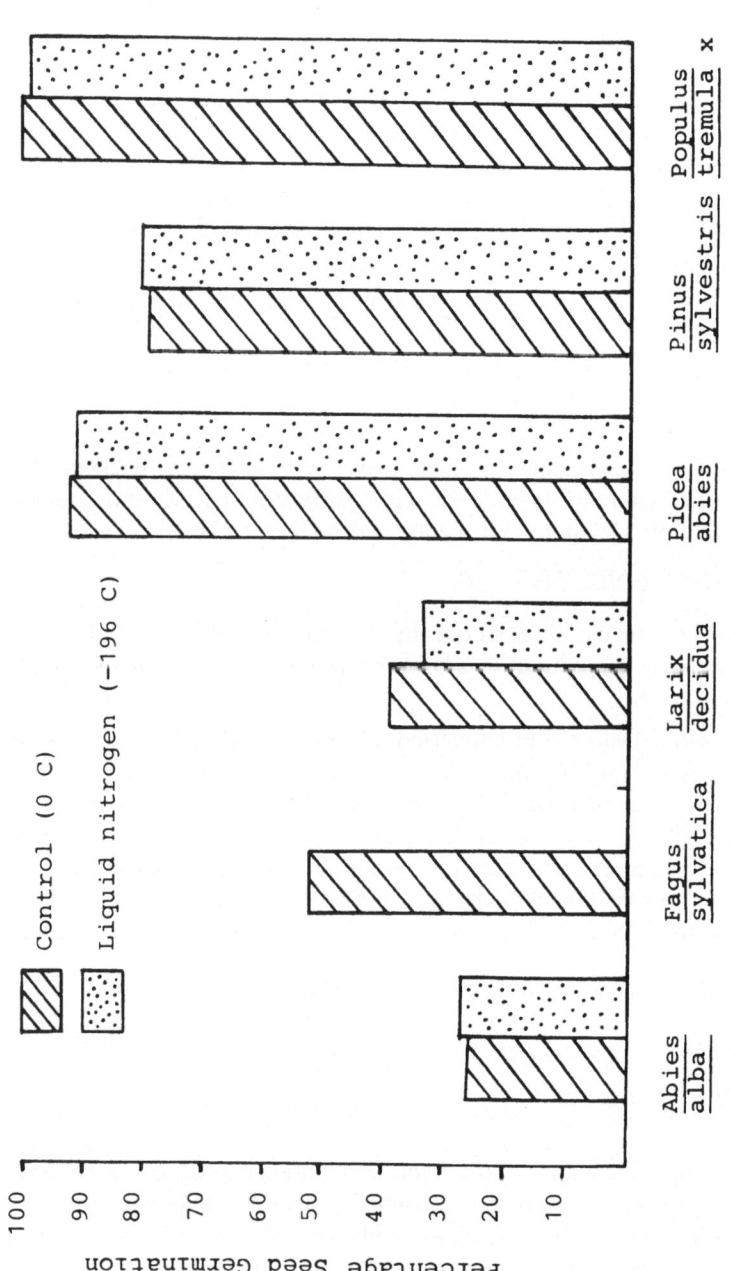

Fig. 1: Germination data on seeds from six forest tree species stored for six days in liquid nitrogen (−196°C) and controls (0°C). Note that beech seeds stored in liquid nitrogen did not germinate.

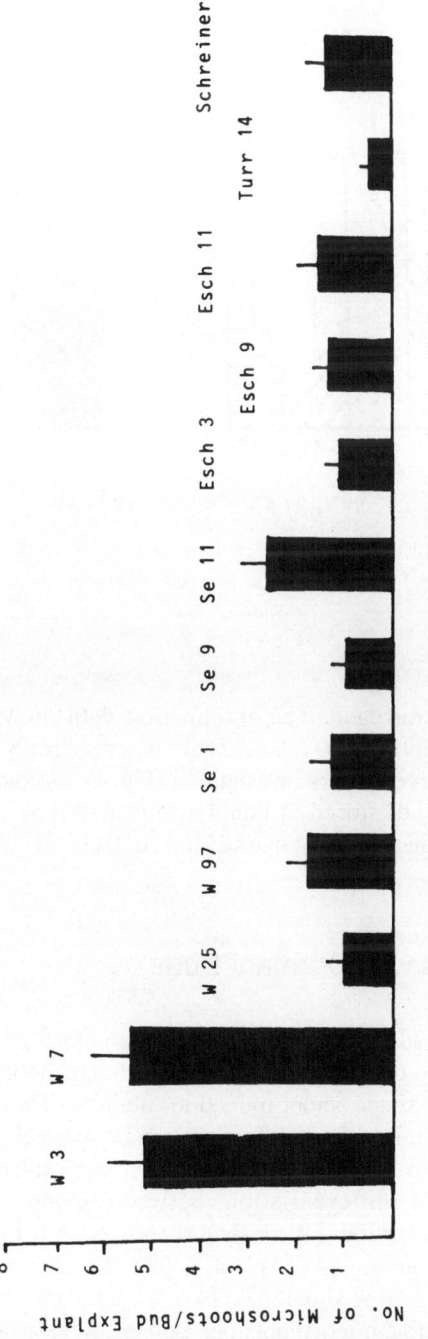

Fig. 2: Degree of microshoot differentiation on explants taken from dormant buds of 12 aspen and hybrid aspen clones stored with twigs at 0°C for one week. These included: *P. tremula* (W 3, W 7, W 25, W 97), *P. tremuloides* (Turr 141, Schreiner), and hybrid aspen (Se 1, Se 9, Se 11, Esch 3, Esch 9, Esch 11).

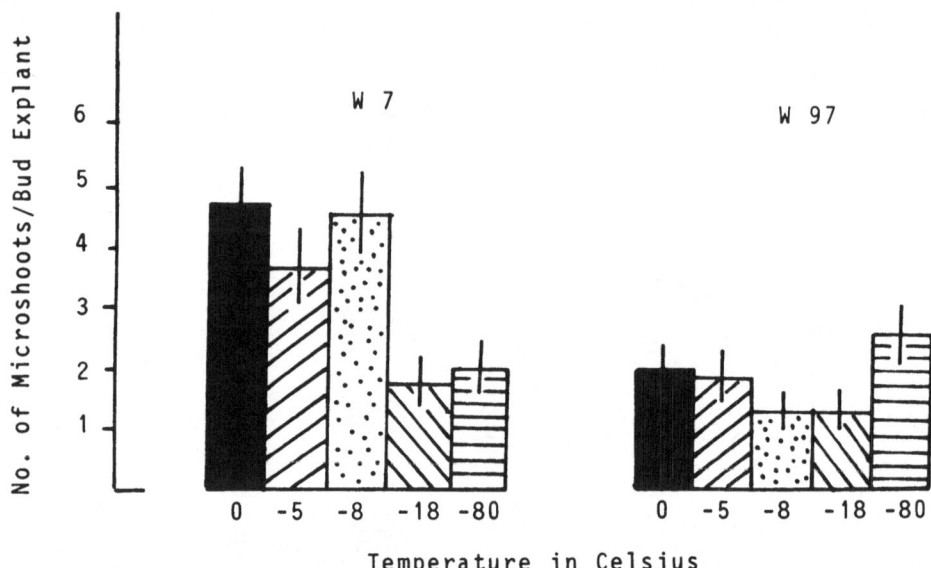

Fig. 3: Degree of microshoot differentiation on explants derived from dormant buds of clones W 7 and W 97 stored with twigs for four weeks at low temperatures.

following storage in liquid nitrogen. The germination data show that in the seeds stored for six days in liquid nitrogen (-196°C), there was practically no loss of viability in the five of the six forest tree species investigated (Fig. 1), as compared to controls. Beech, *(Fagus sylvatica)*, seeds stored in liquid nitrogen for one or six days did not germinate following thawing, as compared to controls at 0°C (51.6 percent germination; Fig. 1)..

Cold Storage of Twigs with Dormant Buds

Before testing the cold storage potential of aspen clones, 12 out of 14 clones were cultured as controls to determine the degree of microshoot differentiation on primary bud explants on a single shoot-induction medium. There were significant differences in the degree of microshoot differentiation between clones (Fig. 2), which ranged from 0.2 (Turr 141) to 5.4 (W 7) microshoots per bud explant. There were also differences in microshoot differentiation between clones within a family (Wedesbuttel) and between families (Wedesbuttel and Seedorf; Fig. 2). On the basis of the number of   microshoots per explant, aspen clones were classified in two groups: "high shooters" and "low shooters". Two *P. tremula* clones, W 3 and W 7, which produced more than five microshoots per explant, were arbitrarily grouped as "high shooters" and the remaining 10 clones, which produced less than three microshoots per explant (Fig. 2), were grouped as "low shooters".

· **1. Cold storage for four weeks:** After four weeks of cold storage, bud explants from two representative clones, a "high shooter" (W 7) and a "low shooter" (W 97), were cultured to determine the effects of five low temperatures on growth and differentiation. Both clones, stored at all the five temperatures, differentiated according to their genetic potentialities (Fig. 3). The clone W 97, a "low shooter", differentiated fewer than three microshoots per explant, ranging from 1.3 at -8°C and -18°C, to 2.6 at -80°C. In the clone W 7, a "high shooter", the range of variation in microshoot number was somewhat larger at different temperatures. The degree of microshoot differentiation was relatively higher in W 7 (approaching five) at 0°C to -8°C, as compared to -18°C and -80°C (fewer than five; Fig. 3).

2. Cold storage for one year : After one year of cold storage, the bud explants from 14 clones exhibited a variable number of microshoots at five different temperatures (Fig. 4). The aspen and hybrid aspen clones responded quite differently to the low temperatures. All six *P. tremula* clones showed extreme sensitivity to storage at -18°C; the bud explants turned brown and showed practically no growth and differentiation of microshoots (Fig. 4). Explants from four (W 3, W 7, W 14, and W 90) of the six clones also did not grow following the storage of their twig-buds (buds stored along with the twigs) at 0°C for one year. All the six *P. tremula* clones showed differentiation of microshoots on explants derived from twig-buds stored at -5°C, -8°C, and -80°C, although there were significant differences between clones and among temperatures (Fig. 4). The maximum number of microshoots differentiated on explants taken from twig-buds stored at -5°C and -8°C for one year.

Explants from two *P. tremuloides* clones (Turr 141 and Schreiner) showed practically no growth and differentiation of microshoots following the storage of twig-buds at the five low temperatures for one year (Fig. 4).

The six hybrid aspen clones exhibited differential response in degree of microshoot differentiation following the storage of twig-buds at low temperatures. Three clones (Se 9, Esch 9, and Esch 11) did not show any growth of the explants taken from twig-buds stored at -18°C; and two clones (Esch 3 and Esch 9) showed sensitivity to 0°C storage (Fig. 4). Of the five low temperatures tested, -80°C seemed to give optimal response in the hybrid aspen clones, and in particular, in Se 1, Se 10, Esch 3, and Esch 11 (Fig. 4).

3. Cold storage for two years: The twig-buds showed extreme sensitivity to low temperatures after two years of cold storage. Explants from only three of the 14 clones survived and exhibited a differentiation of microshoots after the storage of twig-buds at -8°C and -80°C. Clones W 3 and W 7, the "high shooters", showed differentiation of microshoots on explants taken from twig-buds stored at -80°C (Fig. 5). The clone W 7 produced significantly more microshoots per explant (4.1), as compared to W 3 (0.2).

Cold Storage of Dormant Buds With or Without Wax-Vacuum Treatments

1. Cold storage for five weeks: In order to find out the response of buds stored for five weeks at low temperatures, bud explants from two of the four clones were cultured to monitor the induction of microshoots. Of the five temperatures tested,

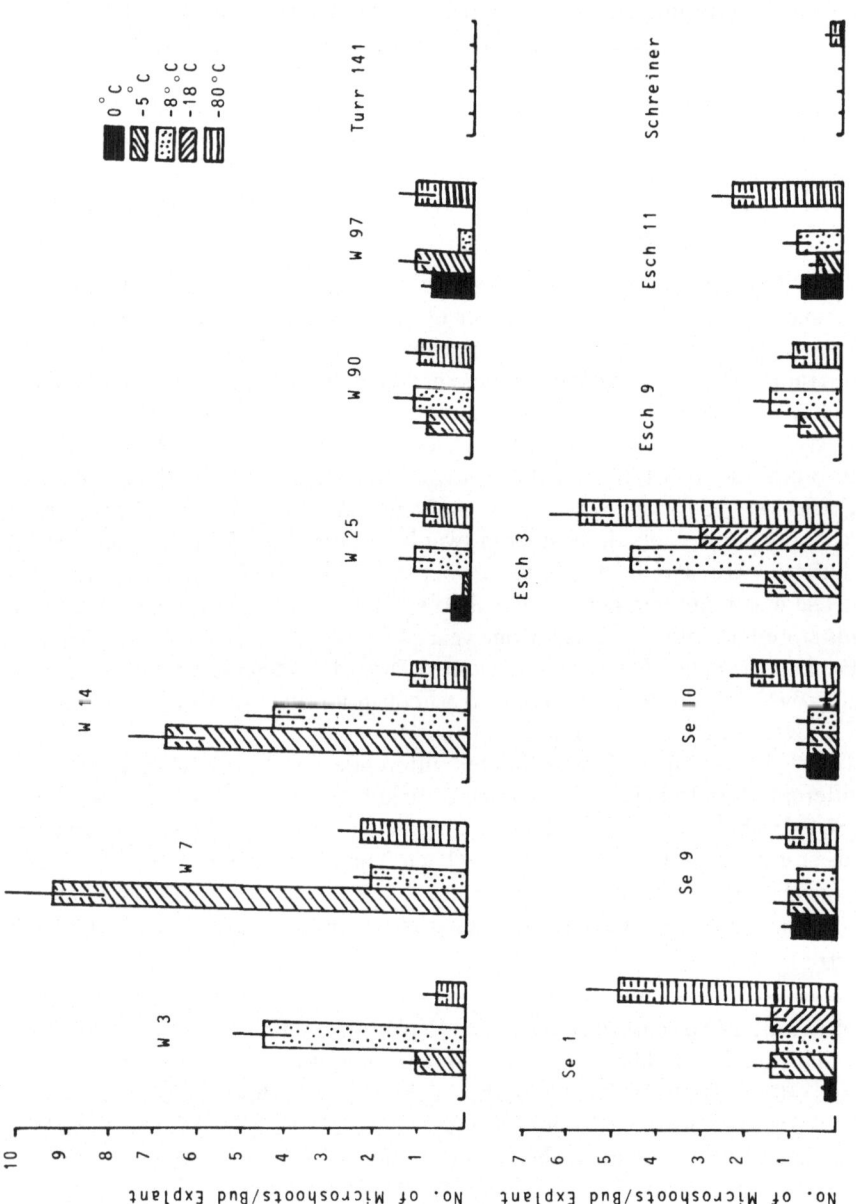

Fig. 4: Degree of microshoot different ation on explants taken from dormant buds of aspen and hybrid aspen clones stored with twigs for one year at low temperatures.

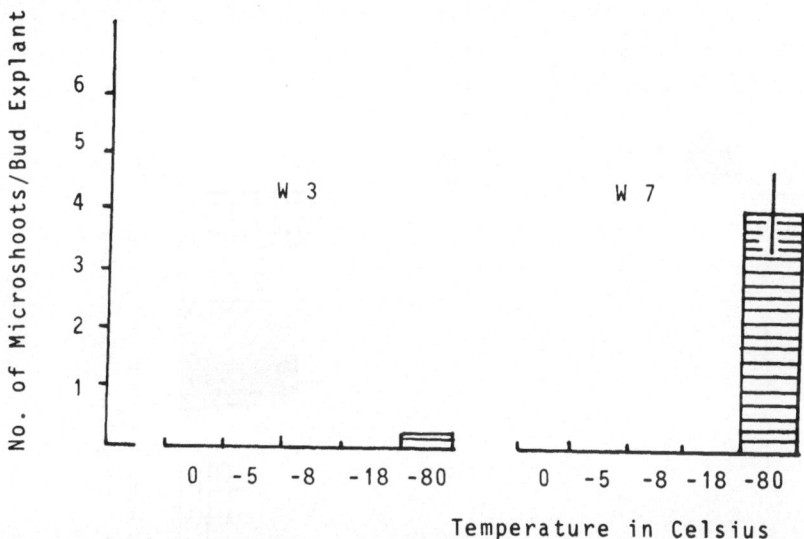

Fig. 5: Degree of microshoot differentiation on explants taken from dormant buds of cones W 3 and W 7 stored with twigs for two years at low temperatures.

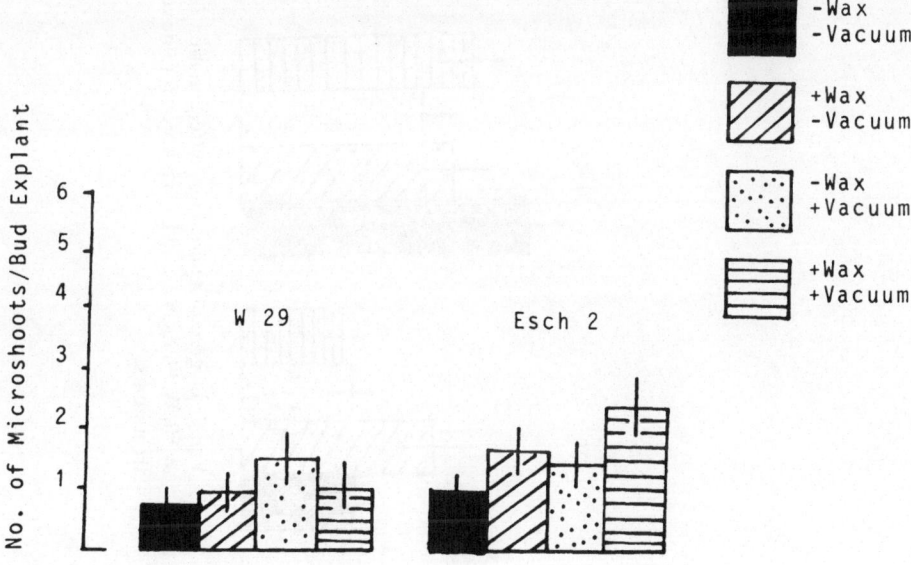

Fig. 6: Degree of microshoot differentiation on explants taken from dormant buds of clones W 29 and Esch 2 stored for five weeks at -80°C, with (+) or without (-) wax-vaccum treatments.

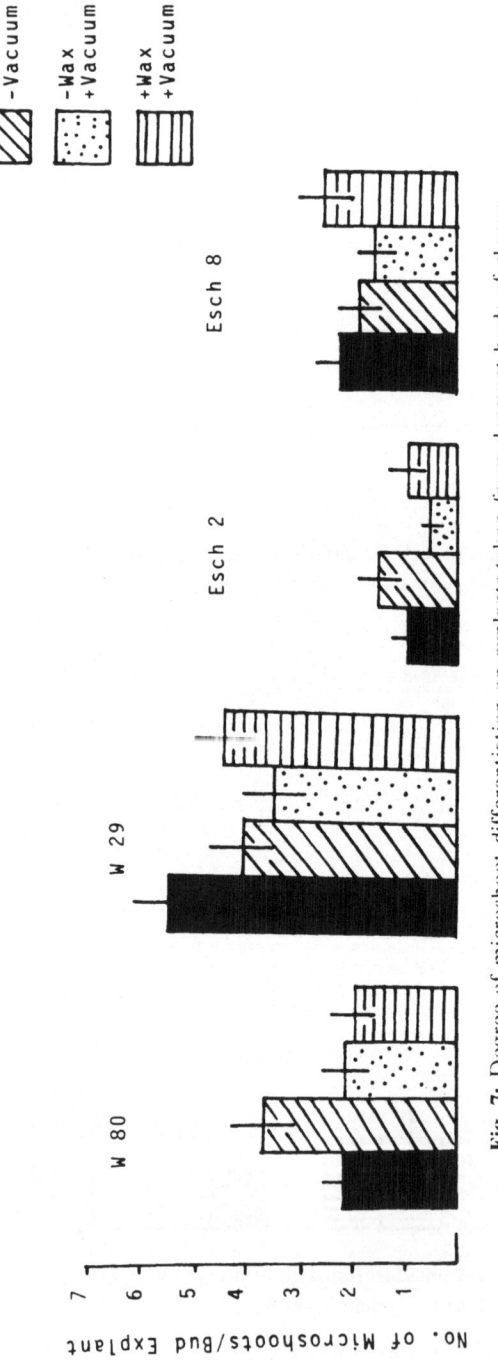

Fig. 7: Degree of microshoot differentiation on explants taken from dormant buds of clones W 80, W 29, Esch 2, and Esch 8 stored for one year at -80°C, with (+) or without (-) wax-vacuum treatments.

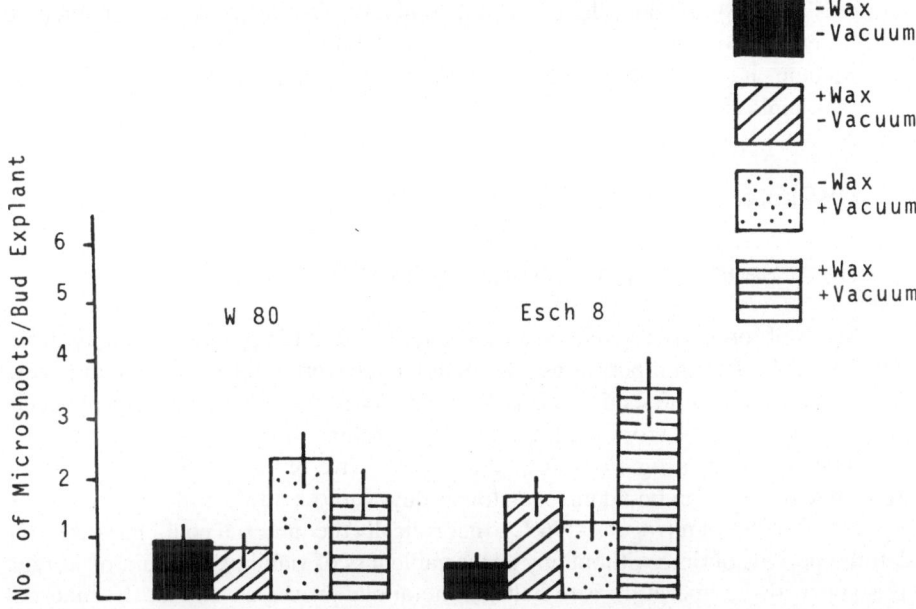

Fig. 8: Degree of microshoot differentiation on explants derived from dormant buds of clones W 80 and Esch 8 stored for two years at -80°C, with (+) or without (-) wax-vacuum treatments.

expands from buds stored at -18°C, with or without wax-vacuum treatments, did not grow or differentiate microshoots. On the other hand, explants from buds stored at 0°C, -5°C, -8°C and -80°C differentiated a low number of microshoots (fewer than three per explant). Since the buds remained viable only at -80°C after prolonged storage, results will be discussed only for -80°C storage. Explants from clone W 29 showed no differences in the degree of microshoot differentiation following the storage of buds at -80°C, with or without wax-vacuum treatments (Fig. 6). On the other hand, in Esch 2, the explants from buds coated with wax and stored under vacuum stored at -80°C differentiated relatively more microshoots as compared to other treatments (Fig. 6).

2. Cold storage for one year: There was no growth and differentiation of microshoots on explants taken from dormant buds of all the four clones (W 29, W 80, Esch 2, and Esch 8) stored at 0°C, -5°C, -8°C and -18°C for one year, with or without wax-vacuum treatments. On the other hand, explants from buds of all the four clones stored for one year at -80°C exhibited growth and differentiation of microshoots. Within each clone, no significant differences in microshoot differentiation were observed among the four treatments (Fig. 7). However, there were differences in the degree of microshoot differentiation between clones and treatments.

3. Cold storage for two years: Only two clones (W 80 and Esch 8) were tested for survival after the storage of dormant buds, with or without wax-vacuum treatments, at -80°C for two years. Both clones showed microshoot differentiation after the four treatments (Fig. 8). In one clone (W 80), buds stored under vacuum seemed to perform better, while in the second clone (Esch 8), buds coated with wax and stored under vacuum at -80°C differentiated relatively more microshoots as compared to other treatments (Fig. 8).

DISCUSSION

Storage of Seed in Liquid Nitrogen (-196°C)

Seeds of forest tree species are generally stored at temperatures ranging from 4°C to -20°C. At these temperatures, the seeds can be stored only for a limited period of time. In most cases, the seeds begin to lose viability after a few years. Seeds of beech, oak and silver fir can be stored at these low temperatures only for one to four years.

The present investigation shows that seeds of five of the six forest tree species tested could be stored in liquid nitrogen for six days without any loss of viability. At the temperature of liquid nitrogen (-196°C), theoretically the material could be stored for indefinite periods of time without any appreciable loss of viability. The aim of storage at this super low temperature is to halt the metabolic processes, so that the material remains genetically stable for a long time.

It seems that certain seed types may present problems for storage in liquid nitrogen (Ahuja, 1986a). Beech seeds did not germinate after storage in liquid nitrogen for one to six days. It appears that relatively small seeds such as those of aspen, scots pine, Norway spruce, larch and silver fir, which exhibited practically no loss of viability following storage in liquid nitrogen, are less prone to freezing-thawing injury, as compared to larger seeds of beech and perhaps oak. We are investigating whether in beech seed, the freezing-thawing injury is caused to the embryo or to the cotyledon or both and how to remedy the problem by tissue culture.

1. Cold Storage of Twigs with Dormant Buds: The present study indicates that twigs with dormant buds (twig-buds) in aspen can be stored for one year at low temperatures. In the *P. tremula* clones, -5°C to -8°C temperatures seem to be optimal for storage for one year. On the other hand, in hybrid aspen, -80°C seemed to be optimal for storage. After storage for two years, only two of the 14 clones survived at the lowest temperature tested (-80°C). Based on their regenerative potential under normal conditions, these two *P. tremula* clones (W 3 and W 7) were classified as "high shooters", as compared to the remaining clones that are "low shooters". It would appear that "high shooters" are able to survive and differentiate microshoots after storage at -80°C for two years, or perhaps longer, while the "low shooters" are unable to survive, when twig-buds are stored at this very low temperature. This might suggest a kind of "Darwinian selection" in operation at -80°C for those clones that are able to regenerate relatively more microshoots (in this case, more than five per explant); those clones that regenerate a low number (less than three) of micro-shoots on primary bud explants seem to be at a selective disadvantage at -80°C for long-term cold storage.

2. Cold Storage of Dormant Buds: When dormant buds are removed from twigs and then stored at low temperatures, they seem to respond somewhat differently to the storage conditions, as compared to twig-buds. Such low alone as 0°C, -5°C, -8°C and -18°C do not appear to be suitable for long-term storage of aspen clones. Under the experimental conditions, dormant buds stored at -80°C temperature were able to retain their viabilitiy and ability to differentiate microshoots. Treatments involving wax and vacuum may be beneficial for different clones in different combinations. These experiments with dormant buds of aspen did not indicate that a single treatment was beneficial for all clones for long-term cold storage.

Thus, it is evident that low temperatures can be effectively employed for the preservation of forest tree germplasm. Preservation of seed is important for maintaining the genetic variability of a population. On the other hand, the storage of vegetative parts, such as dormant buds, would be useful for the preservation of selected genotypes. Winter-hardy twigs from several tree species, e.g. willow, poplar, and apple were stored in liquid nitrogen and were found to be viable following the storage (Sakai, 1986). Bud meristems from aspen clones stored for 24 hours by slow cryopreservation method (Kartha, 1984) exhibited the differentiation of microshoots after retrieval from liquid nitrogen (Ahuja, 1987).

3. Cold Storage of Shoots, Tissues and Pollen: Shoots, plantlets, cell and callus cultures have been stored between -3°C to 15°C, depending upon the species (Aitkin-Christie and Singh, 1987). Shoots of a number of woody species have been stored in cultures for up to two years. In *Pinus radiata*, they have been stored for more than five years at 4°C, and following their return to normal temperatures, the shoots resumed growth (Aitkin-Christie and Singh, 1987).

In addition to seeds, vegetative parts, somatic tissues, and pollen grains of woody plants have also been successfully stored for a number of years at low temperatures (-18°C to -20°C), with or without vacuum (Herrmann, 1976; Akihama and Omura, 1986). Storage of pollen is valuable for hybridisation programs, where the male and female parents mature at different time periods, or are geographically separated. Stored pollen grains may also be employed for the production of haploids.

CONCLUSIONS

In view of the depletion of forest gene resources, it is necessary to preserve the germplasm of forest trees for the present and future use in gene banks. Cold storage of seed, vegetative parts and pollen offers prospects for long-term preservation of forest tree germplasm. However, optimal conditions for cold storage of most woody plants need to be investigated. Efforts should be directed towards developing simplified, cost-effective approaches for cold storage of forest tree germplasm. In this respect, seeds and dormant buds may offer certain advantages.

REFERENCES

Ahuja, M.R., 1983, Somatic cell differentiation and rapid clonal propagation of aspen, *Silvae Genet.,* 32: 131-135.

Ahuja, M.R., 1986a, Storage of forest tree germplasm in liquid nitrogen (-196°C), *Silvae Genet.,* 35: 249-251.

Ahuja, M.R., 1986b, Aspen. In: *Handbook of Plant Cell Culture* (D.A. Evans, W.R. Sharp and P.V. Ammirato, eds.), pp. 626-651, Macmillan Publishing Company, New York.

Ahuja, M.R., 1987, The importance of regenerating systems in forest tree biotechnology, *Ann. Forestales,* 13: 13-23.

Ahuja, M.R., 1988, Differential growth response of aspen clones stored at sub-zero temperatures. In: *Somatic Cell Genetics of Woody Plants* (M.R. Ahuja, ed.), Martinus Nijhoff, Dordrecht (in press).

Aitkin-Christie, J. and Singh, A.P., 1987, Cold storage of tissue cultures. In: *Cell and Tissue Culture in Forestry, Vol.2. Specific Principles and Methods: Growth and Development* (J.M. Bonga and D.J. Durzan eds.), pp. 285-304, Martinus Nijhoff, Dordrecht.

Akihama, T. and Omura, M., 1986, Preservation of fruit tree pollen. In: *Biotechnology in Agriculture and Forestry, Vol 1. Trees I* (Y.P.S. Bajaj, ed.), pp. 101-112, Springer-Verlag, Berlin.

Breloh, P. and Dieterle, G., 1985, Ergebnisse der Waldschaden-serhebung, Allgem, *Forstz,* 40: 1377-1380.

Herrmann, S., 1976, Verfahren zur Konservierung und Erhaltung der Befruchtungsfahigkeit von Waldbaumpollen uber mehrer Jahre, *Silvae Genet.,* 25: 223-229.

Kartha, K.K., 1984, Freeze preservation of meristems. In: *Cell Culture and Somatic Cell Genetics of Plants, Vol 1. Laboratory Procedures and Their Applications* (I.K. Vasil, ed.), pp. 621-628, Academic Press, New York.

Melchior, G.H., Muhs, H.J. and Stephan, B.R., 1986, Tactics for the conservation of forest gene resources in the Federal Republic of Germany, *For. Ecol. Manag.,* 17: 73-81.

Sakai, A., 1986, Cryopreservation of germplasm of woody plants. In: *Biotechnology in Agriculture and Forestry, Vol. 1. Trees I* (Y.P.S. Bajaj, ed.), pp. 113-129, Springer-Verlag, Berlin.

III

COMMERCIAL EXPLOITATION OF TISSUE CULTURE IN HORTICULTURE

17

TISSUE CULTURE OF ORNAMENTAL PLANTS

L.J. Maene

ABSTRACT

This paper reviews the most recent progress made in the *in vitro* studies since the micropropagation scheme developed by Debergh and Maene (1981). According to their scheme, transfer on a fresh medium is no longer needed for elongation and root induction purposes. It is sufficient to add an appropriate liquid medium on top of the old exhausted agar medium (Maene, 1985; Maene and Debergh, 1985a). Special attention has been given to the quality of shoots produced *in vitro*. These are very often physiologically and anatomically abnormal compared to *in vivo* produced plant material and there is very little difference between vitrified and normal looking plants. The reason(s) for these physiological and anatomical abnormalities of tissue-cultured plants have been examined. Different approaches to overcome acclimatisation problems have also been discussed. The most promising results are obtained by lowering

L.J. Maene * Industriele Hogeschool van het Rijk C.T.L.. Voskenslaan 270-9000. Gent. Belgium.

the relative humidity in the *in vitro* culture-container caused, by bottom-cooling. As the transpiration in tissue-cultured plants is comparable to *in vivo* conditions, this new culture technique can be described as *hydroculture under aseptic conditions*. The system has been illustrated for a few ornamental plants.

INTRODUCTION

Micropropagation techniques are available for many ornamental shrubs and trees, cut flowers, fruit and forest trees, vegetables and agricultural plants (Murashige, 1974b; Pierik, 1979; Conger, 1980; Margara, 1982; George and Sherrington, 1984). Irrespective of the group to which a plant belongs, i.e., tree, shrub, herb, ornamental, etc., the methodology used is based on the 3-stage system of Murashige (1974a,b; 1977;1978). The different stages of this system are as follows :

Stage I   : establishment of an aseptic culture;

Stage II   : induction and multiplication of propagules;

Stage III   : preparation of shoots for re-establishment in soil. This includes rooting of the shoots; gradual adaptation to the climatological circumstances *in vivo*; and conversion from the heterotrophic to the autotrophic state.

Each of these stages has its own specific aim. However, this does not necessarily mean that each stage has its own specific demands. For instance, Stages I and II can take place on a culture medium with the same physical and chemical characteristics. However, for a successful transfer of the plants to *in vivo,* the third stage generally differs from the others.

Adaptation of Murashige's Multiplication System (Debergh and Maene, 1981)

The methodology proposed by Murashige (1974a) is successfully used for numerous plants on laboratory scale. However, for commercial micropropagation, it is quite expensive, being more labor-intensive. Therefore, it is obvious that commercial micropropagation of plants is not merely developing suitable culture media, but also developing methods to make the technique more economical. The cost price of tissue-cultured plants, their surplus value included, should always be comparable to the cost of traditionally propagated plants. For the species that can be cloned by conventional vegetative methods, aseptic methods of cloning, if followed according to Murashige's scheme, prove uneconomical and to make them commercially viable, costs need to be reduced considerably.

Increasing the multiplication rate by further improving the composition of the culture medium in Stage II is a possible way to reduce costs. Indeed, for many plants, the multiplication rate increases as the time in culture increases. There are two reasons for this: firstly, the tissue becomes more juvenile, and secondly, the cultural conditions can be improved. Consequently, the cost price could be lowered. However, in Stage II, one never manipulates individual shoots but clusters; the resulting benefits per plant are, therefore, rather small.

The major costs are attributable to those stages in which individual shoots are manipulated, i.e., the rooting stage (Stage III) and in the transfer of the individual shoots to the greenhouse.

An adaptation of Murashige's system for commercial application was proposed by Debergh and Maene (1981), which is as follows :

Stage 0   : preparation of stock plants under hygienic conditions (to get the maximum number of aseptic cultures);

Stage I   : establishment of aseptic cultures;

Stage II   : induction of meristematic centres, their development into buds and their rapid multiplication;

Stage IIIa : elongation of the buds to shoots and the preparation of uniform shoots for Stage IIIb;

Stage IIIb : rooting and the initial growth of the *in vitro* produced shoots under *in vivo* conditions.

Recent Progress - Remaining Problems

The physiological status of cuttings is very important for rooting and for producing shoots of excellent quality at the end of the *in vitro* stages (Stage IIIa).

The use of some co-factors cannot completely change the rooting behavior; they are only supposed to accelerate the rooting process (Reuveni, pers. comm.). The physiological status of cuttings is, however, strongly influenced by the C/N ratio (Letouze and Daguin, 1983) and by the shoot length (Maene and Debergh, 1983). To obtain more than 90 percent rooting in *Cordyline*, the length should vary between 2.5 and 6 cm; smaller shoots are usually thick, succulent and there are fewer normal leaves; shoots longer than six cm have larger leaf surface area and thus water loss from the leaf surface upon transplantation becomes critical.

All the components of the Stage IIIa culture medium and their effect on shoot elongation and rooting *in vivo* were investigated for *Cordyline* (Maene and Debergh, 1983; Maene, 1985; Maene and Debergh, 1985a). None of the non-hormonal ingredients except cytokinin were very critical. The use of BAP in Stage IIIa inhibited rooting to a great extent; Kn, however, was not inhibitory if used in the concentrations which stimulate uniform elongation of clusters in Stage IIIa (Maene and Debergh, 1985a).

The light spectrum in Stage IIIa is of less importance; however, the intensity influences the results : a high light intensity (> 5000 lux or > 100 µmol s^{-1}m^{-2}) stimulates apical dominance in Stage IIIa cultures and is beneficial for the subsequent rooting *in vivo* of these shoots. On the contrary, a low light intensity in Stage IIIa allows the production of uniformly elongated shoots; however, it is less favorable for subsequent rooting *in vivo*. A compromise is optimal : at the beginning of Stage IIIa, cultures are incubated under low light intensities (about 1000 lux or $\pm$ 15µmol s^{-1}m^{-2}) until uniform elongation in the shoot clusters is manifest. Rooting and initial growth are subsequently improved by placing the cultures under higher light intensities for about three weeks, before they are transferred *in vivo*.

As the composition of the culture medium in Stage IIIa, with the exception of the cytokinin, is not very critical, and as the culture medium is quite exhausted at the end of Stage II, we tried to eliminate the labor-intensive transplantation of these Stage II cultures onto a fresh Stage IIIa medium (Maene and Debergh, 1985a). This is possible by supplementing these exhausted cultures with a liquid Stage IIIa medium. This was successfully realised with different species (Maene and Debergh, 1983; 1985a).

Adding liquid media to exhausted cultures offers significant possibilities of rational micropropagation. For instance, with *Cordyline*, a very significant gain in shoot production could be obtained at the end of Stage IIIa (Maene, 1985; Maene and Debergh, 1985a).

Liquid Stage IIIa media should be added before the onset of apical dominance in Stage II clusters. The former contains an especially high salt and sucrose concentration, and Kn is applied in higher amounts as compared to a traditional solid Stage IIIa medium.

Till recently, in all the experiments, the rooting of shoots produced *in vitro* took place on an artificial substrate, such as rockwool, saturated with an auxin solution. After the root induction period, the auxin could be easily removed from the rockwool blocks by simply drenching in water or by abundant irrigation. In this way, the rooting of tissue-cultured shoots was almost perfect. However, problems could arise when the rockwool-rooted shoots were planted in a conventional peat substrate, as rockwool in contact with peat loses water very quickly. This drying could adversely affect the growth of those shoots that do no have their roots spread throughout the volume of rockwool. For this reason, an alternative was tried to root shoots directly in peat. However, the difficulty in such a system was the elimination of the auxin added for root induction.

This problem was resolved exactly as for Stage IIIa. A liquid root-induction medium was added at the end of Stage IIIa. This can be done about a week before the transfer of the shoots to natural conditions (Maene, 1985; Maene and Debergh, 1985a). The composition of this medium is not very critical, so long as the salt concentration is not too high, and even tap water is an excellent pre-treatment.

The above mentioned technique is now being successfully applied in our laboratories for almost all the investigated plants: not only has the time for rooting *in vivo* decreased, but smaller shoots could also be rooted (Maene, 1985). Propagation costs could be considerably lowered as well. However, a fundamental problem still remains.

Tissue culture in general, and the use of liquid media in particular, can give rise to vitrification or related physiological abnormalities. Indeed, the symptoms of vitrified shoots and of shoots or plantlets having difficulties upon transfer from *in vitro* to *in vivo*, show a lot of parallelism (Maene, 1985; Debergh, 1986a,b); and both result in a very poor transplantation success (Maene 1985;1986). These vitrification phenomena or physiological aberrations are thoroughly examined and their possible causes explored.

Many tissue-cultured plants have the following more or less pronounced abnormalities (Grout and Aston, 1977; Sutter and Langhans, 1979; Brainerd et al., 1981; Brainerd and Fuchigami, 1981, 1982; Debergh et al., 1981; Debergh, 1982; Wetzstein and Sommer, 1982; Leshem, 1983; Donnely and Vidaver, 1984):

*   the leaves are thinner and have a different shape: smaller in the case of *Rubus idaeus* and *Dianthus* and larger and narrower in *Cynara scolymus*;
*   the mesophyll cells contain larger vacuoles and are less compact;
*   the palisade cells are thick and have a different shape;
*   there is a reduced quantity of epicuticular wax, or no structural wax layer at all;
*   the number of hydathodes is lower;
*   the number of stomata is lower and they have a higher topography;
*   the stomata do not close when subjected to stress;
*   most of the stomata are located at the periphery of the leaves;
*   besides abaxial stomata, most leaves also have adaxial stomata;
*   the vascular system is distorted;
*   the older leaves have a thick succulent appearance and are friable.

As there are only a few minor differences between *normal* looking and *vitrified* tissue-cultured plants, it is not advisable to assess the quality of tissue-cultured plants on the basis of visual appearance of translucency or vitrification. It should be accepted that the more difficult the transfer of tissue-cultured plants to *in vivo* conditions, the more these plants will be physiologically aberrant from normal plants.

The lack of epicuticular wax, which results in abundant cuticular transpiration, and malfunctioning of stomatal closure mechanism, are the two main reasons for the losses upon transfer of tissue-cultured plants from *in vitro* to *in vivo* conditions (Walkey and Matthews, 1979; Brainerd et al., 1981; Wardle et al., 1981).

According to Leshem (1983) the problem is even more complex : the disorganised structure of the abnormal shoots which develop *in vitro* can prevent the survival of plantlets when transferred to the greenhouse.

The defective vascular system, the reduced or absence of a protective layer and the non-functioning stomata are all quite normal consequences as they are not required *in vitro* ; the humidity in the tissue culture container is almost 100 percent. The only way to avoid these failures is to make *in vitro* conditions more comparable to the *in vivo* growth environment. This can be done by creating a sapstream in the plants under the influence of a gradient in relative humidity (RH). Using this method, the uptake of minerals with the water will also be more comparable to *in vivo* conditions.

It is generally accepted that the atmosphere, in a closed *in vitro* culture container, is nearly saturated with water vapor under normal cultural conditions; there is no pressure deficit and the stomata are open. The temperature in a container is always a few degrees higher than in the culture room because of the greenhouse effect. The cooling system in the culture room creates a cold surface on the container on which condensation takes place. One can often observe condensation water on the closure and on the surface of the container; this proves that the air at these places is saturated with water vapor. Therefore, we can accept that the sapstream of the tissue-cultured plants, at these locations, is reduced to approximately nil. Compared to the *in vivo* circumstances, where more than 90 percent of all the water taken up by the plant is lost through transpiration again, it is a very different situation. Hence, it is quite normal that the transport system of the *in vitro* produced plant is totally unadapted to normal *in vivo* conditions. Possibilities were investigated to activate the sapstream by decreasing the humidity in the container.

Removal of Condensation Water, Induced by Bottom Cooling, on the Culture Medium

In a closed container, the temperature of the coldest area will also determine the humidity of the air. Condensation on the cover means that the RH is 100 percent at this place percent and lower elsewhere. Inversely, condensation on the culture medium means a 100 percent RH in this region while the rest of the container is dryer, depending on its temperature.

Is it possible to create different temperatures in the same container, under normal tissue culture conditions?

In most culture rooms, the culture flasks placed just above a light source (fluorescent tube) show condensation on the cover and on the vertical surface. Those placed away from a heat source do not show this phenomenon. This is a clear indication that the temperature (microclimate) is different in both cases.

The material used as a support for the flasks is also very important in controlling condensation. Condensation is most pronounced on a glass or wooden surface but less obvious on an iron network. With the latter, microventilation is possible due to local convection; also, the conductivity of iron is higher than that of wood or glass. A faster exchange of heat with the cooling system of the culture room is possible in the latter case.

Cooling the containers at their base creates a lower RH in the rest of the container. Table 1 indicates the RH that can be obtained for different temperatures of the culture medium (condensation water) and the air in the container.

Example of a Stage IIIa Culture of *Calathea ornata* at Different Temperature Gradients in the Container

Stage IIIa *Calathea ornata* cultures during the first three weeks were incubated under normal tissue culture conditions. They were subsequently cultured on different cooling plates.

Table 1: Relative humidity (RH) within the container at different temperatures of the culture medium (condensation water)

Temperature of the culture medium (condensation water) (°C)	Temperature of the air within container (°C)	RH of the air within the container
19	19	100
	20	94
	21	88
	22	83
	23	78
	24	74
20	20	100
	21	94
	22	88
	23	83
	24	78
21	21	100
	22	94
	23	88
	24	83
	25	78
22	22	100
	23	94
	24	88
	25	83

Source: Aspirations-Psychrometer-Tafeln. Deutschen Wetterdienst. Braunschweig (1963).

The temperature in the Meli-jars measured through the lid at four levels with thermocouples, the resulting calculated RH at the mean leaf level of the plants and a visual observation of the container lid are presented in Table 2.

It should be noted that the different gradients were not freely chosen but were a result of the technical installation of the experiment. Treatments II and V are controls without bottom-cooling.

The leaf-quality of the cultures, incubated on the bottom-cooling systems, appeared excellent: the leaves did not look wet as in the control treatments and had a waxy look, exactly as that of young leaves of plants developed *in vivo*. Their colour was darker.

Table 2: Microclimate parameters within the container (culture of *Calathea ornata*) bottom cooling experiment

Height level in the container	Treatments (temperatures in °C)				
	I	II	III	IV	V
1.5 cm below the lid	23.2	22.7	24.8	22.9	24.5
Half way between lid and culture medium	23.0	22.9	24.7	22.2	24.5
Just above the culture medium	22.6	23.1	24.5	21.0	24.3
Culture medium (condensation water)	22.6	23.1	24.3	20.4	24.3
Temperature and RH half between (1) and (2)	$\pm$23.1 $\pm$97%	$\pm$22.8 $\pm$99%	$\pm$24.75 $\pm$97%	$\pm$22.6 $\pm$87%	$\pm$24.5 $\pm$99%
Condensation (+) No condensation on the lid (-)	-	+	-	-	+

After four weeks of bottom-cooling at Stage IIIa-culture, all fully developed leaves in five containers were harvested and were air-dried at 40 percent RH. Their loss of water was measured during three hours and compared with *in vivo* obtained leaves of one-month-weaned tissue-cultured plants (Fig. 1).

As observed from Fig. 1, it is certainly remarkable that leaves developed *in vitro* under 87 percent RH (treatment IV) are better protected against water loss than leaves harvested from one-month-old *in vivo* hardened plants; leaves of treatment V (control at 99 percent RH) lost twice as much water and those of treatment II (control at 99 percent RH), even three times more.

The behavior of leaves developed under 97 percent RH (treatments I and III) was almost the same and comparable to *in vivo* grown leaves (treatment VI).

A comparative study of transverse sections of the leaves grown in low and high RH (Maene, 1986) revealed that the cells in the latter case are larger and more elongated; also, the intercellular spaces are bigger so that the leaves are structured in a less compact way.

The improvement in the leaf quality could be proved through *in vivo* rooting experiments : 282 of the 300 shoots (94 percent) arising from the bottom-cooling treatments (I, III and IV) and rooted in peat, survived the transfer; only 40 percent of

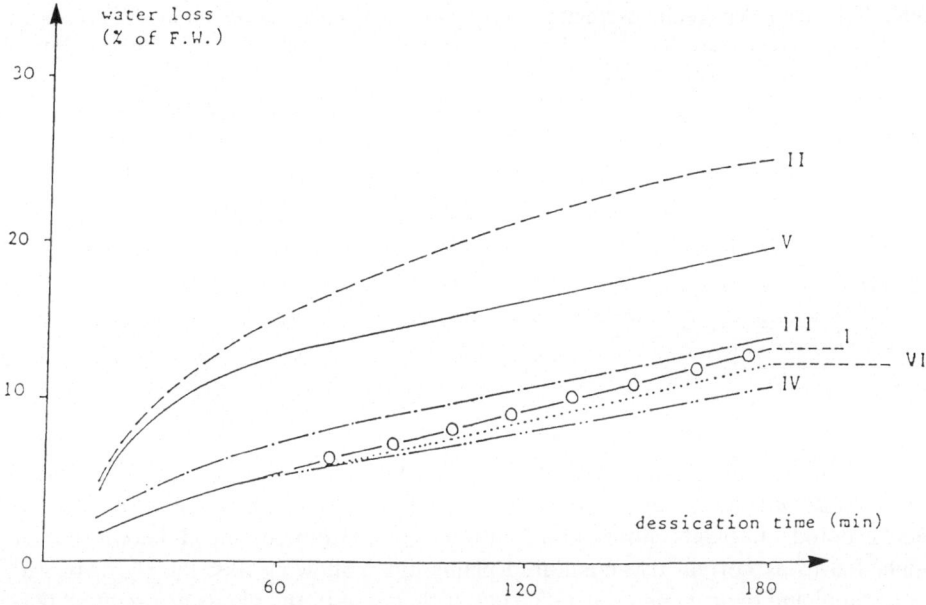

Fig. 1: Amount of water lost by *Calathea ornata* leaves harvested (a) from the *in vitro* microclimate conditions of Table 2 (Treatments I, II, III, IV, V) and (b) from tissue-cultured plants weaned and raised *in vivo* (VI), dried for three hours in the laboratory at 22°C and 40 percent RH.

- Treatments I, II, III, IV, V : See Table 2
- Treatment VI : *in vivo*-developed leaves of one-month-old tissue-cultured plants

the usual tissue-cultured plants (treatments II and V) survived. There was no statistical difference in the rooting between the bottom-cooling treatments I, III and IV.

Additional Physiological Aspects of Bottom-Cooling

Stage IIIa *Ficus beniamina* cultures, treated as in the previous section, revealed a different stomatal behavior of bottom-cooled plants compared to controls (Table 3).

Leaves developed *in vitro* under low RH obviously contain more stomata compared to those grown under normal tissue culture conditions and this could be an indication that intensified transpiration has taken place (Lemeur, personal communication). However, such leaves have a normal-functioning stomatal mechanism comparable to *in vivo* developed leaves (Table 3). In addition, their topography is less superficial compared to the control treatments.

Table 3: Stomatal density and percentage closed stomata of *Ficus beniamina* leaves, developed under two different microclimate conditions *in vitro* and after rooting and acclimatisation *in vivo* (only abaxial stomata were counted; Maene, 1986)

Treatment*	Stomatal density (number mm^{-2})	Percent closed stomata
Ficus II	27,604	41.8
Ficus IV	63,670	87.6

* (see Table 2)

The beneficial effect of bottom-cooling became more apparent when these *Ficus beniamina* plants were rooted in the greenhouse: almost all plants of both treatments rooted, but plants cultured for four weeks on bottom-cooling did not lose their upper leaves as normal tissue-cultured plants did. Following this, their growth was more rapid and their quality was much better than that of the plants grown at 99 percent RH.

Transverse sections of *Ficus beniamina* leaves developed *in vitro* at 87 percent RH (treatment IV; Maene, 1986) showed a more compact mesophyll tissue with few intercellular spaces; certainly the high cell density of the spongy parenchyma tissue was remarkable. Additionally, the upper epidermal and hypodermal cells were smaller compared to those developed at a high RH. In the later case, the cell density was low and there was no clear differentiation of the mesophyll into a well-defined palisade and spongy parenchyma.

CONCLUSIONS

Bottom-cooling is an excellent energy-efficient method to lower the RH in the culture-container without drying out the culture medium. This is a pre-requisite for good rooting of the shoots *in vivo*.

As translocation is intensified, one should be aware of the possible detrimental effect of salt accumulation in the leaves; it is advisable to correct the salt formulation of the culture medium (cf. hydroculture *in vivo*, but under axenic conditions; Maene, 1985, 1986; Maene and Debergh, 1985b).

In our production scheme (Debergh and Maene, 1981; Maene, 1985; Maene and Debergh, 1985b), bottom-cooling can be easily applied after about four weeks of normal Stage IIIa culture. At this time, the salt concentration can be reduced to half. Further research is needed to determine the optimal RH for each plant species. This can be achieved by varying the temperature of the culture medium (bottom-cooling), the temperature of the incubation room and the illumination.

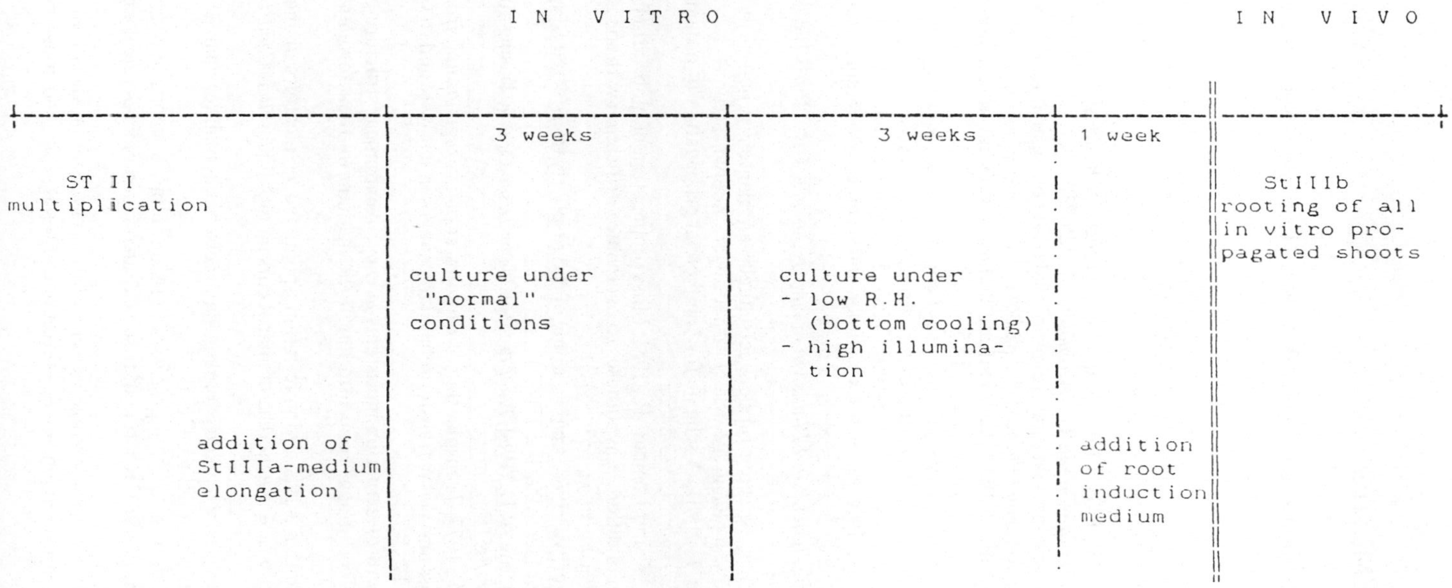

I N   V I T R O   I N   V I V O

ST II multiplication	3 weeks	3 weeks	1 week	StIIb

ST II
multiplication

addition of
StIIIa-medium
elongation

3 weeks

culture under
"normal"
conditions

3 weeks

culture under
- low R.H.
  (bottom cooling)
- high illumina-
  tion

1 week

addition
of root
induction
medium

StIIb
rooting of all
in vitro pro-
pagated shoots

Optimisation of plant micropropagation

Fig. 2: New production scheme for the micropropagation of plants in vitro (Maene, 1985, 1986; Maene and Debergh, 1985a, 1985b).

ACKNOWLEDGEMENTS

This research is sponsored by the I.W.O.N.L. (Institute of Scientific Research in Industry and Agriculture, Belgium).

REFERENCES

Benoit, H. and Ducreux, G., 1981, Etude de quelques aspects de la multiplication vegetative *in vitro* de l'artichaut (*Cynara scolymus* L.), *Agronomie,* 1: 225-230.

Bowen, J., 1972, Sugar transport in immature internodal tissue of sugarcane. I. Mechanism and kinetics of accumulation, *Plant Physiol.,* 49: 82-86.

Brainerd, K. and Fuchigami, L., 1981, Acclimatisation of aseptically cultured apple plants to low relative humidity, *J. Am. Soc. Hortic. Sci.,* 106: 516-518.

Brainerd, K. and Fuchigami, L., 1982, Stomatal functioning of *in vitro* and greenhouse apple leaves in darkness, mannitol, ABA and CO_2, *J. Exp. Bot.,* 33: 388-392.

Brainerd, K., Fuchigami, L., Kwiatowski, S. and Clarck, L., 1981, Leaf anatomy and water stress of aseptically cultured Pixy plum grown under different environments, *Hortscience,* 16: 173-175.

Conger, B., 1980, *Cloning Agricultural Plants via In Vitro Techniques,* CRC Press Inc. Boca Raton, Florida, pp. 272.

Debergh, P., 1982, Nieuwe aspekten bij de weefselteelt van planten, Proefschrift tot het verkrijgen van de graad van geaggregeerde voor het hoger onderwijs, Gent.

Debergh, P. and Maene, L., 1981, A scheme for commercial propagation of ornamental plants by tissue culture, *Sci. Hort.,* 14: 335-345.

Debergh, P., Harbaoui, Y. and Lemeur, R., 1981, Mass propagation of globe artichoke (*Cynara scolymus*) : Evolution of different hypotheses to overcome vitrification with special reference to water potential, *Physiol. Plant.,* 53: 181-187.

Donnelly, D. and Vidaver, W., 1984, Leaf anatomy of red raspberry transferred from culture to soil, *J. Am. Soc. Hortic. Sci.,* 109: 172-176.

George, E. and Sherrington, P.D., 1984, *Plant Propagation by Tissue Culture,* Basingstoke, Exegetics Ltd., Haunts, pp. 709.

Grout, B. and Aston, M., 1977, Transplanting of cauliflower plants regenerated from meristem culture. I : Water loss and water transfer related to changes in leaf-wax and to xylem regeneration, *Hortic. Res.,* 17: 1-7.

Leshem, B., 1983, Growth of carnation meristems *in vitro* : Anatomical structure of abnormal plantlets and the effect of agar concentration in the medium on their formation, *Ann. Bot. (Lond.),* 52: 413-415.

Letouze, R. and Daguin, F., 1983, Manifestation spontanee et aleatoire d'une croissance anormale en culture *in vitro*. Recherche de marqueurs metaboliques, *Rev. Cann. Biol. Exp.,* 42: 23-28.

Maene, L., 1985, Optimalisering van de overgang van weefselteeltoplantjes naar *in vivo* omstandigheden, Doctoraatverhandeling, Gent.

Maene, L., Stimulation of translocation processes in tissue cultured plants, In preparation.

Maene, L. and Debergh, P., 1983, Rooting of tissue culture plants under *in vitro* conditions, *Acta Hortic.,* 131: 210-208.

Maene, L. and Debergh, P., 1984, Problems related to *in vivo* rooting of *in vitro* propagated shoots, In: EEC symposium on *in vitro* techniques-Propagation and long term storage, Braunschweig.

Maene, L. and Debergh, P., 1985a, Liquid additives to established tissue cultures to improve elongation and rooting *in vivo, Plant Cell Tissue Organ Cult.,* 5:23-33.

Maene, L. and Debergh, P., 1985b, Optimization of the transfer of tissue cultured shoots to *in vivo* conditions. In: *Symposium on in vitro problems related to mass propagation of horticultural plants, Acta Hortic.,* 212: 335-348.

Maene, L. and Debergh, P.C., 1987, Optimization of the transfer of tissue culture shoot to *in vivo* conditions, *Acta Hortic.,* 212: 335-342.

Margara, J., 1982, Bases de la multiplication vegetative les meristemes et l'organogenese. INRA, Versailles, pp. 262.

Murashige, T., 1974a, Plant propagation through tissue cultures, *Ann. Rev. Plant Physiol.,* 25: 135-166.

Murashige, T., 1974b, Plant cell and organ culture methods in the establishment of pathogen-free stock, A.W. Dimock Lectures, Cornell Univ., Ithaca, New York, pp. 26.

Murashige, T., 1977, Plant cell and organ cultures as horticultural practices, *Acta Hortic.,* 78:17-30.

Murashige, T., 1978, The impact of plant tissue culture on agriculture. In: *Frontiers of Plant Tissue Culture 1978* (T.A. Thorpe, ed.), pp. 15-26, Univ. Calgary Press, Calagary.

Pierik, R., 1979, *In vitro* Culture of Higher Plants, Bibliographie Kniphorst Sci. Bookshop, Wageningen, pp. 149.

Sarkanen, K.V. and Ludwig, C.H., (Eds.), 1971, *Lignins: Occurrence, formation, structure and reactions,* Wiley Interscience, New York, pp. 916.

Sutter, E. and Langhans, R., 1979, Epicuticular wax formation on carnation plantlets regenerated from shoot tip culture, *J. Am. Soc. Hortic. Sci.,* 104: 493-496.

Walkey, D. and Matthews, K., 1979, Rapid clonal propagation of rhubarb (*Rheum rhaponticum* L.) from meristem tips in tissue culture *Plant. Sci. Lett.,* 14: 287-290.

Wardle, K., Dixon, P. and Simpkins, H., 1981, Sodium accumulation by leaves of cauliflower plantlets and the effect of the mode of plant formation, *Ann. Bot. (Lond.),* 47: 653-659.

Wardle, K., Dobbs, E. and Short, K., 1983, *In vitro* acclimatisation of aseptically cultured plantlets to humidity, *J. Am. Soc. Hortic. Sci.,* 108: 386-389.

Wardle, K., Quinlan, A. and Simpins, I., 1979, Abscisic acid and the regulation of water loss in plantlets of *Brassica oleracea* L. var. *botrytis* regenerated through apical meristem culture, *Ann. Bot. (Lond.),* 43: 745-752.

Wetzstein, H. and Sommer, H., 1982, Leaf anatomy of tissue-cultured *Liquidambar stryraciflua* during acclimatisation, *Am. J. Bot.,* 69: 1579-1586.

18

TISSUE CULTURE OF ORCHIDS IN THAILAND

Uthai Charanasri

Commercial micropropagation of orchids in Thailand started in the year 1972-1973. Since then, business has been very prosperous and expanding rapidly. The industry now produces 30 million plantlets and employs more than 300 workers distributed among 15 laboratories around the country. The estimated gross turnover of tissue culture propagated plants is of the order of US $ 2.4 million. Orchids successfully multiplied on a commercial scale include *Arachnis, Cattleya, Dendrobium, Vanda* and their intergeneric hybrids. The market is basically export-oriented. The hardened plants are exported to Japan, USA and Europe. Few plantlets are grown locally for the cut flower industry.

The export of orchid flowers and planting materials from Thailand has always depended on the results of the extensive hybridisation and asymbiotic embryo culture in aseptic conditions. While there were many orchid seed germination laboratories in the country during 1960 to 1971, the first two commercial tissue culture laboratories, Intuwong Orchid Laboratory and Bangkok Orchids were opened simultaneously in 1972-73. They were established by a graduate and a trainee, respectively, trained at the University of Hawaii by Dr. Y. Sagawa. Due to the high demand for a micropropagation service to multiply new orchid cultivars, which are normally very slow to propagate by conventional methods, more tissue culture laboratories were established. Some of

Uthai Charanasri * Bangkok Flowers Centre Co. Ltd., 34/19 Moo 7 Petchkasem Road, Nongkangplu, Nong Kam, Bangkok, Thailand.

the existing seed germination laboratories were diversified to include tissue culture propagation.

With low labor costs, appropriate technology, good management and the effective use of tissue culture plants in the cut flower and plant exports, the orchid export from Thailand rose steadily to US $ 20 million in 1982. Since then, the increase in cut flower export has slowed down because of limited air cargo space. The micropropagation business, however, still continues to rise steadily. This situation has lured experienced college graduates who have gained their skills through working in well-established commercial firms to leave their employers and start their own business. Boosted by support from capital partners and frustrated customers who could not wait for the slow propagation services of laboratories overloaded with contracts beyond their production capacity, new laboratories have emerged, and with some exceptions, have become firmly established in a short period of time. At present, there are about 15 laboratories in the country and the number is still increasing.

The size of commercial tissue culture laboratories in Thailand varies. Their staff may range from 10 to 100 workers, the number of laminar air flow cabinets from 5 to 50, and the total annual production of tissue culture plants from 500,000 to six million. A laboratory with approximately 300 workers produces about 30 million plants per annum at a gross turnover of approximately US $ 2.4 million. Sales of tissue-cultured plants include additional breeder's royalty for the cultivars and additional contract growing fees for the growers. The genera of orchids successfully multiplied on a commercial scale include *Arachnis, Cattleya, Dendrobium, Vanda* and their intergeneric hybrids. *Phalaenopsis* is the latest type which has been mass propagated for commercial purposes. A majority of the laboratories produce plants on a contract basis with payment in instalments as the work progresses. Cancellation of contracts by clients is frequent, since the market situation makes certain cultivars obsolete in a short time. *Dendrobium* growers are the largest group of customers. Other tropical-orchid growers and planting material exporters (i.e., suppliers to overseas growers) constitute the second biggest group. Exporters who supply potted plants in bloom are the third group of clients. The last, but fast-growing segment is the direct production, on contract, for overseas customers including Japanese, Malaysian, Singaporian, Surinamese and Taiwanese importers. A small proportion of the product is produced and sold by the laboratories themselves. Plantlets are exported to Australia and Malaysia in flasks, where the quarantine laws are strict; to other destinations, they are exported after hardening for two to three months after removing from the cultures. Exports to Japan, USA and Europe are mainly of the mature, ready-to-flower plants.

In general, most of the Thai laboratories specialise in orchid cloning. A few laboratories that have diversified are working on the propagation of other ornamentals such as *Amaryllis, Anthurium, Caladium, Chrysanthemum, Gypsophila, Lilium* and ferns; a few fruit species such as *Musa* and pineapple; vegetable crops such as potato and timber trees such as *Tectona grandis*. Embryo rescue of the naturally lethal homozygous recessive *makapuno* mutant of coconut is being carried out in a few laboratories on a limited scale. Recently, the Science and Technology Development Board of Thailand has made available special loans for investing in biotechnology at an interest rate of five percent per annum. Special privileges from the National Board of Investment, another promoting body, include corporate tax and income tax exemption for a number of years. The prospects of foreign joint venture investment for products for export and self-sufficiency by substitution of existing imports are good. The production of plants or artificial seeds in Thailand for shipment to countries with high labor costs is another area for future development.

19

ACCLIMATIZATION OF TISSUE CULTURE-RAISED PLANTS FOR TRANSPLANTATION TO THE FIELD

**Sant S. Bhojwani and
Vibha Dhawan**

INTRODUCTION

The ultimate success of micropropagation on commercial scale depends on the ability to transfer plants out of culture on a large-scale, at low cost and with high survival rates. Tissue culture conditions that promote rapid growth and multiplication of shoots often result in the formation of structurally and physiologically abnormal plants. The tissue culture plants are often characterised by abnormal leaf morphology and anatomy, poor photosynthetic efficiency, malfunctioning of stomata and a marked decrease in epicuticular waxes. Qualitatively also, the waxes present on the surface of the leaves of *in vivo* and *in vitro* raised plants may vary. The heterotrophic mode of nutrition and poor mechanism to control water loss render micropropagated plants

Sant S. Bhojwani, * Department of Botany, University of Delhi, Delhi - 110 007, India and *Vibha Dhawan,** Tata Energy Research Institute, 90 Jor Bagh, New Delhi - 110 003, India.

vulnerable to the transplantation shocks. Although considerable efforts have been directed to optimise the conditions for the *in vitro* stages of micropropagation, scant attention has been paid to understand the process of acclimatisation of micropropagated plants to the soil environment. Consequently, the transplantation stage continues to be a major bottleneck in the micropropagation of many plants (Earle and Langhans, 1975; Broome and Zimmerman, 1978; Conner and Thomas, 1981; Ziv, 1986).

Successful transplantation of plants to the field requires a sound knowledge of silvicultural practices generally employed in the nursery for the propagation of plants. Sometimes, there is a tendency to use small vials for initial stages of transplantation; this causes root curling. Such plants, if transplanted, may survive but often show drastic reduction in growth at later stages. Further, it should be realised that plant propagation via tissue culture differs from the conventional methods of propagation in several key aspects. Unlike nursery seedlings or cuttings, the tissue culture plants are raised under perfectly controlled conditions. Since, under *in vitro* conditions, plantlets are grown under very high humidity and on a sucrose rich medium, they require gradual acclimatisation to the field conditions. The transfer of individual plantlets to potting mix and their acclimatisation under specified conditions of humidity, temperature and light requires special facilities and is, therefore, very expensive.

This article discusses the major abnormalities in tissue culture plants that require attention at the time of transplantation to field and reviews the methods used to harden the plants for transplantation.

ABNORMALITIES OF THE PLANTS GROWN IN TISSUE CULTURE

Photosynthetic Efficiency

High sucrose-and salt-containing media, often employed for raising cultures, and the poor light conditions seem to restrict the photosynthetic efficiency of the leafy shoots. For *in vitro* growth, a continuous supply of exogenous sucrose (2-3 percent) is required. Although, such plantlets may appear normal, they are unlikely to be actively photosynthesising. This is because of the exogenous supply of sucrose, which does not necessitate the normal development of photosynthetic apparatus. Therefore, tissue culture plants are either poor in chlorophyll content, or the enzymes responsible for photosynthesis are inactive or absent altogether (Grout and Aston, 1977a; Wetzstein and Sommer, 1982; Donnelly and Vidaver, 1984a). Peculiar leaf anatomy, incomplete development of the chloroplasts (Wetzstein and Sommer, 1982) and low net CO_2 uptake (Grout and Aston, 1977a; Donnelly and Vidaver, 1984a) are some of the characteristics of the *in vitro* raised plants.

Grout and Aston (1977b) and Grout and Crisp (1977), examined the photosynthetic efficiency of tissue culture plants of *Brassica oleracea* by measuring net CO_2 uptake in light and in the dark. Even after seven days of transfer from *in vitro* conditions, there was no net CO_2 uptake (i.e. CO_2 released in respiration was greater than that used in photosynthesis). It was only after 14 days of transplantation that the plants became fully autotrophic and could sustain normal growth. In *Leucaena leucocephala,* the leaves from cultured shoots and plantlets exhibited total lack of starch grains normally found in mesophyll cells of field-grown plants, thereby, suggesting the photosynthetic inefficiency of the plants in cultures (Dhawan and Bhojwani, 1987).

Epicuticular Wax

Scant deposition of the protective epicuticular wax on the surface of the leaves of the *in vitro* grown plants, has been regarded as one of the most important factor responsible for excessive loss of water, leading to poor transplantation success (Grout and Aston, 1977a; Sutter and Langhans, 1979, 1982; Fuchigami et al., 1981; Brainerd and Fuchigami, 1981, 1982; Wetzstein and Sommer, 1982). The chemical nature of the wax deposited on the surface of the leaves under *in vitro* conditions is also known to differ from that formed under natural conditions, allowing excessive diffusion of water from *in vitro* formed leaves (Sutter, 1984).

The scanning electron microscope studies in *L. leucocephala* have revealed a definite increase in the amount of epicuticular wax deposited on the leaves following the transfer of plants out of culture. The micropropagated plants attained wax density comparable to that of field-grown plants, with in 6-7 weeks of transplantation (Dhawan and Bhojwani, 1987). These observations corroborate well with observations on the rate of water loss from leaves at different stages of micropropagation and hardening. The decline in the rate of water loss coincided with the increase in the amount of wax deposited on the leaves. However, the efficient water economy of the nature-nurtured plants, could not be matched by the transplanted plants even after five months. This may perhaps be due to the difference in the chemical nature of the wax deposited; hydrophobic wax, typical of the leaves of plants grown *in vitro* (Sutter, 1984), being more abundant on the transplants than the hydrophilic wax predominantly found in the plants growing in the field. However, this hypothetical assertion needs to be confirmed by proper chemical analysis. Differences in the rate of water loss by leaves at different stages of micropropagation have also been reported in *Malus domestica* (Brainerd and Fuchigami, 1981), *Prunus insititia* (Brainerd et al., 1981), *Brassica oleracea* (Sutter and Langhans, 1982) and *Solanum laciniatum* (Conner and Conner, 1984).

Stomatal Functioning

Besides epicuticular wax, poor anatomical differentiation of leaves, greater stomatal frequency (Wetzstein and Sommer, 1982) and impaired stomatal movement (Brainerd and Fuchigami, 1982; Wetzstein and Sommer, 1983; Ziv et al, 1987) have been considered as factors contributing to excessive loss of water by cultured plants.

Role of stomatal number in controlling water loss remains debatable. Published reports on this aspect are contradictory. Whereas Wetzstein and Sommer (1983) observed that leaves of sweetgum plants cultured *in vitro* had significantly more stomata per unit area than the acclimatised or field-grown plants. Fewer stomata were observed in the tissue-cultured plants of *Prunus instititia* (Brainerd et al., 1981), *Malus domestica* (Brainerd and Fuchigami, 1981), *Rubus idaeus* (Donnelly and Vidaver, 1984b), and in *Leucaena leucocephala* (Dhawan and Bhojwani, 1987). Detailed investigations involving more plant species is, therefore, required to have a general idea of the differences in stomatal frequencies and their effect on transplantation success.

In *Malus domestica,* higher rates of water loss from leaves of tissue culture-raised plants are attributed to slow stomatal responses. After acclimatisation at 30-40 percent relative humidity, these leaves regained normal stomatal functioning (Brainerd and Fuchigami, 1981). Similarly, detached leaves of transplanted plants of *Solanum laciniatum* had closed all their stomata within 30 minutes of detachment, whereas half of the stomata from tissue cultured leaves were wide open even after 16 hours of detachment (Conner and Conner, 1984). Thus, the failure of stomata to close in response to environmental stress, darkness and ABA in tissue culture plants is one of the major causes of excessive loss of water (see Ziv et al., 1987) by these plants leading to their wilting and finally death.

Anatomy

The poor mesophyll differentiation and weak vasculature of the leaves formed *in vitro* render the plants highly susceptible to transplantation shock. The cauliflower leaves formed *in vitro* lacked or had poorly developed palisade cells as compared to the transplanted plants (Grout and Aston, 1978). Similarly, in 'Pixy' plum, the layer of palisade cells were shallow and the mesophyll air spaces were greater in cultured plantlets than in greenhouse plantlets (Brainerd et al., 1981). Such dissimilarities in leaf anatomy of *in vivo* and *in vitro* grown plants were also observed in *Liquidambar styraciflua* (Wetzstein and Sommer, 1982) and *Rubus idaeus* (Donnelly and Vidaver, 1984b). In all these studies, comparison was made between the plantlets rooted *in vitro* and the field-grown plants. In the detailed anatomical investigations on *Leucaena leucocephala* (Dhawan and Bhojwani, 1987), it was found that the plants suffer first, and most, during the energy-demanding, rooting stage. The considerably well-defined mesophyll tissue at the shoot multiplication stage, undergoes drastic depletion at the rooting stage. During this period, the size of the palisade cells is reduced, and the spongy parenchyma becomes virtually non-existent, indicating a severe drain of the meager metabolic resources of the plant. The recovery is indicated by the gradual enlargement of palisade cells, re-differentiation of spongy parenchyma cells, and the appearance of starch grains after the pre-transplant hardening stage. Tissue culture plants of *Rubus idaeus* showed very poor development of mechanical tissues (collenchyma and sclerenchyma) in petiole, stem and root (Donnelly et al., 1985). The tissues developed in the culture showed little or no change, when transplanted to the field. They grew minimally, and slight secondary wall deposition occurred. Whereas collenchyma appeared in the leaves formed during the first week of transplanation, sclerenchyma showed a gradual return to normal situation, following transplantation.

Hardening of Tissue Culture Plants

Traditionally, transplantation follows the *in vitro* rooting stage but, *in vivo* rooting is gaining popularity for obvious reasons of economy and the quality of roots. In the latter case, both the rooting and the hardening stages are combined. While transferring out the plants/shoots from the culture, their lower part is gently washed so as to remove the medium sticking to them. The individual plantlets/shoots are then transferred to potting mix and irrigated with inorganic nutrient solution. A variety of potting mixes, such as pumice, peat, vermiculite, soil, sand, etc. or their mixtures in different proportions are used for transplantation. For initial 10-15 days, it is essential to maintain high humidity (90-100 percent) around the plantlets to which they got adapted during culture. Thus, in the initial phase following transplantation, the major environmental shock to the plants is the change from a substrate rich in organic nutrients, to a substrate providing only inorganic nutrients. This probably activates the photosynthetic machinery of the plants, enabling them to withstand the subsequent reduction in the ambient relative humidity and survive under field conditions.

Several methods have been used to build up high humidity around the plants during the early phase following transplantation. The most primitive approach has been to cover the plants with clean, transparent plastic bags and make small holes in them for air circulation. In *Mamillaria elongata,* shoots were transferred to soil and covered with a transparent plastic bag with small holes. In order to reduce humidity inside the bag, the size of the holes was enlarged over a two-week period. By carrying out this exercise, shoots got well adapted to greenhouse conditions and established functional roots (Johnson and Emino, 1979). Partial defoliation of plantlets (Bhojwani, 1980; Tisserat, 1981) and application of anti-transpirants such as one percent v/v 'Acropol' (McCown and Newton, 1981; Tisserat, 1981), in the initial stages of transplantation have been reported to improve the survival frequency, presumably due to reduced water loss by the plantlets. These days, most commercial laboratories have computerised hardening rooms with controlled conditions of light, temperature and humidity.

A simple method of direct transplantation of cultured plants of *Nicotiana tabacum, Oryza sativa* and *Solanum sisymbrifolium* to the field has been reported by Selvapandiyan et al. (1988). Application of a thin film, of 50 percent aqueous glycerol, paraffin (melting point 52-54°C) or grease dissolved in an equal amount of diethylthere on the surface of leaves with the help of a brush, before transplantation, resulted in the survival of transplants with very high frequencies. The treated plants showed cent percent survival for tobacco and *Solanum* and 70 percent for rice. None of the untreated plants of *Solanum* or rice survived and only 16.7 percent of tobacco plants could be transplanted.

Attempts have also been made to harden the shoot system (in terms of inducing autotrophism and develop surface wax on *in vitro* formed leaves) without disturbing the delicate root system. Sutter and Langhans (1982) observed increased wax formation in cabbage plants by exposing cultures to $CaCl_2$ *in vitro*. However, $CaCl_2$ absorbed water from the agar medium to an extent that after three weeks, the medium dried completely, which affected the general morphology of the plants, possibly due to their poor nutritional status resulting from inadequate nutrient supply.

Wardle et al. (1983) tried this approach with chrysanthemum and cauliflower plants. The medium was covered with a layer of lanolin, with in the presence or absence of silica gel. These treatments proved effective in reducing humidity and inducing epicuticular wax deposition in cauliflower, but proved detrimental for growth in both the experimental systems.

For reducing humidity, so as to harden the tissue culture plants of carnation *in vitro,* the culture tubes were opened and placed inside a desiccator with $CaSO_4$ as the desiccant (Ziv, 1986). Surface wax developed after seven days at 50-70 percent relative humidity (RH) but not at 80 percent RH. The survival rate of acclimatised plants rose from 72 percent to 90 and 96 percent after nine days under 70 and 50 percent RH, respectively.

In *Leucaena leucocephala,* over 80 percent survival could be obtained by introducing a pre-transplant hardening stage (Dhawan and Bhojwani, 1987). The plants were initially transferred to screw-cap bottles containing sterilised quartz sand and irrigated with an inorganic salt solution carrying an efficient strain of *Rhizobium* (NGR 8). The bottles were kept closed for two weeks initially. Thereafter, the caps were removed and the plants were maintained under controlled conditions of temperature and light ($25 \pm 2^{\circ}C$ and 18 W m^{-2}) for another two weeks, before their final transfer to the field (Dhawan and Bhojwani, 1987).

Several species produce storage organs *in vitro,* which can be transplanted to the soil with survival frequencies comparable to the structures formed *in vivo*. In such cases, the critical step of transplanting delicate plants can be obviated altogether (Hosoki and Asahira, 1980; Steinitz and Yahel, 1982). It also eliminates the additional step of rooting the shoots. The production of aerial tubers by the cultured shoots of potato under the influence of Chlormequat (CCC) is well known (Wang and Hu, 1984). *In vitro* tuberisation has also been reported in *Dioscorea bulbifera* (Sengupta et al. 1984), *D. alata* (Ammirato, 1984) and *D. rotunda* (Ng, 1988). Bulblet formation in *Muscari armeniacum* (Peck and Cumming, 1986) and *Narcissus tazetta* (Steinitz and Yahel, 1982) was supported by activated charcoal. In *Gladiolus,* where transplantation of micropropagated plants had been a serious problem, corms developed *in vitro* showed 75-80 percent germination under field conditions (Dantu and Bhojwani, 1987). High sucrose concentration was critical for *in vitro* cormlet formation.

CONCLUSIONS

Understanding the physiological and morphological characteristics of tissue culture plants and the changes they undergo during the hardening process should facilitate the development of efficient transplantation protocols. Therefore, additional data must be gathered on these aspects by extending the studies to more species.

REFERENCES

Alderson, P.G. and Rice, R.D., 1986, Propagation of bulbs from floral stem tissue. In: *Plant Tissue Culture and its Agricultural Applications* (L.A. Withers and P.G. Alderson, eds.), pp. 91-97, Butterworths, London.

Ammirato, P.V., 1984, Yams. In: *Handbook of Plant Cell Culture, Vol. 3. Crop Species* (P.V. Ammirato, ed.) pp. 327-354., Macmillan Publishing Company, New York.

Brainerd, K.E. and Fuchigami, L.H., 1981, Acclimatization of aseptically cultured apple plants to low relative humidity, *J. Am. Soc. Hortic. Sci.,* 106: 515-518.

Brainerd, K.E. and Fuchigami, L.H., 1982, Stomatal functioning of *in vitro* and greenhouse apple leaves in darkness, mannitol, ABA and CO_2, *J. Exp. Bot.,* 33: 388-392.

Brainerd, K.E., Fuchigami, L.H., Kwaitkowski, S. and Clark, C.S., 1981, Leaf anatomy and water stress of aseptically cultured 'Pixy' plum grown under different environments. *Hortscience,* 16: 173-175.

Broome, O.C. and Zimmerman, R.H., 1978, *In vitro* propagation of blackberry, *Hortscience,* 13: 151-153.

Conner, A.J. and Thomas, M.B., 1981, Re-establishing plantlets from tissue culture: A review. *Proc. Int. Plant Prop. Soc.,* 31: 342-357.

Conner, L.N. and Conner, A.J., 1984, Comparative water loss from leaves of *Solanum laciniatum* plants cultured *in vitro* and *in vivo*, *Plant Sci. Lett.,* 36: 241-246.

Dantu, P.K. and Bhojwani, S.A., 1987, *In vitro* propagation and corm formation in *Gladiolus, Gartenbauwissenschaft,* 2: 90-93.

Dhawan, V. and Bhojwani, S.S., 1987, Hardening *in vitro* and morpho-physiological changes in the leaves during acclimatization of micropropagated plants of *Leucaena leucocephala* (Lam.) de Wit, *Plant Sci.,* 53: 65-72.

Donnelly, D.J. and Vidaver, W.E., 1984a, Pigment content and gas exchange of red raspberry *in vitro* and *ex vitro*, *J. Am. Soc. Hortic Sci.,* 109: 177-181.

Donnelly, D.J. and Vidaver, W.E., 1984b, Leaf anatomy of red raspberry transferred from culture to soil, *J. Am. Soc. Hortic. Sci.,* 109: 172-176.

Earle, E.D. and Langhans, R.W., 1975, Carnation propagation from shoot tips cultured in liquid medium, *Hortscience,* 10: 608-610.

Fuchigami, L.H., Cheng, T.Y. and Soeldner, A., 1981, Abaxial transpiration and water loss in aseptically cultured plum, *J. Am. Soc. Hortic. Sci.,* 106: 519-522.

Grout, B. and Astron, M.J., 1977a, Transplanting of cauliflower plants regenerated from meristem culture, I. Water loss and water transfer related to changes in leaf wax and to xylem regeneration, *Hortic. Res.,* 17: 1-7.

Grout, B. and Aston, M.J., 1977b, Transplanting of cauliflower plants regenerated from meristem culture, II. Carbon dioxide fixation and the development of photosynthetic ability, *Hortic. Res.,* 17: 65-71.

Grout, B. and Aston, M.J., 1978, Modified leaf anatomy of cauliflower plantlets regenerated from meristem culture *Ann. Bot. (Lond.),* 42: 993-995.

Grout, B.W.W. and Crisp, P., 1977, Practical aspects of the propagation of cauliflower by meristem culture, *Acta Hortic.* 79: 289-296.

Hussey, G., 1977, *In vitro* propagation of *Gladiolus* by precocious axillary shoot formation, *Sci. Hortic.,* 6: 287-296.

Ng, S.Y.C., 1988, *In vitro* tuberization in white yam (*Dioscorea rotundata* Poir), *Plant Cell Tissue Organ Cult,* 14: 121-128.

Peck, D.E. and Cumming, B.G., 1986, Benificial effects of activated charcoal on bulblet production in tissue cultures of *Muscari armeniacum, Plant Cell Tissue Organ Cult.,* 6: 9-14.

Selvapandiyan, A., Subramani, J., Bhatt, P.N. and Mehta, A.R., 1988, A simple method for direct transplantation of cultured plants to the field, *Plant Sci.,* 56: 81-83.

Sutter, E., 1984, Chemical composition of epicuticular wax in cabbage plants grown *in vitro, Can. J. Bot.,* 62: 74-77.

Sutter, E.G., 1986, Micropropagation of *Ixia viridifolia* and a *Gladiolus* x *Homoglossum* hybrid, *Sci. Hortic.,* 29: 181-189.

Sutter, E. and Langhans, R.W., 1979, Epicuticular wax formation on carnation plantlets regenerated from shoot tip culture, *J. Am. Soc. Hortic. Sci.,* 104: 493-496.

Sutter, E. and Langhans, R.W., 1982, Formation of epicuticular wax and its effect on water loss in cabbage plants regenerated from shoot tip culture, *Can. J. Bot.,* 60: 2896-2902.

Uduebo, A.E., 1971, Effect of external supply of growth substances on axillary proliferation and development in *Dioscorea bulbifera, Ann. Bot. (Lond.),* 35: 159-163.

Wang, P.J. and Hu, C.Y., 1984, *In vitro* mass tuberization and virus-free seed potato production in Taiwan, *Ann. Potato. J.,* 59: 33-37.

Wetzstein, H.Y. and Sommer, H.E., 1982, Leaf anatomy of tissuecultured *Liquidambar styraciflua* (Hamamelidaceae) during acclimatization, *Am. J. Bot.,* 69: 1579-1586.

Wetzstein, H.Y. and Sommer, H.E., 1983, Scanning electron microscopy of *in vitro* cultured *Liquidambar styraciflua* plantlets during acclimatization, *J. Am. Soc. Hortic. Sci.,* 108: 475-480.

Ziv, M., 1979, Transplanting *Gladiolus* plants propagated *in vitro, Sci. Hortic.,* 11: 257-260.

Ziv, M., 1986, *In vitro* hardening and acclimatization of tissue culture plants, In: *Plant Tissue Culture and its Agricultural Applications* (L.A. Withers and P.G. Alderson, eds.), pp. 187-196. Butterworths, London.

Ziv, M., Schwartz, A. and Fleminger, D., 1987, Malfunctioning stomata in vitreous leaves of carnation *(Dianthus caryophyllus)* plants propagated *in vitro;* Implications for hardening, *Plant Sci.,* 52: 127-134.

20

PILOT PLANT TISSUE CULTURE UNIT

R.D. Lai and
P. Mohan Kumar

ABSTRACT

As a part of national goals in the spheres of energy conservation and afforestation, Tata Tea Limited has embarked upon a biotechnology program for the generation of fuelwood planting material. In some parts of India, the tea industry is replacing the use of fossil fuels for the process of tea drying, and extensive energy plantations are required to meet this objective. To increase biomass production, a Pilot Plant Tissue Culture Unit for the propagation of the superior planting material has been set up. Relying on indigenous equipments, the micropropagation protocols for three *Eucalyptus* species have been standardised. The plants have been transferred to soil in the nursery and thence to the field. Survival rate on transplantation to soil was found to be about 65 percent and from nursery to field about 90 percent. The laboratory is also undertaking fundamental *in vitro* studies on the propagation of tea (which is conventionally propagated vegetatively) and on breeding programs. Some aspects of costing are also discussed. It is recommended that plantation companies be encouraged to participate in industrial forestry.

R.D. Lai and *P. Mohan Kumar* * Tata Tea Limited, 1 Bishop Lefroy Road, Calcutta - 700 020, India.

INTRODUCTION

The current perception of the Government of India and several other organisations that the reforestation of India is the need of the hour, indicates a growing national interest in modern techniques of large-scale production of superior planting material. There should, in fact, be a new and increasing demand for elite plantlets of various woody species. This coincides with a growing awareness in the tea industry that continuing dependence on fossil fuels is not altogether desirable. Besides rising prices and questionable quality, the fact that tea gardens tend to be situated in extremely remote areas leads to diverse problems of transportation and supplies. Nevertheless, the requirement of thermal energy by the industry remains indispensable and enormous: to produce 1 kg of tea, about 3.5 kg of water has to be evaporated. The obvious alternative to irreplaceable fossil fuels, in this case, is fuelwood.

Commercial forestry for the purpose of producing fuel to dry tea is, in fact, already a reality in some tea-growing areas. For example, the entire thermal requirements of processing the tea leaves plucked from about 10,000 ha of tea gardens in the High Range of Kerala are now being met by captive energy plantations. Based on trials which commenced in the 1950s, currently favored trees include some species of *Acacia, Alnus, Eucalyptus, Grevillea* and *Thuja*.

To explore the potential of biotechnology in the propagation of certain plantation crops, Tata Tea Limited set up a pilot plant tissue culture laboratory in 1986. Initial efforts have been directed at the micropropagation of Eucalyptus and tea. It is felt that the project can, in the short-term, contribute towards fulfilling the internal demand for fresh planting material of superior stock, and, in the long-term, form the basis of a modern breeding program by applying biotechnological methods.

THE PILOT PLANT TISSUE CULTURE UNIT

Sufficient land was made available in a tea estate in southern India, and allocated for the construction of a laboratory, a nursery for hardening of plants, and suitable area for field trials. The building has four separate rooms, totalling to a floor space of about 130 m², which serve the following functions:

* Storage, washing and sterilisation of glassware
* Media preparation
* Inoculation
* Controlled-environment culture room

The design of the laboratory building was based on recommendations by a consultant (*vide infra*), but for the nursery, the techniques practised by the Company in the 50-odd tea and shade nurseries all over the country were retained. For example, subsequent experiments proved that there was no need for a greenhouse, and that sheds built of locally available materials were better suited. All the equipment and instruments are indigenous.

Tea gardens are located in remote districts of the country, where, quite often - as in this case -various infrastructural necessities are lacking. Besides the road to the tissue culture facility, the transmission lines, voltage transformers, switchboards and associated electrical hardware and telephone lines had to be set up. New accommodation for the staff had to be built, there being no housing available.

A judicious balance had to be established between the following factors:

* The need to have adequate equipment and facilities upto a certain minimum standard and yet maintain control over costs
* sophisticated laboratory techniques and relatively practicable industrial methods, and
* dependence on well-qualified scientists, as in university/government research institutions, and the need to employ unskilled labor.

The experience of the past 18 months since the laboratory was commissioned has resulted in a review of some of the early assumptions in order to optimise the economic evaluation of the plants.

MICROPROPAGATION OF EUCALYPTUS

Existing *Eucalyptus* stands in the private man-made forests of the High Range comprise three main species, *viz. E. grandis, E. globulus, E. robusta*, and small pockets of *E. citriodora*. Whereas *E. grandis* has been found to be generally well-suited to the region, in that it grows the fastest (an important consideration from the point of view of commercial firewood production), *E. globulus* is attractive because of its higher calorific value. *E. robusta* is used in the water-logged soils of valleys. Both *E. globulus* and *E. citriodora* are sources of saleable oil, thus conferring economic value on the foliage as well as the wood.

Propagation is commonly carried out from seed, which results in tremendous variation owing to heterozygosity. The clonal selection program has so far been concentrated on the measurement of girth at breast height and growth rate, but additional criteria will be used to identify elites from the 500 trees chosen to date. Attempts of vegetative propagation through cuttings did not yield reliable results, and it was decided to investigate *in vitro* techniques for the rapid clonal propagation of *Eucalyptus* species.

A number of research institutions in India were contacted, and finally the National Chemical Laboratory, Pune, was selected to provide consultancy services for technology transfer. Many reports in the existing literature on *in vitro* production protocols of *E. grandis* (Lakshmi Sita and Shobha Rani, 1985), *E. globulus* and *E. citriodora* (Mascarenhas et al., 1982) did not clearly establish good survival rates in large-scale operations under field conditions. However, it was felt that in-house methodologies developed for tea could, perhaps, be successfully adapted for this purpose. Thus did one hope to overcome the bottleneck of transplantation to the field that seemed to be the bane of many laboratories.

The explants chosen for multiplication were axillary buds from young coppice shoots generated by the felling of selected trees. After washing with detergent, surface sterilisation was carried out by treating explants with 0.05 percent mercuric chloride for 15-20 minutes. This gives 50-60 percent aseptic cultures. Bud opening and initial multiplication were achieved in two to three sub-cultures of 20-25 days each, using a basal MS medium supplemented with high content of cytokinins. Different media were required at this stage for different species of *Eucalyptus,* but a common medium with low cytokinins was found suitable for all three to induce subsequent shoot multiplication. The cultures were sub-cultured four to five times to produce a sizeable number of shoots. A method for obtaining healthy rooted plants in about 20-25 days was established after trying numerous different treatments. Once again, a single medium (hormone-depleted MS salts of half concentration with 0.25 percent charcoal) applicable to all three species was developed. The rooted plantlets were removed to disinfected soil in sleeves under polythene tents, as is usual in plantation nurseries. Once established, the sheets were removed and the plants maintained under partial shade until they were ready for transplanting to the field. It was determined that this technique of acclimatisation afforded 65 percent survival, which compared very favorably with the one that had less than 40 percent chances under glasshouse conditions. Mortality of the first lot of plants, four months after transplanting to the field, is less than 10 percent to date. These are now being compared with seedlings planted at the same time.

In addition to the laboratory and nursery work mentioned above, forestry research is continuing in the form of spacing and species trials. Future efforts will be directed towards cross-fertilisation of these activities to improve the productivity of the energy plantations.

MICROPROPAGATION OF TEA

The technique of vegetative propagation of *Camellia sinensis* for the production of new clonal planting material was standardised in India over two decades ago. This is now the primary and most economic method of generating tea plants, and *in vitro* micropropagation can become competitive only in certain special situations. One of these would be the opening up of virgin lands to tea plantation, when very large number of new plants (7,000-14,000 per ha) would be required. A second might be the discovery of a new elite with a yield potential considerably higher than the present maximum (4000-6000 kg per ha, depending on agro-climatic conditions). A discovery of this nature would result in an immediate, but short-lived, demand for a large number of plants. Neither situation is very likely, however, since land in this country is a scarce commodity and it takes 10 years of patient evaluation before a new clone can be released (time required to create a bank of mother bushes from where, cuttings could be obtained).

Clonal selection is the mainstay of the tea industry. Major programs are under way, using numerous well-established criteria to identify possible elites. Close to the Pilot Plant Tissue Culture Unit is a germplasm bank comprising over 200 selections. Many tea gardens have seed orchards. Selfed and crossed progeny are being investigated. Classical methods of plant breeding continue to be important for the industry.

The tissue culture of tea may be an indispensable pre-requisite for biotechnological breeding programs, and it is accordingly a subject of intensive study in numerous organisations in India. However, it is only in those laboratories which have commercial interests in developing the protocols that top priority has been accorded to *Camellia sinensis*: these are the Tea Research Institute in Tamil Nadu; the Tea Research Association in Assam, and Tata Tea Limited. Research is at different stages of progress in areas such as embryo and cotyledon culture, bud multiplication, callus induction (somatic embryogenesis) and anther culture.

The need to increase productivity in the tea industry cannot be over-emphasised. The beverage is ubiquitous, being the cheapest drink next to water. Its consumption rate in this country is growing faster than the rate of production, which can not keep up with the demand. India is the largest producer and exporter of tea in the world today (1984-85 foreign exchange earnings: Rs. 771 crores, equivalent to US $ 600 million). However, the rate at which our domestic consumption is increasing, it has been predicted that India may, by the turn of the century, drop to become a net importer of tea. The application of biotechnology to tea is, therefore, an area of crucial importance and enormous potential for this country.

ECONOMICS

It is evident from the foregoing sections that the protocols have only just been standardised. Minor modifications to the existing techniques may be required to overcome certain location-specific problems and optimise success-rates at each stage before scaling up to a commercial production level. Thus, it is too early to present any conclusions about the cost of production at the Pilot Plant Tissue Culture Unit. Nevertheless, it is worth drawing attention to certain aspects, which have had a bearing on the subject.

The capital investment required to set up the unit *per se* was limited to the extent possible by creating a well-appointed, but not extravagant, laboratory. Foreign equipment was eschewed not merely on account of the high customs duty on imports, but because after-sales service is practically impossible to obtain in remote tea garden districts. In fact, this factor of remoteness was responsible for the greater part of the expenditure, new housing for the staff being its most capital-intensive element.

Among operating costs, the role of time becomes important. Direct costs, including labor, utilities and consumables such as glassware and nursery materials are influenced by multiplication and casualty rates. The number of sub-cultures carried out and the time spent therein governs, to a large extent, the number of plantlets produced. One of the specific objectives of the research on the standardisation of *Eucalyptus* micropropagation was the development of a common medium for certain stages in different species, as this saves time and labor during production.

Additional problems arise from the allocation of indirect costs, such as the proportion of effort devoted by supervisory scientists and administrators to *Eucalyptus* plantlet production and fundamental tea research. There are various hidden costs when a laboratory is part of a large, established organisation, and finally, the question of whether past R&D costs should be written off when a pilot plant enters a purely production mode.

From a financial standpoint, this sort of analysis is of little value without some indication of the prices, the market can bear. On this point, however, there is insufficient information available.

RECOMMENDATIONS

Although the Pilot Plant Tissue Culture Unit was set up primarily with internal desiderata in mind, it should also be viewed in the national context since it is one of the pioneering tissue culture laboratories in the Indian private sector. In order to clearly define the role of commercial organisations in future, a few suggestions can be made for consideration by policy-making authorities :

(1) **Conduct Market Surveys:** Much has been written about the priorities for basic and applied research in plant biotechnology in India, and the government accords great importance to this field for its potential *vis-a-vis* the nation's future development. What is required is need-based planning using data gathered from market surveys, and forecasts according to different price structures. This is particularly necessary for the unorganised farming sector *vis-a-vis* the anticipated shift towards intensive horticulture. So, too, the requirement of superior planting material for afforestation programs.

(2) **Demonstration Plots:** While it is unlikely that, in the foreseeable future, tissue-cultured plants will be cheaper than those propagated by conventional methods, it would not be unreasonable to expect that they may be superior in terms of quality, yield, etc. To convince rural buyers of the greater benefits associated with the higher price of better planting material, it is vital to establish demonstration plots in appropriate areas as soon as possible.

(3) **Undertake Industrial Forestry:** It has already been pointed out that the social forestry program cannot meet the needs of industrial forestry in this country, nor should they be expected to. Rather, those industries whose raw materials or energy needs are derived from wood should be encouraged to raise captive plantations. In fact, voluntary and government programs for the creation of fuel reserves to supply future firewood needs could be supported by plantation companies who understand well the economics and operation of cultivating vast tracts of land. The case in favor of

this (Lappin, 1985) takes into account the fact that the model developed for plantations (tea, coffee, rubber, etc.) would be very suitable for afforestation. The plantation industry, which has been in existence in India for the past century and a half, is familiar with the entire gamut of operations: R&D, nursery work, planting, cultural practices, nutrition, pest and disease control, harvesting, processing, packing, transport and marketing. In addition, the requisite management systems are already in existence for administration and accounts, labor employment and welfare, engineering and infrastructural maintenance in remote parts of the country. Thus, the application of modern biotechnological methods in forestry, utilising the vast experience of the plantation industry, will genuinely contribute to the resolution of the problem of firewood shortage and the reforestation of India.

REFERENCES

Lakshmi Sita, G. and Shobha Rani, B., 1985, *In vitro* propagation of *Eucalyptus grandis* L. by tissue culture, *Plant Cell Rep.,* 4: 63-65.
Lappin, M.R.P., 1985, Adoption of plantation model for afforestation, *Planters' Chronicle,* 80: 145-147.
Mascarenhas, A.F., Hazara, S., Potdar, U., Kulkarni, D.K. and Gupta, P.K., 1982, Rapid clonal multiplication of mature forest trees through tissue culture. In: *Proc. 5th International Congress: Plant Tissue and Cell Culture* (A. Fujiwara, ed.), pp. 719-720, Jpn. Assoc. Plant Tissue Culture, Tokyo.

21

LARGE-SCALE PRODUCTION OF PLANTS THROUGH MICROPROPAGATION: PROBLEMS, PROSPECTS AND OPPORTUNITIES FOR INDIA

Jitendra Prakash

ABSTRACT

Current techniques for the *in vitro* propagation of plants and the ready acceptance of tissue culture transplants by the commercial sector have allowed for strong and continued growth within the micropropagation industry. Some companies in the USA, the Netherlands and UK, are producing up to five million plants per year and major laboratories are thinking of expanding the production to twenty million plants per year. This trend will soon saturate certain markets if product diversity does not take

Jitendra Prakash * A.V. Thomas Group of Companies, P.B. No. 520, Willingdon Island, Cochin - 682 003, India.

place. The major difficulties in expanding propagation techniques to new crops are the high production cost and the problems encountered in developing efficient systems to deal with a large number of plants. Most of these crops have seasonal peaks of demand. Better production planning with good crop mix, improved *in vitro* storage techniques, reduced labor costs and a better grip on contamination control are the critical factors for the successful expansion of micropropagation technology. These are discussed in detail with special reference to India for the development and expansion of this technology.

INTRODUCTION

During the past 20 years, there has been considerable progress in the development of plant tissue culture (PTC) technology and it is very clear that this technology has become a powerful tool for studying the basic and applied problems in plant science. The potential value of this technology is being commercially exploited by various organisations all over the world. Currently, the most practical application is the mass propagation of selected genotypes. This is evident by the large number of commercial firms engaged in propagating a variety of plants through tissue culture. In the Netherlands itself, the total number of firms propagating plants through PTC methods has increased considerably in a short span of time. The number of firms producing up to 10,000 plants have increased from 12 to 14; 10,000 to 1,00,000 plants have gone up from 14 to 18; 500,000 to 1,000,000 plants from one to two and more than five million plants from two to three. This progress has been observed during the period 1985-86(Table 1).

Table 1 : Comparative survey of *in vitro* propagation in the Netherlands *

Production (No. of plants produced per year)	Number of firms	
	in 1985	in 1986
10,000 or less	12	14
10,000 - 100,000	14	18
100,000 - 500,000	6	6
500,000 - 1,000,000	1	2
1,000,000 - 5,000,000	7	7
More than 5,000,000	2	3
Total	42	50

* (Personal communications from R.L.M. Pierik)

However, there are several major factors that limit the production of herbaceous horticultural crops beyond one million plants per annum and the widespread application of the latest advances in this field for the production of other high-value crop species. The present paper attempts to critically evaluate the basic areas which require attention for development and the prospects involved in the large-scale production of plants through micropropagation. Recent advances that can be implemented in the present technology to suit Indian conditions are also highlighted.

GROWTH IN MICROPROPAGATION INDUSTRY vs MARKET SATURATION

One of the major aspects of plant biotechnology is the ability to produce large number of identical individuals via *in vitro* cloning and the recent advances have paved the way for effective and continued growth within the commercial sector of the micropropagation industry. For example, the total number of plants cloned *in vitro* in the Netherlands in 1985 was 35,981,960 and in 1986, this figure rose to 42,753,600, which shows an increase of 19 percent within one year (Prof. R.L.M. Pierik, personal communication). In general, tissue culture methods have greatly increased the scope and potential of propagation by exploiting regenerative behavior more efficiently and in a wider range of plants than is possible with conventional procedures. For example, in Dutch plant tissue culture laboratories, a remarkable growth was noticed from 1980 to 1986 (Table 2). The species that are grown as potted plants or for their flowers, in general, have received more attention as compared to other crops such as ornamental bulbs and corms, vegetables, field crops and trees.

Table 2: Number of plants vegetatively propagated in Dutch plant tissue culture laboratories *

Category	1980	1986
Pot-plants	3,118,150	19,822,274
Cut-flowers	910,500	12,639,758
Orchids	2,700,000	1,449,190
Ornamental bulbs and corms	563,000	8,085,920
Vegetables	22,300	68,825
Field crops (potato, sugarbeet)	117,000	311,828
Trees	-	1,040
Miscellaneous (shrubs, garden plants, etc.)	800	374,765
Total	7,431,750	42,753,600

* (Personal communications from R.L.M. Pierik)

At present, the demand for most of these plants exceeds the production capacity of the commercial laboratories. This trend is very marked in the Netherlands: in 1986, only 50 million plants were produced against a demand of 85 million plants. In 1987, the demand for plants had gone upto 135 million (Mr. Andrew Brown, personal communication). This sort of trend generally initiates growth within the commercial industry. Currently, some commercial plant tissue culture laboratories have been producing five to eight million plants per year and efforts are in progress to expand to

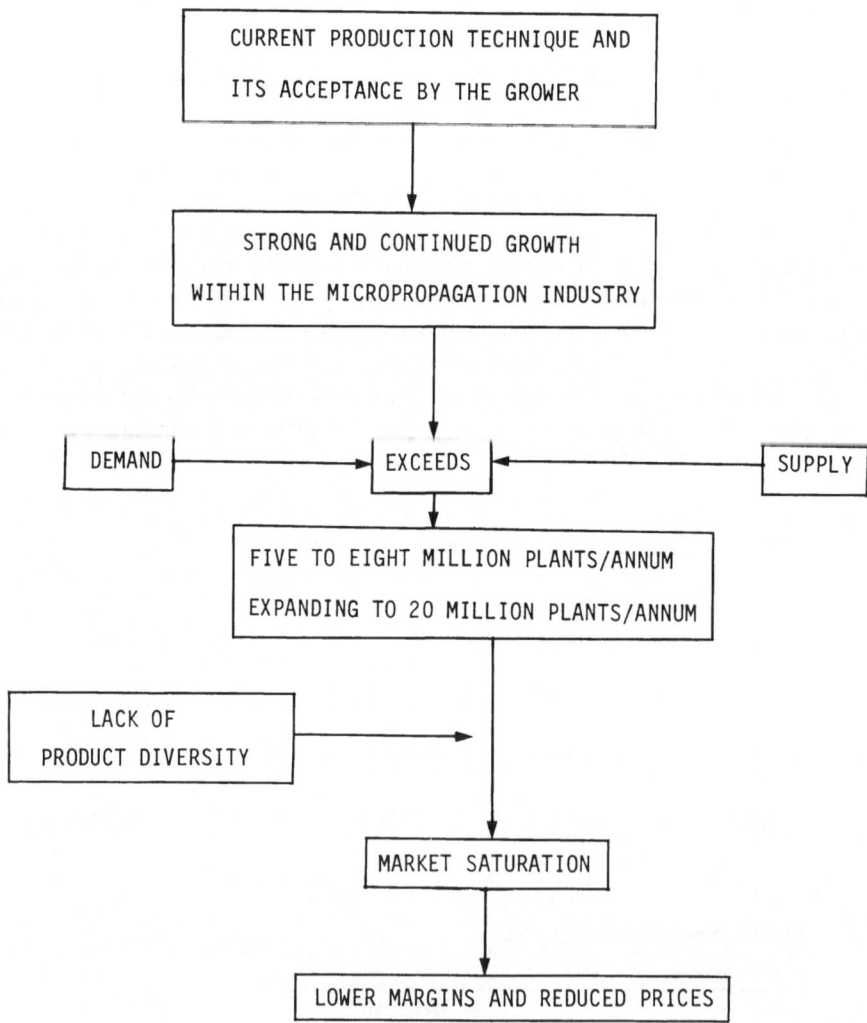

Fig. 1: Growth in micropropagation industry and market saturation

nearly 20 million plants per annum (Rowe, 1986). Unless and until there is a balance between product diversity and quantity, this kind of effort will saturate, and thereby bring down, the price in some of the commercial markets. For instance, in the Netherlands, pot-plants such as *Anthurium, Nephrolepis* and *Saintpaulia;* cut-flowers *Alstroemeria, Anthurium andreanum* and *Gerbera;* vegetables such as cauliflower and white cabbage; and ornamental bulbs and corms such as *Lilium* and *Nerine* are being propagated in very large numbers as compared to *Kalanchoe; Yucca* in the pot-plants category; carnation, matricaria in the cut-flowers group; kohlrabi and fennel in the vegetables group; and *Gladiolus* and *Watsonia* in ornamental bulbs and corms category, which are being propagated in very few numbers.

Continued lack of product diversity and subsequent competition in well-defined markets would force the commercial laboratories to sell their plants at lower cost and with less profits. This will then lead to limited cash resources and thereby, the inability to stay in business (Fig. 1). Therefore, more and more commercial firms have to initiate and extend their efforts towards other economically important plants, so that this problem could be overcome to a certain extent.

PROBLEMS ASSOCIATED WITH THE EXPANSION OF MICROPROPAGATION TECHNOLOGY

When the economics of an industry is under consideration, it becomes necessary to scrutinise the factors responsible for its performance over the past years, in order to set priority of its objectives. Though an extensive range of techniques has been developed during the last 20 years, there are still some major difficulties, especially those associated with the expansion of this technology to a variety of crops. Some of these problems can be discussed under the following headings :

Production Costs

Initially, in the case of selected crops, conventional planting materials were being sold at a much lower price as compared to tissue cultured/transplants/propagules. However, later, when it was proved that through tissue culture technology, uniform and disease-free plants could be produced in large numbers all the year round, there was no problem for the commercial growers in accepting a tissue culture-transplant at a higher price. The present trend in the micropropagation industry gives us a picture that more and more companies are producing only a few selected varieties of plants in large numbers, thereby generating stiff competition in the market. For example, Vietnamese farmers are producing tissue cultured potato plants at a total cost of one US cent each, as compared to eight cents per plant in the USA. The reason is that the current procedure of plant tissue culture is labor-intensive. Consequently, the labor inputs (in the form of salaries) form a significant proportion of the total production cost. In order to extend the technology to a variety of high value crops, one must be able to cut down production costs to the minimum by implementing the technical advances that are made in the research and developmental field. Given technical support and low transportation costs, future production centres will shift to countries where labor is less expensive. The standard of living in every country is increasing day by day and the same trend, as can be seen in Vietnam, may not continue in the future. Hence, the future of the micropropagation industry clearly lies in the ability to develop new technologies such as the application of computer controlled handling systems. Deleplanque et al. (1985), presented a robotic vision system which locates plantlets on their medium, picks them up and transfers them into containers. To reduce labor costs, automated plant culture systems are being developed where they duplicate the features found in manual plant tissue culture procedures but without the necessity for frequent manual transferring of cultures to a fresh medium. This would occur through computer control via pre-programmed instructions (Tisserat and VanderCook, 1985). However, the use of computers in this technology is under debate because the whole system cannot be automated. Wherever and whenever a precise cutting is to be done again, automation is difficult. In order to automate these features and computerise quality control at each step of micropropagation, heavy expenditure in the research and development would be required, thereby increasing the cost of each unit considerably. Hence, big multinational companies in the USA and Europe are not sure whether to go for automated technology or to shift the production unit to a place where the production costs would be less due to the availability of cheap labor.

Seasonal Nature and Heavy Peak Demands

Most of the crops are seasonal and are in heavy demand only at certain times of the year. To overcome this problem and to meet the requirements of extremely large number of plants, efficient methods must be developed. Also, most commercial laboratories produce a mixture of plants required at different times in order to maximise the use of the facilities throughout the year. These definitely cut down the production cost.

Personnel/Organisational Problems

Since large-scale micropropagation is a labor-intensive process, one must correspondingly increase the number of employees to increase the number of plants produced. As the activities in any commercial laboratory expand, organisational problems will arise and, at times, become critical. To reduce these problems, the management structure must be well planned. Supporting information systems such as inventory control, production scheduling, space utilisation, daily targets and actuals should be well defined for better management. For the successful large-scale micropropagation of quality products, there must be perfect coordination among the technicians, supporting staff, supervisors and the researchers. The job description and reporting systems should be very clear. In addition to these, personnel training is a critical component for successful large-scale production, which needs to be periodically reviewed on the basis of the customers' needs. The inventory control must be computerised. All these features will add considerably to the management responsibilities.

Contamination Control

Close attention has to be given to this aspect, to avoid much of the industrial effort going waste. For example, if in an Indian commercial laboratory, 10 percent of the one million plants produced through micropropagation methods perish due to contamination, then the loss in direct income would amount to approximately Rs. 6,00,000. This loss of cultures during the culture period would increase the production cost of the remaining plants to a large extent.

Contamination in cultures could be caused mainly due to the following:

(a) *Natural contamination:* Occurs from any of the following agents - Dust, air-borne salts, metal filings, chippings, spares, seeds, vegetable matter, fibers, animal hair, etc.

(b) *Man-made contamination:* Occurs mainly from body and clothing. It was calculated that each operator will generate a minimum of 500,000 to 1,000,000 particles (greater than 0.5 μm diameter) per minute.

For reducing contamination rates and increasing productivity, a clean indoor room environment for culture work is essential. In the work-area (where aseptic manipulations are carried out), three sources contribute to work-area particle contamination of the room air. They are the outside air, the liberation of particles by equipment machinery and processes, and the liberation of particles by operator. The objective of clean room is to establish and maintain an internal atmosphere devoid of particulate contamination in the working environment. Air cleanliness is measured as particle counts such as Class 1, Class 10, Class 100, Class 1,000, Class 10,000 etc., i.e., on a particle concentration (number of particles per unit volume). For the high cleanliness area of Class 100 or better, low turbulence displacement flow, also called laminar flow, has to be employed where the air moves along parallel flow paths and permits the particles liberated in the room to be carried quickly and as directly as possible. This sort of facility may be established by introducing the supply air into the room through walls or ceilings built up from HEPA or ULPA filters or through special clean air distributing elements. Since the installation and running cost of laminar flow cabinets is high, it is advisable to design it according to the plant species to be worked on, so that the working space is utilised effectively. To achieve this, a perfect understanding is needed between researchers, engineers and architects. Open-minded collaboration between the supplier of the clean room system and the end-user is absolutely essential, if optimum balance between quality and flexibility on the one hand and investment and running cost on the other, are to be achieved (Bruderere et al., 1983; Schicht, 1985).

Modern high quality clean rooms have environmental control systems in the widest sense. Besides cleanliness, temperature and humidity control, they encompass advanced air handling technology, vibration protection and many other aspects. In addition to this, correct and well-defined working procedures are necessary. All this should be backed by performance monitoring of the operation as a whole. Wearing gowns, which is mandatory in Class 100 clean rooms, reduces the contamination from clothing, skin and hair.

Maintaining cleanliness in the working area is equally important. Apart from contaminants appearing in the cultures and those carried along with the explant, there is a continuous entry of contaminants along with the worker and through air entering the room. A definite protocol should be developed in order to reduce contamination in the laboratory and this varies with the location of the laboratory.

OPPORTUNITIES FOR DEVELOPMENT IN INDIA

Among all the countries in the world, India is rich in natural resources; its diverse agroclimatic zones have the potential to produce a wide variety of agricultural and commercially valuable crops. It also has a germplasm of rare ornamental and medicinal species that are native to the tropical forests of different states. As a whole, to develop and extend this technology to the maximum, India has many advantages over other countries in the world.

The native forests all over the country are well known for their richness and variety in foliage and ornamental plants. These plants can be propagated on a large-scale by tissue culture methods and can be used either as pot-plants or for their cut-flowers. Since commercial growers, breeders and producers are interested in new varieties of plants, these plants can be introduced as new products in horticultural markets.

Plants obtained from anywhere in the world can be maintained, propagated and transplanted into soil in India, as the country has diverse agroclimatic regions. To economise the procedure, it is advisable to cut down costs on the purchase and maintenance of the equipment (air conditioning systems) needed for high rates of multiplication and survival, if the appropriate site is located within the country. However, before selecting a site for the micropropagation industry, one must decide which products are going to be produced.

In India, the scientific talent and the operational skills of the people are at par with the best in the world. The cost of skilled personnel is also low as compared to other countries. As a result, a significant portion of production costs comes down, whereas the reverse is true in the case of most advanced countries of Europe and the USA. In fact, some of the important discoveries in PTC have been made by Indian scientists. For example, anther culture technology, widely practised throughout the world today, was introduced by Indian scientists from Delhi University in 1964. The discovery of *in vitro* pollination to effectively overcome the incompatibility mechanism in many plants and ensuring proper fertilisation, good seed-set and the possibility of obtaining rare hybrids, are also from the same school.

Embryo culture is yet another novel *in vitro* technology that originated in India and is being used extensively by plant breeders the world over. This technology can help to overcome the hurdles of endosperm-embryo incompatibility and poor endosperm development encountered in traditional breeding methodologies; rare hybrids can also be obtained by this method. Interspecific as well as intergeneric hybrids have been successfully regenerated through embryo rescue and culture. Similarly, Indian universities and institutes have pioneered micropropagation procedures for various crops.

However, one drawback of the Indian research program has been that, although the best technology is available here, the implementation/commercialisation is being done in other parts of the world because of lack of proper direction.

Blessed with rich flora and scientific talents, India is in a better position to commercialise micropropagation technology in comparison to other so-called cheap labor areas of the world, as it does not have to import the technology. What it needs is to scale up, protect and patent the technology for the benefit of the nation.

It will not be out of place to mention that India has trained a large number of bright and exceptionally enthusiastic plant tissue culturists, who have gone abroad to see whether their basic knowledge can be applied in the real sense. Now, since the country has begun to realise the importance of the application of tissue culture technology, it is high time that adequate incentives are provided to Indian scientists working abroad to come back and provide an effective solution to the brain drain. The management of the A.V. Thomas Group of companies has been working intensively towards this national cause. They have been successful in attracting several key personnel in the field of biotechnology from abroad. The new laboratory in the Cochin Export Processing Zone, the largest in Asia and the first of its kind in India, is expected to commence production from June, 1988. This modern laboratory, equipped with top class facilities and at par with the best in the world, is designed to provide an ideal working environment and facilities for Indian tissue culturists returning from abroad, and is of great importance. One of the many advantages of having a large-scale tissue culture facility in the export processing zone is that the capital equipment/consumables etc., can be imported into the zone without any duty and custom formalities. All our research and development programs are directed towards the basic aspects of application and commercialisation in the real sense.

ACKNOWLEDGEMENTS

I am grateful to Prof. R.L.M. Pierik, Department of Horticulture, Agricultural University, 6700, AA Wageningen, The Netherlands, and Mr. Andrew Brown, Managing Director, Advanced Bio Centre, 2 Lancaster House, Cox's Yard, Somerton, Somerset, UK, for providing some of the information that has been presented in this paper.

REFERENCES

Bruderere, J., Luwa, A.G. and Anemonenstr, 1983, An economic and efficient Air Distribution Method for the establishment of clean working environment, *Swiss Pharma,* 5: 17-21.

Deleplanque, H., Bonnet, P. and Postaire, J.G., 1985, An intelligent robotic system for *in vitro* plantlet production. In: *Proc. Int. Conf. Robotic Vision & Sensory Controls,* pp. 305-314, Amsterdam.

Rowe, J.W., 1986, New technologies in plant tissue culture. In: *Tissue Culture as a Plant Production System for Horticultural Crops* (R.H. Zimmerman R.J. Griesbach, F.A. Hannerschlag and R.H. Lawson eds.), pp. 33-51, Martinus Nijhoff, Dordrecht.

Schicht, H.H., 1985, Clean room technology: The concept of total environmental control for advanced industries, *Technol. Appl. Ion Phy.,* 485-491.

Tisserat, B. and Vandercook, C.E., 1985, Development of an automated plant culture system *Plant Cell Tissue Organ Cult.,* 5: 107-118.

22

COST ANALYSIS OF MICROPROPAGATED PLANTS

R.K. Pachauri and
Vibha Dhawan

INTRODUCTION

Clonal propagation of plants through tissue culture is popularly called 'micropropagation'. In developed countries, this is being done in the commercial laboratories, some of which produce up to five million plantlets every year. The technique has so far been applied to ornamentals and a few horticultural species, as tissue culture plantlets are generally 5-10 times more expensive than seed-raised plants and therefore, there are very few buyers for plants of low economic value or forest trees from which returns could be expected only after long periods. The market for tissue culture plants would expand dramatically, if their selling price is reduced. The growth of this industry will, therefore, depend on the technological advancements made, especially in the field of automation and mechanisation of the process, in order to reduce the cost of propagules.

R.K. Pachauri and *Vibha Dhawan* * Tata Energy Research Institute, 90 Jor Bagh, New Delhi 110 003, India.

In this article, an attempt has been made to highlight the factors that contribute towards the cost of tissue culture plants. A model, defining the role of each component, has been suggested, the analysis of which may lead to steps that can contribute in reducing the cost of plantlets.

Before developing protocols for the micropropagation of a given species, it is necessary to evaluate the need for developing such method of cloning. Aseptic cloning is a technique employed for mass propagation of a selected plant. Micropropagation can be used in the following cases:

* Plants that can be conveniently propagated by rooting of cuttings but either the number of mother trees are limited or for which large areas of land are required for maintaining the mother trees to supply material for substantial planting programs. Also, elite trees might be dispersed over large areas and collection of material from them may prove to be expensive.

* When the species are dioecious but plants of a particular sex are more desirable commercially (such as male plants of *Asparagus officinalis* and female plants of *Carica papaya*). In such cases, plants of the desired sex could be obtained in large numbers through tissue culture.

* With increasing awareness of seed-borne diseases and the danger of introducing pathogens while importing plant material, quarantine procedures all over the world are being followed with strictness. Tissue-cultured plants, being disease-free and maintained under aseptic conditions, prove to be ideal for international exchange. Further, in the species that are conventionally propagated by vegetative techniques, there is also a danger of accumulating diseases which, over a period of time, may affect productivity. However, such risk could be eliminated if these plants are multiplied through tissue culture (e.g. seed potatoes).

* Tissue culture propagated plants can be used to raise seed orchards. Since this is an expensive technique, small numbers of plants of 10-15 selected clones could be planted. The seeds thus produced are of superior quality (overcoming the danger inherent in stands of genetically identical plants) and can then be used for large-scale plantation.

While all attempts should be made to reduce the cost of test-tube plants, it should not be at the cost of clonal uniformity of the plants. One has to be very careful while selecting the method of shoot multiplication. Regeneration from the callus might result in plants of different ploidy. The production of somatic embryos is a desirable method, as studies conducted so far, indicated that only diploid cells form somatic embryos and plants produced through somatic embryogenesis are clonally uniform (I.K. Vasil, personal communication). Other advantages of somatic embryogenesis are:

- somatic embryos, being bipolar structures, reduce one step of aseptic manipulation (viz. rooting),

- very fast rates of multiplication can be achieved, and

- somatic embryos can be encapsulated in a jelly-like substance for the production of artificial seeds (see Chapter 6).

However, not all plant species form somatic embryos *in vitro*. For those species, the axillary shoot proliferation has so far proved to be the safest method of shoot multiplication. Although initially slower, astronomical figures of production can be achieved within a year's time, since the number of shoots increases logarithmically with each passage. Theoretically, a three-fold increase in shoot number every four weeks or a six-fold increase every six weeks would allow more than a million shoots to be produced from a single-node segment within a year. Some examples of tree species, where an annual rate of multiplication of up to 1,000,000-fold is possible, are as follows:

Dalbergia sissoo (Suwal et al, 1988)

Eucalyptus citriodora (Gupta et al., 1981)

Leucaena leucocephala (Dhawan and Bhojwani, 1985)

Populus alba (Christie, 1978)

Populus tremula (Christie, 1978)

Salix matsudana x *alba* (Bhojwani, 1980)

Santalum album (Lakshmi Sita, 1986)

High rates of propagation can be attained only if there are no restrictions on the supply of skilled labor for shoot separation and sub-culture, various other resources and incubation space. Since sub-culturing is carried out manually, the production of one million plants within a short time would require a work force of several hundred trained persons.

Apart from manpower, another limiting factor is the incubation space. Although, the shoots can be multiplied and rooted *in vitro* throughout the year, their final transfer to the field is restricted to a particular time of the year. To obtain maximum transplantation success, the plantlets should be of optimal size and this can be achieved only if the multiplied shoots are put in for rooting at certain fixed times of the year so that plantlets of the desired size are transferred to the field.

The cost of propagating plants varies from species to species, depending upon the techniques followed, rates of multiplication achieved and the total skill of operation. The cost of tissue culture-propagated plants is inversely proportional to the number of plants produced. Also, if a variety of plant species are propagated in the same laboratory, with different times of transplantation, the costs are reduced significantly by fully utilising the facilities available. No generalisations can be made as the labor cost is different at different places and so is the cost of materials, overhead expenditure and salaries. A break-up could be provided in two ways, e.g. costs of individual components or expenditures incurred at each stage. Sluis and Walker (1986) estimated that for most ornamental plants, the cost of different components is as follows: 40 percent labor, 10 percent material, 20 percent overheads, and 30 percent as sales, general and administration, etc. Depending on the plant species and wages, the share of labor costs might be much higher. Anderson et al. (1977) estimated that 76 percent of the cost of producing broccoli plants by tissue culture would be taken up in wages and salaries (37 percent for explanting, 8 percent for media preparation and clean-up; and 31 percent for supervision and administration). With such a high proportion of the total related to wages, the actual cost of tissue-cultured plants will necessarily differ between countries.. For instance, plants produced in the USA will cost somewhat more in international currency units than those produced in Europe. Goh (1982) mentioned that the price of mericloned orchid plants raised in Singapore was four times higher than that of plants produced by the same technique in Thailand.

Another significant component that contributes to the cost of the micropropagated plants is the overhead expenses related to supervision, research facilities and utilities. Tissue culture laboratories consume large quantities of energy. The utility consumption rates for lights, air conditioning and autoclaving can be significant. The capital equipment components of a tissue culture facility contribute to overheads both as variable costs, in terms of consumption, of utilities and as fixed costs, in terms of capitalisation and depreciation. A typical facility for about one million plants per year requires a space of about 230 m^3. Also, the plants must be grown to a suitable size and hardened off for planting in the field. The cost of land, building the facility, depreciation cost of equipment/building and the cost of maintaining the plants in greenhouse, should all be taken into consideration while calculating the cost of tissue culture-plants.

Cost Savings

Again, the cost of land, utilities, labor, etc. vary from country to country and no generalisations can be made. The greatest proportion of the cost of micropropagation is associated with labor wages, as all manipulations are to be carried out manually. While lack of automation is generally regarded as the bottleneck to advancement, progress in this area has been slow. This is basically because of the complex physical arrangement of nodes and shoots on a microplant and the resultant difficulties in mechanical access to the material being separated. For example, a cluster of shoots or individual nodes on stem represents an interconnected framework of tissues and living systems, which shows much variability in the configuration of these units, e.g. the distance between buds or shoots and their spatial orientations are not consistent and accurately predictable in a tissue culture environment. Another impediment to the automation of tissue culture systems is the shape of the culture vials. These tend to be small so as to limit contamination, are normally deep and therefore, do not provide easy mechanical access to the shoots being separated.

Attempts have, however, been made to develop labor-saving approaches. The significant among these are: adding liquid nutrient medium to exhausted agar media (Maene and Debergh, 1985), computerised changing of nutrient solutions at different stages of growth (Tisserat and Vendercook, 1985), use of industrial robots (DeBry, 1986) and nutrient-mist generation in special culture enclosure (Weathers and Giles, 1987).

Levin et al (1988) are reported to have developed an automated propagated system which has substantially eliminated manual tasks, resulting in as much as 85 percent savings in labor costs. The cost of production is reduced to as much as 60 percent of the conventional, compared to conventional methods of tissue culture propagation. This technique is being tested at a commercial laboratory in Israel.

The system supports the growth of both organogenic and embryogenic cultures in bioreactors which are sized with the aid of a bioprocessor. The technique, however, can only be applied for plants that can be grown in liquid cultures and which regenerate either shoot-buds or somatic embryos. Further, since the process involves a callusing phase, there is no certainty about the clonal uniformity of the regenerated plants.

Costs at the shoot multiplication and rooting stages can be reduced by transferring shoot clumps, rather than individual shoots. The eventual separation of shoots at stage III inevitably takes time and labor. The introduction of some new techniques for the plants for which protocols are available, can increase efficiency (in terms of the multiplication rate or the increasing survival percentage), so as to reduce the cost and thus broaden the range of plants that can be propagated economically *in vitro* (Debergh and Maene, 1981).

For some plants, it is very difficult to free explants from microbial contaminants. Growing mother plants in a greenhouse at high humidity gives rise to fresh growths which are less contaminated. By this method, 100 percent healthy cultures of *Ficus lyrata* were obtained in contrast to the failure to obtain a healthy culture from field-grown material.

For some plants, it has been possible to treat the shoots formed in cultures as mini-cuttings and to root them *in vivo*. A number of methods have been developed for rooting *in vivo*. Usually the shoots are treated with a high concentration of auxins (either by initially saturating the potting mix with auxin solution or dipping the basal cut-end in the rooting powder (for details, see Debergh and Maene, 1981). Some species root equally well both *in vivo* and *in vitro* as in *Rhododendron* (Kyte and Briggs, 1979) and *Rubus ideaus* (Welander, 1985). However, in other species such as *Vaccinium corymbosum* (Cohen and Elliott, 1979) and *Feijoa sellowiana* (Bhojwani et al., 1987), rooting is much better *in vivo*. However, *in vivo* rooting is not always successful. For example, attempts to root *in vitro* multiplied *Leucaena* shoots *in vivo* resulted in a hundred percent loss of shoots due to microbial infection (Dhawan, unpublished).

Market Structure

The size of market for micropropagated plants is a subject of some debate. There has been a steady rise in the total number of plants produced annually from tissue culture (both in terms of number of species and plants of a particular species). This is especially true in the case of ornamentals, where a new variety or morphotype appears for which alternative methods of propagation are either not available or are equally expensive. In case of some crop plants such as potato, tissue culture techniques are commercially exploited. The crop is planted as a vegetative unit, known as seed-piece or tuber-section. The potato crop is plagued by a large number of pests which can spread through seed; therefore, seed-piece quality standards have been set by the certification agencies. This has now been overtaken by tissue culture to mass-produce the early stock of potato seed. The field is currently expanding through the addition of a greenhouse tuber production phase, as a means of providing the grower with a field-plantable unit that does not require special handling.

On a commercial scale, plant tissue culture is being most widely used for the clonal propagation of ornamental plants. Over 250 commercial laboratories exist in the USA and Canada alone, of which some 10 are producing over five million plants every year. The market for tissue culture-propagated plants is largely determined by their cost. Since the process is labor-intensive, the test-tube plants are about ten times more expensive than seed-raised plants. It is apparent that the market for plants from tissue culture could expand dramatically if the prices are reduced. The growth of the industry will be largely dependent on the technological advancements, especially in the field of automation and mechanisation.

The structure of total annual costs of producing plants by tissue culture can be depicted by the following expression:

$$TC = (K_c + K_b)(i+d) + E + S + L_k$$

where TC = total annual cost of production

$\quad K_c$ = total cost of equipment

$\quad K_b$ = total cost of buildings and infrastructure

$\quad i$ = annual rate of interest applicable to investments

$\quad d$ = annual rate of depreciation

$\quad E$ = annual cost of power consumed

$\quad S$ = annual cost of consumables and supplies

$\quad L_k$ = annual cost of labor in the kth process activity

$\quad k$ = an integer denoting the number of activities covered in the process, such as collection and preparation of explants, media preparation, etc.

There are obvious economies of scale in the production of tissue culture plants, but there is no empirical evidence on which to base any estimates in the Indian context. The Tata Energy Research Institute has collected some information on a proposed Tissue Culture Pilot Plant facility, which can be treated as a basis for estimating the economics of tissue culture plant production. The facility is designed for a total annual production volume of one million plants.

The cost-relationship defining the operation of this plant would be:

$$TC = (9+2.5)(0.012+0.01) + 0.9 + 0.4 + (0.225 + 0.525 + 0.225 + 0.375 + 0.150)$$

where K_c = Rs. 9.000 million

$\quad K_b$ = Rs. 2.500 million

$\quad i$ = Rs. 0.012 million

d = Rs. 0.010 million

E = Rs. 0.900 million

S = Rs. 0.400 million

L_1 = Rs. 0.225 million (initiation of cultures)

L_2 = Rs. 0.525 million (shoot multiplication)

L_3 = Rs. 0.225 million (rooting)

L_4 = Rs. 0.375 million (hardening)

L_5 = Rs. 0.150 million (transplantation)

Hence, TC = Rs 5.33 million. For a total annual production of one million plants, the cost per plant would then work out to Rs. 5.33. At this level, the operation may not be viable, and therefore, if in a country like India, large-scale production of plants by tissue culture has to compete with conventional methods, much engineering efforts would be required to evolve designs and methods to develop the commercialisation of tissue culture as a process. Also much larger units, perhaps for producing five to 10 million plants per year, would have to be designed for deriving the full benefits of economies of scale. Thus, priorities can be laid down for further work in this area in a developing country such as India.

Capabilities in engineering need to be developed, which would enable the design and construction of tissue culture facilities that optimise costs under Indian conditions. For instance, in the pilot plant discussed earlier, almost half the cost of production would be related to equipment and building costs. Processes and techniques which minimise these costs would be desirable.

Tissue culture techniques should be developed for those plants where:

* planting material produced by conventional methods is either scarce or expensive (which may be the case with plants such as most species of bamboo);

* there is a large variability due to genetic factors related to yields or resistance to diseases and pests, etc.; in such cases, there would then be a great benefit in the selection of plus trees and their large-scale multiplication by tissue culture, a process that is not likely to come about through natural methods;

* there appears to be some potential for applying genetic engineering or molecular biology techniques that may provide plants with certain desirable characteristics, which could then be multiplied by tissue culture.

While labor may appear to be a small component of costs in India in comparison to other countries, there is need for developing methods by which labor costs can be reduced, particularly during rooting, acclimatisation and hardening.

The present level and structure of tissue culture costs should not act as a barrier in further efforts in this field. A strategy, as detailed above, would be in the interests of rapid and large-scale afforestation and raising biomass yields in India.

REFERENCES

Anderson, W.C., Meagher, C.W. and Nelson, A.G., 1977, Cost of propagating broccoli plants through tissue culture, *Hortscience,* 12: 543-544.

Bhojwani, S.S., 1980, Micropropagation method for a hybrid willow (*Salix matsudara olba* N2 - 1002), *N.Z.J. Bot.,* 18: 209-214.

Christie, C.B., 1978, Rapid propagation of aspens and silver poplars using tissue culture techniques, *Proc. Int. Plant Prop. Soc.,* 28: 255-260

Debry, L., 1986, Robots in plant tissue culture, *IAPTC Newslett.,* 49: 2-22.

Dhawan, V. and Bhojwani, S.S., 1985, *In vitro* vegetation propagation of *Leucaena leucocephala* (Lam.) de Wit, *Plant Cell Rep.,* 4: 315-318

Goh, C.J., 1982, Development of orchid tissue culture in southeast Asian Countries. In: *Tissue Culture of Economically Important Plants* (A.N. Rao, ed.) pp. 205-209, COSTED and ANBS, Nat. Univ. Singapore.

Gupta, P.K., Mascarenhas, A.F. and Jagannathan, V., 1981, Tissue culture of forest trees - Clonal propagation of mature trees of *Eucalyptus citriodora* Hook, by tissue culture, *Plant Sci. Lett.,* 20: 195-201.

Hasnain, S., Pigeon, R. and Overend, R.P., 1986, Economic analysis of the use of tissue culture for rapid forest improvement. *The Forestry Chronicle,* Aug. 1986: 240-245.

Lakshmi Sita, G., 1986, Sandalwood (*Santalum album* L.), In: *Biotechnology in Agriculture and Forestry 1. Trees I* (Y.P.S. Bajaj, ed.) pp. 363-374, Springer-Verlag, Berlin.

Levin, R., Gaba, V., Tal, B., Hirsch, S., De-Nola, D. and Vasil, I.K., 1988, Automated plant tissue culture for mass propagation, *Biotechnology,* 6: 1035-1040.

Maene, L. and Debergh, P. 1985, Liquid medium additions to established tissue culture to improve elongation and rooting *in vitro, Plant Cell Tissue Organ Cult.,* 5: 23-33.

Sluis, C.J. and Wolker, K.A., 1985, Commercialization of plant tissue culture propagation, *IAPTC Newslett.,* 47: 2-12.

Smith, R., 1986, Radiata pine (*Pinus radiata* D. Don), In: *Biotechnology in Agriculture and Forestry 1. Trees I* (Y.P.S. Bajaj, ed.) pp. 274-290, Springer-Verlag, Berlin.

Suwal, B., Karki, A. and Rajbhandary, S.B., 1988, The *in vitro* proliferation of forest trees, *Silvae Genet.,* 37: 26-28.

Tisserat, B. and Vandercook, C.E., 1985, Development of an automated plant tissue culture system, *Plant Cell Tissue Organ Cult.,* 5: 107-117.

Weathers, and Giles, K.L., 1987, Regenaration of many varieties of plants using a novel method of *in vitro* cell culture, *In Vitro Cell Dev. Biol.,* 23: 69 (Abstract)

IV

NITROGEN FIXATION STUDIES IN FORESTRY

23

MICROPROPAGATION AND NODULATION OF TREE LEGUMES

Vibha Dhawan

ABSTRACT

Considerable progress has been made over the last decade in the development of aseptic techniques for multiplying superior and novel plants. While micropropagation methods are being widely adopted for ornamental and horticultural species, their application in forestry continues to be meagre. This paper describes the constraints and potential of tree tissue culture with special emphasis on tree legumes. Further, microbial associates, which can considerably improve the productivity of trees, have not been well-identified. The initial studies on nodulation of micropropagated plants of *Leucaena leucocephala* with *Rhizobium* strain NGR8 are described with a speculation of potential increase in transplantation success and biomass production.

Vibha Dhawan, Tata Energy Research Institute, 90 Jor Bagh, New Delhi - 110 003, India.

INTRODUCTION

Forestry has not been in the mainstream of research and development of plant sciences and therefore, could not take advantage of the recent advances made in the biological sciences in general.

However, with increasing per capita energy consumption all over the world and increasing population in developing countries, it is clear that considerable efforts are required to increase the biomass production per unit area. The over-exploitation of our natural resources has resulted in vicious cycles of floods and droughts. This has further accentuated the crisis of renewable energy resources. Now there is a general awareness all over the world of the need to afforesting the existing wastelands and also increase biomass production per unit area.

In forestry, cloning of trees through tissue culture is all the more important because most forest species are cross-pollinated and, therefore, the progenies raised through seeds represent half sibs and thus are heterogeneous. Furthermore, most tree species lose the ability to form roots with age and as a result, the time by which trees are evaluated for the desirable characteristics, it is too late to propagate them vegetatively by conventional methods. To increase the biomass production without increasing the forest land, the only alternative is to plant superior genotypes provide efficient microbial associates wherever required and follow good nursery and management practices. Tree legumes, with inherent capacity to fix atmospheric nitrogen, are the prime candidates for afforesting the degraded lands because apart from being hardy and robust, they also improve the soil nutrient status over a period of time.

TISSUE CULTURE OF TREE LEGUMES

There are very few reports of successful propagation of leguminous trees through tissue culture. Also, most of the publications deal with root and/or shoot differentiation from seedling-derived explants and their calli (Table 1). *In vitro* studies on *Albizzia lebbeck* (Gharyal and Maheshwari, 1981; Upadhyaya and Chandra, 1983; Lakshmana Rao and De, 1987), *Dalbergia lanceolaria* (Anand and Bir, 1983), *Dalbergia latifolia* (Sankara Rao, 1986; Ravishankar Rai and Chandra, 1988), *Robinia pseudo-acacia* (Chalupa, 1983), *Sesbania sesban* (Khattar and Mohan Ram, 1982) and *Sesbania grandiflora* (Khattar and Mohan Ram, 1983) were aimed at micropropagation but the objective was not achieved. The leguminous species where micropropagation from adult tissues has been reported are: *Acacia koa* (Skolmen and Mapes, 1976,78); *Dalbergia latifolia* (Lakshmi Sita et al., 1986), *D. sissoo* (Datta et al., 1983; Dawra et al., 1984; Mukhopadhyay and Bhojwani, unpublished) and *Leucaena leucocephala* (Dhawan and Bhojwani, 1985,1987; Goyal et al., 1985),

Table 1: Current status of tissue culture of tree legumes

Species	Seedling/ Adult	Explant	Shoot Multi-plication*	Rooting	Transplan-tation	Reference
Acacia koa	Seedling	Shoot-tip	AxB	+	+	Skolmen and Mapes, 1978; Skolmen, 1986
Acacia mangium	Seedling	Cotyledonary node	AxB	+	+	Crawford and Hartney, 1986
Acacia nilotica	Adult	Stem segment	Callus	+	-	Mathur and Chandra, 1983
Acacia stenophylla	Seedling	Cotyledonary node	AxB	-	-	Crawford and Hartney, 1986
Albizzia lebbeck	Seedling	Root, hypo-cotyl, cotyle-don and leaflet	SE	+	+	Gharyal and Maheshwari, 1981
Albizzia lebbeck	Seedling	Root and hypo-cotyl	Callus	+	-	Upadhyaya and Chandra, 1983
Albizzia lebbeck	Seedling	Hypocotyl, leaf, stem segment	Callus	+	+	Lakshmana Rao and De, 1987
Dalbergia lanceolaria	Seedling	Hypocotyl, leaf etc.	Callus	+	-	Anand and Bir, 1984
Dalbergia latifolia	Seedling	Cotyledonary leaf, hypocotyl, stem and root	Callus	+	-	Nataraja and Sudhadevi, 1985
Dalbergia latifolia	Adult	Shoot segments	Callus	+	+	Lakshmi Sita, et al., 1986
Dalbergia latifolia	Adult	Shoot-tip	Callus	+	+	Ravishankar Rai and Chandra, 1988
Dalbergia sisoo	Seedling	Root segment	AdB	+	-	Mukhopadhyay and Mohan Ram, 1981
Dalbergia sissoo	Adult	Nodal segment	Callus	+	-	Datta and Datta, 1983
Dalbergia sissoo	Adult	Nodal segment		+	-	Datta et al., 1983
Dalbergia sissoo	Seedling	Nodal segment	AxB	+	+	Suwal et al., 1988
Leucaena leucocephala	Adult	Nodal segment	AxB	+	+	Goyal et al., 1985

Species	Seedling/ Adult	Explant	Shoot Multi- plication	Rooting	Transplan- tation	Reference
Leucaena leucocephala	Adult	Nodal segment	AxB	+	+	Dhawan and Bhojwani, 1985, 1987
Leucaena leucocephala	Adult	Nodal segment	AxB	+	+	Datta and Datta, 1985
Mimosa pudica	Seedling	Cotyledon, hypocotyl, leaflets and shoot apices	Callus	+	-	Gharyal and Maheshwari, 1982
Prosopis alba	Seedling	Nodal segment	AxB	-	-	Tabone et al., 1986
Prosopis chilensis	Seedling	Nodal segment and shoot-tip	Callus	+	-	Jordan, 1987
Prosopis cineraria	Seedling	Hypocotyl	Callus	+	+	Goyal and Arya, 1981
Prosopis cineraria	Adult	Nodal segment	AxB	+	+	Goyal and Arya, 1984
Prosopis juliflora	Adult	Nodal segment	AxB	-	-	Wainwright and England, 1987
Prosopis tamarugo	Seedling	Nodal segment and shoot-tip	Callus	+	-	Jordan, 1987
Sesbania grandiflora	Seedling	Cotyledon and hypocotyl	Callus	+	+	Khatter and Mohan Ram, 1983
Sesbania sesban	Seedling	Cotyledon and hypocotyl	Callus	-	-	Khattar and Mohan Ram 1982
Tamarindus indica	Seedling	Various seedling explants	Callus	+	-	Mascarenhas et al., 1987

Abbreviation[*]: AdB : Adventitious branching AxB - Axillary branching, SE - Somatic embryogenesis

Except for Skolmen and Mapes (1978) and Dhawan and Bhojwani (1987), none of the publications refer to the nodulation of plants raised *in vitro*. In *Acacia koa*, Skolmen and Mapes (1978) reported that out of the 82 micropropagated plants that were transferred to the field, 16 developed nodules after three months of transplantation. We had earlier reported that inoculation of micropropagated plants of *Leucaena leucocephala* with an efficient strain of *Rhizobium* prior to their transfer to the field not only resulted in nodule formation but also enhanced transplantation success (Dhawan and Bhojwani, 1987).

In our institute, we are developing aseptic methods for cloning of species that are suitable for semi-arid regions. Tissue culture techniques being expensive, cannot be applied for large-scale plantations on wastelands. Also, tissue culture plants are more delicate than the seedling plants and thus, cannot withstand the harsh conditions prevailing in wastelands. Therefore, the protocols developed will be applied on a limited-scale to clone few hundred copies which will then be utilized for raising seed orchards.

Tissue Culture of *Acacia* Species

In developing countries, the acute shortage of fuelwood in rural communities and the need of integrating multipurpose trees into agricultural systems, have forced a reappraisal of tree and shrub species. Most acacias produce excellent firewood and have the potential to become a component of farming systems providing wood, shade and shelter, and improving soil. Also, they are adapted to a wide range of tropical soils and with inherent capacity to fix biological nitrogen, stand as prime candidates in afforestation.

The germplasm of ten *Acacia* species (viz. *Acacia ampliceps, A. auriculiformis, A. bivenosa, A. holosericea, A. ligulata, A. maconochieana, A. salicina, A. sclerosperma, A. stenophylla* and *A. victoriae*) was obtained from CSIRO (Commonwealth Scientific and Industrial Research Organization), Australia. A field experiment for selecting the suitable species for semi-arid regions of India was initiated and simultaneously, attempts were made to develop protocols for micropropagation with seedling material. Aseptic seedlings were raised on MS basal medium, explants with pre-formed meristems (*viz.* cotyledonary node, nodal segments and terminal portion) were excised from three-week old seedlings and multiplied on medium supplemented with BAP, Kn, 2-ip, etc. Basal media used were: MS (Murashige and Skoog, 1962), MS half strength (inorganic salts reduced to half strength) and B_5 (Gamborg et al, 1968).

Tissue Culture of *Leucaena* Species

Leucaena leucocephala (Lam.) de Wit is perhaps the most talked about leguminous tree of the tropics as it offers a wide range of uses. It produces nutritious forage, firewood, timber and rich organic fertiliser. It coppices readily, producing vigorous sprouts, thus constituting a reliable source of renewable energy. Indeed, the total annual yields of *Leucaena* are among the highest ever recorded. The species, however, is beset with certain problems. The tree is susceptible to frost and *jumping plant lice (Heteropsylla cubana)* which has severely damaged *Leucaena* plantations in several parts of the world. Fortunately, *L. leucocephala* hybridises readily with other *Leucaena* species including those which are resistant to frost and psyllids.

Through conventional breeding techniques, several *Leucaena* hybrids have been developed by NFTA (Nitrogen Fixing Tree Association). These hybrids still await commercial exploitation because they are semi-sterile and cannot be propagated by vegetative means. Some of the hybrids for which protocol up to shoot multiplication stage has been standardised are: *L. leucocephala* x *L. diversifolia*; *L. diversifolia* x *L. leucocephala*; *L. (diversifolia* x *pallida)* x *L. leucocephala*; *L. leucocephala* x *L. esculenta*; *L. leucocephala* x *L.pallida* and *L. pulverulenta* x *L. leucocephala.*

The protocol for *L. leucocephala* has been described in detail and modifications for hybrids are given in the following paragraphs.

Single-node segments either excised from two-week-old aseptically raised seedlings or from field-grown adult trees were used as the explants. For raising healthy cultures from mature trees, the explants were first rinsed in ethanol, air-dried and treated with one percent solution of Cetavlon for five minutes followed by thorough washing under running tap water. The explants were surface sterilised with one percent (v/v) sodium hypochlorite solution for 15 min. After three to four rinses in sterile distilled water, the cut ends were trimmed and inoculated on LSM (*Leucaena* shoot multiplication medium, Table 2). The cultures were incubated at 30°C under diffused light (12 Wm^{-2}).

The bud-break was observed in a week's time and within three weeks, the solitary shoot attained a height of six to seven cm. The single node and terminal cuttings of the solitary shoot that developed in primary cultures were used for further *in vitro* multiplication of shoots. In order to determine the optimum medium for shoot multiplication, MS medium was modified by varying its inorganic constituents, testing different sources of carbon skeleton at different concentrations, supplementing the medium with several auxins (IAA, IBA and NAA) and cytokinins (BAP, Kn, 2-ip and AdSO$_4$) singly or in combinations, and adding some other growth adjuvants (adenine, amino acids, phloroglucinol and polyamines). The presence of cytokinin was found to be essential for shoot multiplication and BAP (3×10^{-6}M) proved superior to all other cytokinins tested.

The addition of auxins inhibited shoot growth. Macronutrient composition of the MS medium was adjudged superior to some other popular tissue culture media. However, all the micronutrients and organics included in the MS medium were not essential for sustained growth and multiplication of *Leucaena* shoots. From the micronutrients, the deletion of CuSO$_4$.5H$_2$O, Na$_2$MoO$_4$.2H$_2$O and KI did not adversely affect the rate of shoot multiplication. However, ZnSO$_4$.7H$_2$O and H$_3$BO$_3$ were absolutely essential. Of the organic nutrients, thiamine accounted for the full effect of all the vitamins recommended in the MS medium. Myo-inositol was not necessary. Optimum level of sucrose for *Leucaena* shoot cultures was four percent as against three percent present in the original MS medium. AR grade sucrose was successfully replaced by a commercial form of table sugar (sugar cubes) of sufficient purity, which is ten times cheaper than the AR grade sucrose.

To improve shoot growth and prevent premature abscission of leaves, various amino acids, polyamines and phloroglucinol were added to the medium individually. Of these, addition of glutamine, adenine and putrescine, in the presence of BAP, prevented leaf abscission. Glutamine also slightly enhanced the multiplication rate. Pholoroglucinol was inhibitory.

Table 2: Composition of LSM (*Leucaena* shoot multiplication)
medium derived for MS medium

Constituents	Amount (mg l⁻¹)
Macronutrients	As in MS
Micronutrients	
$CoCl_2.6H_2O$	0.025
H_3BO_3	6.200
$MnSO_4. 4H_2O$	22.300
$ZnSO_4.7H_2O$	8.600
Organics	
Nicotinic acid	0.5
Pyridoxine HCl	0.5
Thiamine HCl	0.1
Glycine	2.0
Glutamine	58.4
Sugar cubes	40,000
BAP	$3 \times 10^{-6}M^*$

* The hybrids responded better on slightly higher doses of BAP ($6 \times 10^{-6}M$).

Raising the incubation temperature from 25°C to 30°C resulted in increased shoot growth and multiplication rate. Still higher temperature (35°C) was detrimental.

Shoots multiplied *in vitro* could not be rooted *in vivo*. However, they readily formed roots in cultures. Auxin was essential for rooting. On the basis of the rooting percentage and morphology of the roots, IAA ($5 \times 10^{-6}M$) proved to be the best. Reducing the concentration of major and minor salts of MS medium to half was beneficial for rooting. In order to confer an autotrophic character to the plant at the rooting stage, effect of deleting the organic nutrients of MS medium, individually and in different combinations, was studied. However, full complement of the organics was essential for optimal rooting. Of all the vitamins and amino acids, the deletion of glycine was most inhibitory.

Attempts to transplant micropropagated plants directly to the pots or polythene bags were unsuccessful, probably because of inadequate facilities to maintain high humidity around the plants immediately following transplantation. This necessitated the introduction of pre-transplant hardening *in vitro* in the micropropagation protocol.

Various methods were tried for pre-transplant hardening. The one which gave better results involved the transfer of rooted plantlets to 250 ml conical flasks or feeding bottles containing quartz sand and irrigating them with quarter-strength MHB (Modified Hely and Brockwell's) salt solution. The flasks were capped with aluminium foil/lid. After 10 days, the foil/lid was removed and the flasks/bottles were transferred to comparatively high light intensity (18 Watts m^{-2}). After two weeks, the plantlets were transferred to sand in large polythene bags and maintained in shade under natural conditions. Following this protocol, 63 percent plantlets of adult tree origin and 83 percent plantlets of seedling origin survived transplantation.

Morpho-physiological studies of the cultured plants and their comparison with the field-grown trees and intermediate stages of acclimatisation revealed that the low survival of plants grown *in vitro* following transplantation is due to rapid loss of water. This has been correlated with poor epiculticular wax deposition under culture conditions. During acclimatization period, the amount of epicuticular wax gradually increased and the rate of water loss decreased. The starch grains normally found in the mesophyll cells of *Leucaena* leaves were absent in the leaves formed *in vitro*, suggesting that such plants are photosynthetically inefficient. Interestingly, the mesophyll tissue at the shoot multiplication stage appeared as in normal leaves, but the leaves formed at the rooting stage exhibited poor differentiation of leaf tissues with large intercellular spaces. These abnormal features persisted until the hardening stage. Normalisation of leaf tissue, which started after the hardening stage, was preceded by the appearance of starch grains in their cells.

In Vitro Nodulation

The objective of this study was to nodulate the micropropagated plants before transplantation, with the assumption that nodulated plants would have better chances of survival upon transplantation than non-nodulated plants. They were also presumed to perform well, even in the soils devoid of an effective *Rhizobium* strain. Since micropropagated plants have low photosynthetic capacity, they were not regarded as ideal material for exploratory work on nodulation. Therefore, experiments to standardise the conditions for *in vitro* nodulation and to select the most efficient of the available *Rhizobium* strains were done with seedlings. The knowledge gained from these studies was applied to nodulate micropropagated plants.

The nodulation responses of K-8 and K-28 seedlings to the seven local strains of *Rhizobium* (Lcn 1, Lcn 2, Lcn 3, Lcn 7, Lcn 8, Lcn 9 and K-28-2) were found to be similar. Symbiotic effectivity of the different strains was determined on the basis of morphological (number, size and distribution of nodules), growth (length of root and shoot and their fresh and dry weights) and biochemical (nitrogenase activity of the nodules) parameters. Considering all the parameters, Lcn 8 and Lcn 9 were found to be the most effective among local strains. However, when these strains were compared with two exotic strains of *Rhizobium* (NGR 8 and CB 81), widely used for *Leucaena*, NGR 8 proved superior to all others. NGR 8 was, therefore, used for the nodulation of micropropagated plants.

Two methods were followed for the nodulation of micropropagated plants : (i) Gibson's method, and (ii) the pre-transplant hardening method described earlier. In the latter case, *Rhizobium* was suspended in the liquid medium used for irrigating the plants soon after their transfer from the rooting medium to sand. The initial mortality in the first method was very high (50 percent) but all the surviving plants nodulated. In the second protocol, which ensured higher survival frequencies, 80-90 percent plants formed nodules. Unlike seedlings which formed nodules within two weeks after inoculation, the mircropropagated plants took almost five weeks for the nodules to be visible. All the nodulated plants, irrespective of the nodulation method, survived when transferred to soil.

On the basis of above experiments, a medium was formulated for multiplication of *Leucaena leucocephala* (Table 2). This essentially is a modification of the MS medium.

Tissue Culture of *Ougeinia dalbergioides*

It is a moderate sized deciduous tree often with a crooked stem, seldom exceeding 0.9-1.2 m in girth and 9-12m in height. A few individuals are superior in growth with girth of 2.1 m or more and a height of 18 m. The species is cross-pollinated and natural reproduction is by seeds. However, the tree does not set-seed abundantly every year. Also, being an excellent fodder, it is heavily lopped resulting in poor seed-set. Although, vegetative propagation by root suckers is possible, such plants attain much smaller height than the seed-raised plants.

Seeds from an elite tree growing in Tehri Garhwal region of India were collected and aseptic seedlings were raised from them. Explants with pre-formed meristem were excised and cultured on MS (Murashige and Skoog, 1962) and B₅ (Gamborg et al. 1968) medium supplemented with different cytokinins (BAP, 2-ip and Kn) either alone or in combinations. On MS + BAP $(3 \times 10^{-6}M)$ + Kn $(3 \times 10^{-6}M)$, four-to five-fold multiplication was achieved every four weeks. Further experiments are in progress to optimise shoot multiplication and induce rooting of shoots developed *in vitro*.

CONCLUSION

The potential uses of micropropagation in forestry have long been recognised but, as compared to herbaceous plants, the progress in this area has lagged behind (Winton, 1978; Mott, 1981; Sommer and Wetzstein, 1984; Thorpe and Biondi, 1984). With tree species, in nature, one is confronted with a serious problem that by the time the elite nature of a tree is established, it has past the phase when it could be propagated vegetatively. This problem is also encountered in tissue cultures. Consequently, tissue culture of most of the tree species is initiated with juvenile explants. Of the 100 reports on tree tissue culture for which information is available (Sommer and Wetzstein, 1984), only 14 deal with adult explants, and it is only during the past decade that some success has been achieved with the micropropagation of adult trees (Dhawan and Bhojwani, 1986).

In spite of the realisation that leguminous trees, with natural capacity to fix atmospheric nitrogen, protein-rich foliage, pods and seeds, and general robustness, are likely to benefit the developing countries enormously, there is a general lack of research in this field (Table 1). A program on conventional breeding for higher biomass yield and increased rates of nitrogen fixation, coupled with *in vitro* techniques of cloning, would result in exploitation of hybrids developed by breeding techniques. Superior clones, if raised in seed orchards, would cross-breed and produce superior quality seeds also overcoming the danger of monoculture. The selection of efficient strains of *Rhizobium* could further enhance biomass production.

REFERENCES

Anand, M. and Bir, S.S., 1984, Organogenetic differentiation in tissue cultures of *Dalbergia lanceolaria, Curr. Sci.,* 53: 1305-1307.

Crawford, D.F. and Hartney, V.J., 1986, Micropropagation of *Acacia mangium* and *Acacia stenophylla.* In: *Australian Acacias in Developing Countries, ACIAR Proc. No. 16* (J.W. Turnbull, ed.), pp. 64-65, Proc. Intl. Workshop held at the Forestry Training Centre, Gympie, Queensland., Australia.

Datta, K. and Datta, S.K., 1985, Auxin + KNO_3 induced regeneration of leguminous trees - *Leucaena leucocephala* through tissue culture, *Curr. Sci.,* 54: 248-250.

Datta, S.K. and Datta, K., 1983, Auxin induced regeneration of forest tree - *Dalbergia sissoo* Roxb. through tissue culture, *Curr. Sci.,* 52: 434-436.

Datta, S.K., Datta, K. and Pramanik, T., 1983, *In vitro* clonal multiplication of mature trees of *Dalbergia sissoo* Roxb., *Plant Cell Tissue Organ Cult.,* 2: 15-20.

Dhawan, V. and Bhojwani, S.S., 1985, *In vitro* vegetative propagation of *Leucaena leucocephala* (Lam.) de Wit, *Plant Cell Rep.,* 4: 315-318.

Dhawan, V. and Bhojwani, S.S., 1986, Micropropagation in crop plants, *Glimpses in Plant Res.,* 7: 1-98.

Dhawan, V. and Bhojwani, S.S., 1987, *In vitro* nodulation of micropropagated plants of *Leucaena leucocephala* by *Rhizobium, Plant Soil,* 103: 214-216.

Gamborg, O.L., Miller, R.A. and Ojima, K., 1968, Nutrient requirements of suspension cultures of soybean root cells, *Exp. Cell Res.,* 50: 151-158.

Gharyal, P.K. and Maheshwari, S.C., 1981, *In vitro* differentiation of somatic embryoids in a leguminous tree - *Albizzia lebbeck* L., *Naturwissenschaften,* 67: 379.

Gharyal, P.K. and Maheshwari, S.C., 1982, Plantlet formation in tissue cultures of sensitive plant *Mimosa pudica* L., *Z. Pflanzenphysiol.,* 105: 179-182.

Gibson, A.H., 1963, Physical environment and symbiotic nitrogen fixation. I. The effect of root temperature on recently nodulated *Trifolium subterranieun* L. plants, *Aust. J. Biol. Sci.,* 16: 37-49.

Goyal, Y. and Arya, H.C., 1981, Differentiation in cultures of *Prosopis cineraria* Linn., *Curr. Sci.,* 50: 468-469.

Goyal, Y. and Arya, H.C., 1984, Tissue culture of desert trees: I. Clonal multiplication of *Prosopis cineraria* by bud culture, *J. Plant Physiol.,* 115: 183-189.

Goyal, Y., Binghan, R.L. and Felker, P., 1985, Propagation of the tropical tree, *Leucaena leucocephala* K67, by *in vitro* bud culture, *Plant Cell Tissue Organ Cult.,* 4: 3-10.

Hely, F.W. and Brockwell, J., 1962, An exploratory study of the ecology of *Rhizobium meliloti* in inland South Wales and Queensland, *Aust. J. Agric. Res.,* 13: 864-879.

Jordan, M., 1987, *In vitro* culture of *Prosopis* species. In: *Cell and Tissue Culture in Forestry, Vol. 3, Case Histories: Gymnosperms, Angiosperms and Palms* (J.M. Bonga and D.J. Durzan, eds.), pp. 70-384, Martinus Nijhoff, Dordrecht.

Khattar, S. and Mohan Ram, H.Y., 1982, Organogenesis in the cultured tissues of *Sesbania sesban,* a leguminous shrub, *Indian J. Exp. Biol.,* 20: 216-219.

Khattar, S. and Mohan Ram, H.Y., 1983, Organogenesis and plantlet formation *in vitro* in *Sesbania grandiflora* (L.) Pers., *Indian J. Exp. Biol.,* 21: 251-253.

Lakshmana Rao, P.V. and De, D.N., 1987, Tissue culture propagation of tree legume *Albizzia lebbeck* (L.) Benth., *Plant Sci.,* 51: 263-267.

Lakshmi Sita, G., Chattopadhyay, S. and Tejovathi, D.H., 1986, Plant regeneration from shoot callus of rosewood, *Plant Cell Rep.,* 5: 266-268.

Mascarenhas, A.F., Nair, S., Kulkarni, V.M., Agrawal, D.C., Khuspe, S.S. and Mehta, U.J., 1987, Tamarind. In: *Cell and Tissue Culture in Forestry, Vol. 3, Case Histories: Gymnosperms, Angiosperms and Palms* (J.M. Bonga and D.J. Durzan, eds.), pp. 316-325, Martinus Nijhoff, Dordrecht.

Mathur, I. and Chandra, N., 1983, Induced regeneration in stem explants of *Acacia nilotica, Curr. Sci.,* 52: 882-883.

Mott, R.L., 1981, Trees. In: *Cloning Agricultural Plants via In Vitro Techniques* (B.V. Conger, ed.), pp. 217-256, CRC Press, Boca Raton, Florida.

Mukhopadhyay, A. and Mohan Ram, H.Y., 1981, Regeneration of plantlets from excised roots of *Dalbergia sissoo, Indian J. Exp. Biol.,* 19: 1113-1115.

Murashige, T. and Skoog, F., 1962, A revised medium for rapid growth and bioassays with tobacco tissue cultures, *Physiol. Plant.,* 15: 473-497.

Nataraja, K. and Sudhadevi, A.M., 1985, Induction of plantlets from seedling explants of *Dalbergia latifolia* Roxb. *in vitro, Beitr. Biol. Pflanz.,* 59: 341-350.

Ravishankar Rai, V. and Jagadish Chandra, K.S., 1988, *In vitro* regeneration of plantlets from shoot callus of mature trees of *Dalbergia latifolia, Plant Cell Tissue Organ Cult.,* 18: 77-83.

Skolmen, R.G., 1986, Acacia (*Acacia koa* Gray). In: *Biotechnology in Agriculture and Forestry, Vol. 1, Trees I* (Y.P.S. Bajaj, ed.), pp. 375-384, Springer-Verlag, Berlin.

Skolmen, R.G. and Mapes, M.O., 1976, *Acacia koa* Gray plantlets from somatic callus tissue, *J. Hered.,* 67: 114-115.

Skolmen, R.G. and Mapes, M.O., 1978, After care procedures required for field survival of tissue culture propagated *Acacia koa, Proc. Int. Plant Prop. Soc.,* 28: 156-164.

Sommer, H.E. and Wetzstein, H.Y., 1984, Hardwoods. In: *Handbook of Plant Cell Culture, Vol. 3, Crop Species* (P.V. Ammirato, D.A. Evans, W.R. Sharp and Y. Yamada, eds.), pp. 511-540, Macmillan Publishing Company, New York.

Suwal, B., Karki, A. and Rajbhandary, S.B., 1988, The *in vitro* proliferation of forest trees, *Silvae Genet.,* 37: 26-28.

Thorpe, T.A. and Biondi, S., 1984, Conifers. In: *Handbook of Plant Cell Culture, Vol. 2, Crop Species* (W.R. Sharp, D.A. Evans, P.V. Ammirato and Y. Yamada, eds.), pp. 435-470. Macmillan Publishing Company, New York.

Tobone, T.J., Felker, P., Bingham, R.L., Reyes, I. and Loughrey, S., 1986, Techniques in the shoot multiplication of the leguminous tree *Prosopis alba* and clones B_2V50. In: *Tree Planting in Semi-Arid Regions* (P. Felker, ed.), pp. 191-200, Elsevier Science Publishers, Amsterdam.

Upadhyaya, S. and Chandra, N., 1983, *Shoot and plantlet formation in organ and callus cultures of Albizzia lebbeck* Benth., *Ann. Bot. (Lond.),* 52: 421-424.

Wainwright, H. and England, N., 1987, The micropropagation of *Prosopis juliflora* (Swartz) DC: Establishment *in vitro, Acta Hortic.,* 212: 49-53.

Winton, L.L., 1978, Morphogenesis in clonal propagation of woody plants. In: *Frontiers of Plant Tissue Culture* (T.A. Thorpe, ed.), pp. 419-426, Univ. Calgary Press, Calgary.

24

ROLE OF MYCORRHIZAE IN FORESTRY

H.S. Thapar

ABSTRACT

This paper highlights the importance of mycorrhizal inoculation, in afforestation program in forest tree species and discusses the experimental evidence on growth improvement by mycorrhizal inoculation with pure cultures, enhanced nitrogen fixation by ectomycorrhizal fungi/seedlings, host/fungus specificity and rhizobia-VAM interactions. It deals briefly with the potential of mycorrhizal technology as an aid to produce quality planting stocks with efficient mycorrhizal root systems for critical planting sites. Priority areas in mycorrhizal research are also outlined for further investigation.

INTRODUCTION

Mycorrhiza constitutes the most striking example of symbiosis in the plant kingdom. Mycotrophy represents a specialised mode of tree nutrition, the significance of which has been realised during the last few decades. Mycorrhizae help in the faster uptake and translocation of water and nutrients, particularly phosphorus, nitrogen and

H.S. Thapar * Forest Research Institute and Colleges, Dehra Dun, India.

potassium, besides other elements like zinc, calcium, etc. Plants equipped with mycorrhizae are better adapted to withstand drought and invasion by pathogenic organisms in nurseries and in the field. The benefits to the hosts in forest species, chiefly beech, Douglas-fir, pines, yellow-poplar, sweet gum, hoop pine, acer, maple and green ash etc., from deficient soils are documented in excellent reviews published during the past few decades (Harley, 1959; Marks and Kozlowski, 1973; Tinker, 1982; Marx and Schneck, 1983). Mycorrhizal technology has assumed greater relevance in the planting of exotics on critical sites and averting planting failures which are well known from experiences in countries the world over.

IMPORTANCE OF MYCORRHIZAE IN NURSERIES

Raising vigorous and healthy stocks is of prime concern to nursery persons and managers all over the world. Numerous studies show that initial inoculation of nursery soil is necessary, prior to raising planting stocks. In nurseries established on agricultural sites which receive high doses of fertiliser and irrigation, the mycorrhizal fungi may be lacking or present in low densities, or the species may be less effective. Therefore, the inoculation of such soils becomes important and the introduction of a new species desirable (Mikola, 1973). In these circumstances, inoculation alone is not often an effective means of improving the growth of seedlings and the establishment of mycorrhiza, unless it is preceded by the correction of soil conditions. Soil properties such as acidity, fertility or organic matter content tend to favor the growth and multiplication of the introduced fungi. In alkaline soils, as in Chernozems, the inoculation of the larix with mycorrhizal fungi has been found to be successful only with additional acidification of soil followed by foliar sprays of micronutrients like molybdenum and zinc. These induce better mycorrhizal development, besides better plant growth (Zerling, 1958, cited by Meyer, 1973).

It has been found that the scale at which weedicides and pesticides are applied in forestry, exerts a harmful influence on mycorrhizal fungi. In such cases, mycorrhizal inoculation becomes particularly important in improving the quality of planting stocks. Lack of mycorrhizal associations in nursery seedlings results in the retardation of growth and chlorotic appearance of seedlings which are unfit for planting. The potential benefits of mycorrhiza in nurseries are in increase in height and collar diameter and fewer seedling cells at lifting time. In various nurseries in the USA, a substantial increase in terms of height and weight was observed in seedlings inoculated with mycorrhizal fungi. Inoculation with *Pisolithus tinctorius* fungi resulted in a seven percent increase in plantable seedlings and 38 percent fewer seedling culls. The potential annual saving is estimated to be US $ 847 (USDA Gen. Rept., 1980).

IMPORTANCE OF MYCORRHIZAE IN FORESTRY

Mycorrhizal infection of forest trees is essential for their healthy growth in some ecological conditions. This fact is illustrated by the failure of several attempts to grow trees in barren areas. Several researchers such as Hatch (1936), Rayner (1938), Gilmour (1958), Briscoe (1959) and Harley (1959), have summarised the experiences in Australia, Rhodesia, the Philippines, South America, New Zealand and Puerto Rico. In Puerto Rico, all attempts to introduce pines by conventional cultural practices like fertilisation, failed. However, when mycorrhizal fungi were introduced from the USA by transferring soils from successful pine stands, the results were striking: all treatments produced excellent survival and growth (Vozzo and Hacskaylo, 1971). Survival of inoculated slash pine was 85 percent and that of the control 36 percent. Further, inoculated trees grew 13 times taller than uninoculated ones in four years. Similarly, certain tree species (such as *P. sylvestris*) in Calluna heath lands in southern Britain, which had hitherto failed, successfully established, with the inoculation of mycorrhizal strains, and became tolerant to the toxic factors in Calluna soils (Handley, 1963).

Whereas the introduction of mycorrhizal fungi has led to the success of many plantations, there have been instances wherein exotics have been successfully planted without such introductions. *Eucalyptus* has been planted extensively on poor soils and grassland sites for afforestation in southern India. Chir pine (*Pinus roxburgii*) has been successfully established as an exotic on a site previously under Sal (*Shorea robusta*) forests in central India at a distance of over 1,000 km from its indigenous home. However, in both eucalyptus and chir pine, mycorrhizae developed on their own. In such cases, mycorrhizal fungi may have been present on the sites afforested (grasslands in the former case and original forests in the latter) and the trees were able to develop symbiotic associations with these fungi. It is also possible that the spores of mycorrhizal fungi were carried by the wind to the sites and eventually established mycorrhizal associations.

A plausible explanation for the poor survival and growth of seedlings without mycorrhiza on certain virgin soils, is the inadequacy of the non-mycorrhizal root system to absorb minerals in short supply, particularly phosphorus. Not only are mycorrhizae more efficient in this regard than uninfected short roots (Kramer and Wilbur, 1949), but they expose a much larger total absorbing surface (Hatch, 1937). Improved nutrition of mycorrhizal *vis-a-vis* non-mycorrhizal trees has been demonstrated by Hatch (1937), Mitchell et al. (1937), Rosendahl (1943), McComb and Griffith (1946) and others. The physiology of salt absorption by ectotrophic mycorrhiza is further examined by Harley (1959). By using radio-isotope techniques, Melin (1963) demonstrated that the lower symbionts translocate various nutrients from soil to tree and in turn take carbohydrates from the host.

There is evidence of increased phosphorus absorption in endotrophic infection (Daft and Nicolson, 1966). Mosse (1957) found that the growth of infected apple cuttings was significantly more in terms of height and weight than the uninfected ones. Baylis (1959) obtained similar results with *Griselinia littoralis*, especially on poor soils. Mosse (1957) found increased absorption of copper, iron and calcium by infected plants, while Baylis (1959) found stimulated phosphorus uptake.

Trees may be successfully grown without mycorrhiza under certain limiting conditions such as highly fertile soils (Olson, 1944; Harley, 1959). However, success could not be achieved in many trials in non-forested areas, in spite of fertiliser application and cultural practices. Success was achieved only when mycorrhiza was introduced. Briscoe (1959) found that in Puerto Rico, enhanced growth of the plant was not obtained on the application of nitrogen alone, minor elements alone, NPK or complete fertilisers, but was observed only when mycorrhizae were introduced. It has been suggested (White, 1941) that the fungal symbionts probably secrete a growth stimulant which is essential for the healthy growth of the tree. More work is needed on these lines for a better understanding of such situations.

Besides normal mycorrhizae which are beneficial to both symbionts, some abnormal structures are also encountered. These are characterised by deficient or excessive development of fungal sheath or cortical penetration and termed as pseudomycorrhiza. These have been found to be detrimental to plant growth and the fungi involved (e.g. *Rhizoctonia sylvestris*, *Mycelium radicisatrovirens*) are parasites on the host.

AFFORESTATION OF ADVERSE SITES

The superior performance of loblolly pine, *Pinus taeda* with *Pisolithus tinctorius* (PT) on coal spoils has been reported by Marx (1975, 1976, 1977, 1980). Marx and Artman (1979) reported that PT significantly improved survival and growth of both loblolly and short leaf pine on coal spoils in Kentucky and Georgia, USA. Berry (1982) confirmed the value of PT in a recent study on two strip-mined coal spoils in the southern region of the USA. On both sites, seedlings with PT mycorrhizae had greater survival, height and root collar diameter than naturally inoculated seedlings. Volume on plots with PT ectomycorrhizae were 200 percent greater in Tennessee and 380 percent greater in Alabama than the indexes of control seedlings with *Thelophora terrestris* (TT) ectomycorrhizae (Ruehle et al., 1981). The performance of loblolly pine seedlings colonised by PT and TT ectomycorrhizae, two years after planting on an amended borrow pit in the pied mont of South Carolina, revealed that containerised loblolly pine seedlings tailored with *Pisolithus tinctorius* can be successfully established and rapid growth obtained on a subsoil borrow pit amended with sewage sludge (Ruehle, 1980).

In broad-leaved species, growth improvement following inoculations with VAM fungi have been demonstrated in yellow poplar (*Liriodendron tulipifera* L; Clark, 1963); hoop pine (*Araucaria cunninghamii* Sweet; Thapar and Khan, 1985); sweet gum (*Liquidamber styracifera* L.; sugar maple (*Acer saccharum* Marsh.); green ash (*Fraxinus pennsylvanica*); box alder (*Acer negundo* L.); sycamore (*Plantanus occidentalis* L.); black cherry (*Prunus serotina* Ehrh.); and black walnut (*Juglans nigra* L.). In hoop pine, the inoculated seedlings showed significant increase in shoot height and oven dry weight in soils, sterilised by autoclaving and formalin treatment. However, such differences were not significant when compost was added to the soil. No correlation was observed between the spore population and growth improvement. In green ash, black cherry, red maple and sweet gum, the improvement in height following inoculation with *Glomus fasciculatus* was over 400 percent. In uninoculated seedlings, similar enhancement was observed in other parameters such as diameter, root weight, stem weight and leaf weight (Kormanik et al., 1982).

NITROGEN FIXATION

Taking into account other microbial symbiosis in plants such as legumes, alder and casuarina, atmospheric nitrogen fixation by mycorrhizal symbiosis was proposed in earlier research in mycorrhiza. Extensive plant growth studies by Melin (1959), however, failed to show any nitrogen fixation by mycorrhizal seedlings or fungi. However, with the advent of more sensitive assays, recent reports (Richards et al., 1971) have shown ^{15}N enrichment of mycorrhiza of *Pinus radiata*, *P. elliottii* and *P. caribaea*. The nitrogen fixation capability of some cultures of ectomycorrhizal fungi (*Pisolithus tinctorius*, *Laccaria laccata*, *Rhizopogon luteolus*, *Cenococcum graniforme*, etc.) has been detected in assays conducted at the Forest Research Institute (FRI) in India, but the results are not consistent and no claim or generalisation can be made at this stage. Nitrogen fixing ability has been reported in *Rhizopogon* mycorrhiza with *P. radiata* (Giles and Whitehead, 1977), but this claim awaits further confirmation.

RHIZOBIA-VAM INTERACTIONS

Studies on the interaction of VAM fungi and *Rhizobium* species in forestry are very few and confined only to nodulating species such as black locust (*Robinia pseudoacacia*). In these, inoculation with both *Rhizobium* resulted in increased length, higher biomass production, phosphorus uptake efficiency and acetylene reduction activity. The trees inoculated with both *Rhizobium* species and *Glomus fasciculatus* responded similarly to those inoculated with *Rhizobium* species alone but in trees inoculated with VAM fungi alone, such a response was lacking. The symbiotic efficacy of various strains of rhizobia *vis-a-vis* growth, nodulation, yield and nitrogen content of different tree species has been demonstrated in recent field experiments (Dutt et al., 1983). Results of the investigations revealed that all the inoculated trees registered a significant increase in growth over control; 13-14.3 percent in height, 13.7-19 percent in collar diameter, and 15.4-18 percent in breast-height diameter after two years. Similar improvements were registered in respect to all parameters after 30 months of inoculation. The effect of different strains of *Rhizobium* cultures on nodulation, yield and nitrogen content of subabul (*Leucaena leucocephala*) cultivars has been studied by Shinde and Relwani (1981), and the significant effect on nodulation (number of nodules and grain yield) was obtained when two strains (NGR 8 and CB 81) were applied together. The symbiotic efficacy of different strains of VAM has been reported in several studies which shows that most species respond better to indigenous strains, as compared to those isolated from a widely separated habitat and genera of an unrelated tree species. Success in the planting of tree species inoculated with VAM fungi on critical sites has been reported from the USA (Daft and Hacskaylo, 1977).

This establishes that under certain situations, the mycorrhizal dependency of some species is absolute for growth responses to mycorrhizal fungi to fungicide under controlled conditions (Cudlin et al., 1981; Thapar, 1987). The results of laboratory assays and field trials are not always in agreement, which suggests that parallel trials, *in vitro* and in the field, are needed to find safer chemicals and dosages for the treatment of nursery soil against pests.

HOST SPECIFICITY

Specificity of many mycorrhizal fungi for certain host genera has long been recognised, but has been reported only for some species, such as *Boletus elegans* for larch and *Cortinarius hemitrichus* for birch (Trappe, 1977). This relationship is regarded as an important species characteristic by certain mycologists.

Although many mycorrhizal fungi such as *Amanita muscaria* (L.ex.Fr.) Pers.ex.Hook., *Boletus edulis* Bull.ex.Fr., *Laccaria laccata* (Scop.ex.Fr.) BK., *Pisolithus tinctorius* (Pers) Coker and Cough and *Cenococcum geophilum* Fr. commonly associate with diverse hosts (Trappe, 1962) and are designated as broad host ranging fungi, *Suillus* and species of *Rhizopogon* fruit ably in association with a particular host species. Molina and Trappe (1982) tested 27 species of fungi with diverse sporocarp association for ectomycorrhiza formation with seven North East Pacific conifers: *Pseudotsuga menziesii, Tsuga heterophylla, Larix occidentalis, Picea sitchesis, P. contorta, P. pondorosa* and *P. monticola*. It was found that fungi which were host-specific in nature were able to form well developed mycorrhiza with one or more non associated hosts, thereby suggesting that specific sporocarp host associations do not necessarily limit mycorrhizal association with other hosts. However, fungi with known sporocarp and specific hosts showed best mycorrhiza development with that particular host, suggesting further specification in these associations. Among the conifers, the three species of *Pinus* showed little intra-generic differences in their ability to form ectomycorrhiza with the various fungi.

Experimental evidence on the lack of mycorrhizal specificity by Ericaceus hosts *Arbutus menziesii* and *Arctostaphylos uva-ursi* has been reported by Molina and Trappe (1982). Twenty eight cultures of ectomycorrhizal fungi were isolated from diverse ectomycorrhizal hosts (Malajczuk et al., 1982). Few differences *Eucalyptus* species differed in their ability to form ectomycorrhiza with several fungi, thus indicating no evidence of host specificity. Within *Eucalyptus* species and *Pinus radiata*, ectomycorrhizal formation is common with several broad host-ranging fungi. However, it was observed that fungi known to associate exclusively with the members of pinaceae did not form ectomycorrhiza with *Eucalyptus* and vice versa. Intensive studies of host specificity relationships are, therefore, required for better understanding of the degrees and processes of ectomycorrhiza, host-fungus specificity and compatibility to practical consideration in the reforestation of exotic plantations. However, carefully designed experiments are needed to clarify the implication of this phenomenon. The physiological processes which determine ectomycorrhizal specificity and compatibility, have been explored only to a limited extent. It is now recognised that phenolics play an important role in incompatible host fungus interactions. If this is true, we must characterise them and use histochemical techniques to find their cellular origin. Also, the various interactions must be looked at, in relation to the complexity of the natural environment and the microbiological factors which are known to affect ectomycorrhiza formation (Molina and Trappe, 1982).

STATUS IN INDIA

Research on ectomycorrhiza has become increasingly significant in India during the past few decades, in view of the large-scale introduction of tropical pines in the dome areas of the grasslands of central and southern India. Prescribed burning during site preparations may kill beneficial fungi, including mycorrhizal fungi. Soil compaction, erosion, drainage and fertiliser application drastically change the ecosystem and may affect symbiotic relationships, leading to unsuccessful attempts to afforest such marginal sites. Studies have been conducted at the FRI, India, on basic and applied aspects of ecto- and endomycorrhiza of forest trees: chir pine, caribbean pine and silver fir among conifers; *Eucalyptus*, *Agathis* and *Araucaria* among broad-leaved species. The studies have thrown light on the morphological, physiological and ecological aspects of mycorrhizal associations (Bakshi, 1974). Among significant findings, two new species of ectomycorrhizal fungi, viz. *Xerocomus bakshii sing. Singh* and *Pulberoboletus shoreae sing.* Singh and the occurrence of *Gigaspora margarita* Becker and Hall and *Gigaspora aurigloba* Hall, have been reported for the first time from this region. Application of growth hormones and mycorrhizal inoculations (soil inoculation) at low altitude silver fir nurseries in Himachal Pradesh accelerated the height gain in seedlings, thereby reducing the nursery period from six growing seasons (4 1/2 years in the natural zones) to four growing seasons (e.g. three years) (Bakshi et al., 1972). Trimming of roots at the time of field planting removed mycorrhiza considerably. A survey of pine nurseries in different parts of the country revealed acute mycorrhizal deficiency in seedlings (Thapar and Paliwal, 1982). The state-of-the-art on this subject has also been reviewed by Thapar (1982) and Bagyaraj (1987).

CONCLUSION

Despite the spurt in ectomycorrhiza research in the recent past, there is a pressing need for more studies, specially on the physiological and biochemical aspects of mycorrhizal fungi and the factors affecting spore germination, hyphal growth in soil, inoculum density, and screening and selection of fungi for the introduction of specific mycorrhiza in deficient areas. While the efficacy of pure culture inoculation has been established beyond question, the inoculation of nursery seedlings with a mixture of mycorrhizal fungi needs to be given a fair trial, and the performance of such seedlings should be tested under field conditions. The mass cultivation of mycorrhizal inoculum has been attempted with success with only one species, i.e., *Pisolithus tinctorius* amongst many species that have been established as mycorrhizal formers. Search for similar universal fungal symbionts is desirable for developing package programs for field application, in areas with high rainfall and salt concentrations. Furthermore, the uptake efficiency of fungi for various ions such as nitrogen, potassium and calcium, is yet to be established for varying site conditions representing nutrient deficiency. Few studies are available to throw light on the toxic effect of soil fumigants, pesticides and nematicides on mycorrhizal development in potted seedlings and nurseries receiving these treatments. More experimental evidences are required from parallel studies *in vitro* and in the field, for the selection of safer chemicals and minimal acceptable dosages. The establishment of fungus gardens (mycorrhizal banks) of specific fungi or a mixture of fungi has immense scope, as inoculum from rhisozphere of these trees could be used as an effective and economic means to introduce selective fungi on the new crops. This concept holds great promise for the supply of inoculum to the nurseries to be established for afforestation programs on barren areas in future.

Inoculation of seeds with spores of selective mycorrhizal fungi, could be an effective method of introducing inoculum in nurseries. Although, methods of adding spores to naked seeds and to the pelletising matrix of encapsulated seeds have been developed (Marx and Schneck, 1983), extensive field testing of these techniques is advocated.

The mycorrhizal technology in the USA has played a vital role in tailoring tree roots with efficient mycorrhizal fungi adapted to the ecological requirements of various soils. For large plantation programs on wastelands, usarlands, semi-arid and degraded lands in India, superior planting stocks are required to meet future requirements. In such programs, the inoculation of seedlings with specific strains or species of ECM and VAM fungi tolerant to a high concentration of salts, soil toxicity and other extreme conditions can help, to a very large extent, in the success of these programs. There is a pressing need for the selection and mass multiplication of superior mycorrhizal strains, which could be taken up in the development of package programs for inoculation purposes, as in the case of *Rhizobium*. Unfortunately, large quantities of inoculum (about 2-3 lm^{-2} in the case of ectomycorrhizal and 1.5 t ha^{-1} in the case of VAM fungi) limit the exploitation of the potential benefits of mycorrhiza in extensive plantation programs in our country.

REFERENCES

Bagyaraj, D.J., 1987, Current status of mycorrhiza research in India. In: *Proceedings of the National Workshop on Mycorrhizae*. pp. 95-102, Jawaharlal Nehru University, New Delhi.
Bakshi, B.K., Reddy, M.A.R., Thapar, H.S. and Khan, S.N., 1972, Studies in Silver fir regeneration, *Indian For.*,88: 135-144.
Bakshi, B.K., 1974, Mycorrhizae and its role in Forestry, *PL-480 Project Report*. FRI & Colleges, Dehra Dun.
Baylis, G.T.S., 1959, Effect of vesicular arbuscular mycorrhizae on the growth of *Griselinia littoralis* (Cornaceae), *New Phytol.*. 58: 274.

Berry, C.R., 1982, Survival and growth of Pine hybrid seedlings with *Pisolithus tinctorius* ectomy-corrhizae on coal spoils in Alabama and Tennessee, *Rep. of Environ. Qual.,* 11: 709-715.

Bevege, D.J., 1970, Vesicular-arbuscular mycorrhizae of *Araucaria cunninghamii* and their role in nitrogen and phosphorus nutrition, 42nd Congr. WNZ Assoc., Adv. Sci., Section 12.

Briscoe, C.B., 1959, Early results of mycorrhizal inoculation of Pines in Puerto Rico, *Caribbean For.,* 20: 73-77.

Clark, B., 1963, Endotrophic mycorrhizae influence Yellow Poplar seedlings growth, *Science,* 140: 1220-1221.

Cudlin, P., Majstrik, V. and Vaclav, S., 1981, Growth inhibition of mycorrhizal fungi by pesticides, Fifth North American Conference on Mycorrhiza, pp. 65, Quebec.

Daft, M.J. and Nicolson, T.H., 1966, Effect of endogone mycorrhizae on plant growth, *New Phytol.,* 65: 343-350.

Daft, M.J. and Hacskaylo, E., 1977, Growth of endomycorrhizae and non-mycorrhizae seedlings in sand and anthracite spoil, *For. Sci.,* 23: 207-217.

Dutt, A.K., Kumar, V. and Pathnia, U., 1983, Effect of different rhizobial inoculation on the growth of *Leucaena leucocephala.* In: *Proc. National Symposium on Advances in Tree Sciences.* Solan.

Giles, K.L. and Whitehead, R.C.M., 1977, Reassociation of a modified mycorrhiza with the host plant roots (*Pinus radiata*) and the transfer of acetylene reduction activities, *Plant Soil,* 48: 143.

Gilmour, J.W., 1958, Chlorosis of Douglas-fir, *N.Z. For.,* 7: 94-106.

Handley, W.R.C., 1963, Mycorrhizal associations and Calluna heathland afforestation, *For. Comm. Bull.,* 36: 1-70.

Harley, J.L., 1959, *The Biology of Mycorrhiza,* Leonard-Hill Ltd., London.

Hatch, A.B., 1936, The role of mycorrhizae in afforestation, *J. For.,* 34: 22-29.

Hatch, A.B., 1937, The physical basis of mycotrophy in *Pinus. Black Rock Forest Bull.,* 6: 168.

Ho, I. and Trappe, J.M., 1980, Nitrate reductase activity of nonmycorrhizal Douglas-fir rootlets and of some associated mycorrhizal fungi, *Plant Soil,* 54: 395-398.

Kormanik, P.P., Schultz, R.V. and Bryan, C., 1982, The influence of vesicular arbuscular mycorrhizae on the growth and development of eight hard wood tree species, *For. Sci.,* 28: 531-539.

Kramer, P.J. and Wilbur, K.M., 1949, Absorption of radioactive phosphorus by mycorrhizal roots of pine, *Science,* 110: 8-9.

Malajczuk, N., Molina, R. and Trappe, J.M., 1982, Ectomycorrhiza formation in *Eucalyptus. New Phytol.,* 91: 467-482.

Marks, G.C. and Kozlowski, T.T., 1973, *Ectomycorrhizae: Their Ecology and Physiology,* Academic Press, New York.

Marx, D.H., 1975, Mycorrhizae and establishment of trees on strip-mined land, *Ohio J. Sci.,* 75: 288.

Marx, D.H., 1976, Synthesis of Ectomycorrhizae on loblolly pine seedlings with basidiospores of *Pisolithus tinctorius, For. Sci.,* 22: 13-20.

Marx, D.H., 1976, Use of specific mycorrhizal fungi on tree roots of forestation of disturbed bands. In: *Proc. Conf. on Forestation of Disturbed Surface Areas.* pp. 47-65, Birmingham.

Marx, D.H., 1977, The role of mycorrhiza in forest production. In: *Conf. Pap. Ann. Meeting,* Atlanta, Georgia, pp. 1-11.

Marx, D.H., 1980, Role of mycorrhiza in forestation of surface mines. In: *Proc. Trees for Reclamation Symposium,* USDA For. Serv. Gen. Tech. Rep., NE- 61.

Marx, D.H. and Artman, J.D., 1979, *Pisolithus tinctorius* ectomycorrhizae improve survival and growth of pine seedlings on acid coal spoils in Kentucky and Virginia, *Reclamation Review,* 22: 23-31.

Marx, D.H. and Schneck, N.C., 1983, Potential of mycorrhizal symbiosis in agricultural and Forest productivity. In: *Challenging Problems in Plant Health* (T. Kommendhal and P.H. Williams, eds.), pp. 334-347.

McComb, A.L. and Griffith, J.E., 1946, Growth stimulation and phosphorus absorption of mycorrhizal and non-mycorrhizal northern White pine and Douglas-fir seedlings in relation to fertilizer treatment, *Plant Physiol.,* 21: 11-17.

Melin, E., 1959, Mykorrihza. In: *Handbuch der Pflanzenphysiologie, Vol. 11,* (W. Ruhland, ed.), pp. 605-638, Springer-Verlag, Berlin.

Melin, E., 1963, Some effects of forest trees root on mycorrhizal basidiomycetes. In: *Symbiotic Associations: The 13th Symposium of the Society. Fo. General Microbiology* (P.S. Nutman and B. Mosse, eds.), pp. 356, Cambridge Univ. Press.

Meyer, F.H., 1973, Distribution of ectomycorrhizae in native and man-made forests. In: *Ectomycorrhizae: Their Ecology and Physiology* (G.C. Marks and T.T. Kozlowski, eds.), pp. 79-106, Academic Press, New York.

Mikola, P., 1973, Application of mycorrhizal symbiosis in forestry practice. In: *Ectomycorrhizae: Their Ecology and Physiology* (G.C. Marks and T.T. Kozlowski, eds.), pp. 383-411, Academic Press, New York.

Mitchell, H.L., Finn, R.F. and Rosendahl, R.O., 1937, The relation between mycorrhizae and the growth of nutrient absorption of coniferous seedlings in nursery beds, *Black Rock For. Bull.* 1: 58-73.

Molina, R. and Trappe, J.M., 1982, Patterns of ectomycorrhizal host specificity and potential among Pacific Northeast conifers and fungi, *For. Sci..* 28: 425-458.

Mossee, B., 1957, Growth and chemical composition of mycorrhizal and non-mycorrhizal apples, *Nature (Lond.),* 179: 922.

Olson, R.V., 1944, The use of hydroponics in the practice of forestry, *J. For.,* 42: 236-243.

Rayner, M.C., 1938, The use of soil or humus inocula in nurseries and plantation, *Emp. For. J.,* 17: 236-243.

Rhodes, D. and Pope, P.E., 1981, Interactions of *Glomus fasciculatus* and a *Rhizobium* species on the growth, phosphorus uptake and nitrogen fixation in *Robinia pseudoacacia,* Fifth North American Conf. on Mycorrhizae, pp. 26, Quebec.

Richards, B.N., Bevege, D.J. and Lamb, R.J., 1971, Personal communication cited in *Mineral Nutrition of Ectomycorrhizae* (G.D. Bowen, ed.) and *Ectomycorrhizae: Their Ecology and Physiology* (G.C. Marks and T.T. Kozlowski, eds.), Academic Press, New York.

Rosendahl, R.O., 1943, The effect of mycorrhizal and non-mycorrhizal fungi on the availability of phosphorus, *Proc. Am. Soil Sci. Soc.,* 7: 477.

Ruehle, J.L., 1980, Growth of containerized loblolly pine with specific ectomycorrhiza after 22 years on amended borrow pit, *Reclamation Rev.,* 3: 95-101.

Ruehle, J.L., 1982, Mycorrhizal inoculation improves performance of container grown pines planted on adverse sites, In : Proc. of Southern Containerized Forest Tree Seedling Conference, Savannah, Georgia, pp. 133-135.

Ruehle, J.L., Marx, D.H., Barnett, J.P. and Pawuk, W.H., 1981, Survival and growth of container-grown and bare-root short leaf pine seedlings with *Pisolithus* and *Thelephora* ectomycorrhizae, *J. Appl. For.,* 5: 20-24.

Shinde, D.B. and Relwani, L.L., 1981, Effect of different *Rhizobium* cultures on the nodulation and nitrogen content of subabul (*Leucaena leucocephala*) cultivars. In: *Proc. National Seminar on Leucaena leucocephala* in India (R.N. Kaul, M.G. Gogte and N.K. Mathur, eds.), pp. 225-227, Univ. Pune, Pune.

Thapar, H.S., 1982, The state of art on mycorrhiza in India, Tropical pine Seminar-cum-Workshop, Koraput.

Thapar, H.S., 1987, Effect of fungicides on ectomycorrhizal fungi *in vitro.* In : *Proceedings of the National Workshop on Mycorrhizae.* pp. 150-166, Jawaharlal Nehru University, New Delhi.

Thapar, H.S. and Khan, S.N., 1985, Effect of VA mycorrhiza on the growth of Hoop Pine (*Araucaria cunninghamii* Sweet.), *J. Tree Sci.,* 4: 39-43.

Thapar, H.S. and Paliwal, D.P., 1982, Studies on pine mycorrhiza in nursery seedlings, *Indian For..* 108: 51-59.

Thapar, H.S., Khan, S.N. and Uniyal, K., 1985, Endogonaceae of forest soils: New Records, *J. Tree Sci.*, 4: 47-52.

Tinker, P.B., 1982, Mycorrhizae: The present position, Whither Soil Research, Panel Discussion Papers, ISSS, AISS.

Trappe, J.M., 1962, Fungus associates of ectotrophic mycorrhiza, *Bot. Rev.*, 28: 538-606.

Trappe, J.M., 1977, Selection of fungi for ectomycorrhizal inoculation in nurseries, *Ann. Rev. Phytopathol.*, 15: 203-222.

Vozzo, J.A. and Hacskaylo, E., 1971, Inoculation of *Pinus caribaea* with ectomycorrhizal fungi in Puerto Rico, *For. Sci.*, 17: 239-245.

White, D.P., 1941, Prairie soil as a medium for tree growth, *Ecology*, 22: 398-401.

Zerling, G.I., 1958, Die mycorrhizae der larche undihre wirkung uf das wac bustun undden sustand der sampling in Koraput Tschernosem Bodan des Transwolga-Gebietes, *Microbiologia*, 27: 450.

25

ECTOMYCORRHIZAL EFFECTS ON NODULATION, NITROGEN FIXATION AND GROWTH OF *ALNUS GLUTINOSA* AS AFFECTED BY GLYPHOSATE

L. Chatarpaul,
P. Chakravarty and
P. Subramaniam

ABSTRACT

Actinorhizal plants such as *Alnus* and *Casuarina* offer good potential for increasing the productivity of impoverished soils because of their ability to fix atmospheric nitrogen in a symbiotic relationship with the actinomycete, *Frankia*. The occurrence of mycorrhizae on actinorhizal plants is seen as a hypersymbiosis which could be of added nutritional significance. To test this hypothesis, the effects of an ectomycorrhizal fungus, *Paxillus involutus*, on *Alnus glutinosa* seedlings nodulated with *Frankia* were

L. Chatarpaul, P. Chakravarty and *P. Subramaniam* * Canadian Forestry Service, Petawawa National Forestry Institute, Chalk River, Ontario, Canada KOJ 1JO.

investigated in the greenhouse and in the field both in the absence and presence of the herbicide, glyphosate. Under greenhouse conditions, simultaneous inoculation with *Frankia* and *Paxillus involutus* stimulated biomass production, nodulation and acetylene reduction in the absence of glyphosate. Glyphosate at 1.5 l ha^{-1} stimulated biomass production and acetylene reduction, but tempered the influence of the tripartite association. The 9 l ha^{-1} glyphosate treatment reduced shoot growth, stimulated root growth, and further reduced the impact of the tripartite association. In the field, at the end of one season, about 45 percent of the seedlings failed to survive in the non-herbicide treated plots, whereas all survived the 1.5 l ha^{-1} and 9 l ha^{-1} treatments. Trends were similar to those observed in the greenhouse. In *in vitro* studies, pure cultures of *Frankia* and *Paxillus involutus* were treated with different concentrations of glyphosate. At concentrations of up to 10 ppm, neither of the two organisms were affected by glyphosate but their growth was significantly reduced by the 50-1000 ppm treatments. Our results indicate significant enhancement in the performance of *Alnus glutinosa* seedlings when inoculated with both *Frankia* and *Paxillus involutus*. Recommended application rates of glyphosate may not have any negative effects on growth.

INTRODUCTION

Actinorhizal plants that possess root nodules formed by nitrogen fixing actinomycetes of the genus *Frankia,* are increasingly being used in forestry throughout the world. There are more than 20 known genera of actinorhizal plants (Bond, 1983) and many of these have been put to a variety of forestry and land reclamation uses (Tarrant and Trappe, 1971). Actinorhizal trees can be used for sawlogs, pulp and fuelwood and can be inter-planted with other tree crops or used in rotation with them to improve productivity on nitrogen-deficient sites (Tarrant and Trappe, 1971). They are especially useful in reforesting difficult sites such as mine spoils and coastal dunes. From an energy perspective, both rapid production of biomass and reduced input of costly chemical fertilisers are possible with actinorhizal plants (Chatarpaul and Carlisle, 1983).

The presence of ectomycorrhizal fungi on several species of actinorhizal plants has been reported (Daft et al., 1985). This tripartite association of the host with two microbial symbionts represents a possible case of hypersymbiosis, which could be of nutritional significance (Godbout and Fortin, 1983). Mycorrhizae have been shown to improve the uptake of phosphorus, water and other minerals in many plants (Marks and Kozlowski, 1973; Harley and Smith, 1983). Mycorrhizal association on the roots of actinorhizal plants would be advantageous because improved overall plant nutrition would enhance nodulation and nitrogen fixation and result in a more rapid establishment on poor sites. Mycorrhizal roots of *Alnus viridis* absorbed phosphate five times more rapidly than non-mycorrhizal roots and this was directly related to the degree of mycorrhizal infection; seedling height was greater in *Alnus glutinosa* inoculated with both *Frankia* and an ectomycorrhizal fungus than those inoculated with either symbiont alone (Green et al., 1979).

Herbicides are increasingly being used by foresters as a means of controlling weeds (Malik and Vandenborn, 1986). Often, these weeds include actinorhizal plants, which are an important source of nitrogen and organic matter in diverse environmental conditions. There is concern that some herbicides may have detrimental effects on non-target micro-organisms including highly beneficial nitrogen-fixers and mycorrhizal fungi. Such effects could lead to a long-term decline in productivity and negate the benefits of using herbicides. 'Roundup' [N(Phosphonomethyl)glycine] is one of the few herbicides registered for forestry use.

Alnus (alder) is the actinorhizal genus showing the greatest potential for temperate areas (Silvester, 1977). However, there is very little information on the nutritional significance of the tripartite symbiosis. Similarly, very little work has been done on the impact of chemicals such as herbicides on these highly beneficial natural processes. Thus, for optimum utilisation of these plants, a thorough understanding of the symbionts and their interactions with the host and environment is essential. The aim of this study was to investigate the effects of inoculating *Alnus glutinosa* with a *Frankia* strain and an ectomycorrhizal fungus, *Paxillus involutus,* and to evaluate the effects of the herbicide, glyphosate, *in vitro* and *in vivo*.

MATERIALS AND METHODS

In Vitro Growth of *Frankia* and *Paxillus involutus*

The effects of glyphosate on the *in vitro* growth of two micro-organisms forming symbiotic associations with alders were studied. A *Frankia* strain, ACN1[AG], (ULQ 0102001007) and a common ectomycorrhizal fungus *Paxillus involutus* (Batsch. ex Fr.) were used in the experiment. The tests were conducted in 125 ml serum bottles containing 50 ml of sterile liquid QMod (Lalonde and Calvert, 1979) for *Frankia,* and MMN (Marx, 1969) for *P. involutus*. A 10 ml suspension of glyphosate (35.9 percent a.i.) was added to each bottle at concentrations of 0.1, 1, 10, 50, 100, 500 and 1,000 ppm (v/v). A non-herbicide control received 10 ml of sterile distilled water. Each treatment was replicated five times. Following seeding with actively growing cultures of the organisms, the bottles were shaken for 30 seconds and then incubated at 25°C in the dark for 30 days. At the end of the incubation period, the hyphae were harvested by filtration and the biomass of the organisms determined after drying for 48 hours at 80°C.

Greenhouse Test

Seeds of *Alnus glutinosa* (L.) Gaertn. were surface sterilised with 30 percent H_2O_2 for 10 min, washed several times with sterile distilled water and then sown in Spencer-Lemaire containers (1.9 x 1.9 x 10.1 cm: length x width x height) in 22 x 37 cm trays containing sterile growth medium (peat:vermiculite 3:1, w/w) with 25 percent of the recommended dose of slow-release fertiliser (Nutricote, Micromax, dolomitic lime and gypsum). Two-week-old seedlings were simultaneously inoculated with *P. involutus* and the *Frankia* strain used in the *in vitro* studies. Control seedlings were inoculated with *Frankia* only. The seedlings were kept in the greenhouse at 16 hours of daylight at $25°C \pm 5°C : 20°C \pm 2°C$ (day:night) and watered twice a day. At two months, half of the seedlings were transplanted to pots that had been subjected to glyphosate treatment eight weeks earlier. These 12.7 cm plastic pots were filled with peat-vermiculite and slow release fertiliser in the same ratio as mentioned above. Glyphosate at 1.5 and 9 l ha^{-1} was applied on the surface of the pots. Control pots without glyphosate were treated with distilled water. Eight replicates per treatment were kept in the greenhouse under similar conditions, as mentioned earlier. The plants were harvested after two months when soil and debris were carefully washed from the roots. Intact root systems were used to determine nitrogenase activity by the acetylene reduction method using gas chromatography. Mycorrhizal infection was noted and the weight of shoots, roots and nodules determined, following drying for 48 hours at 80°C. The length of the shoots and roots was also measured.

Field Test

The remainder of the *Frankia*-and *P. involutus*-inoculated seedlings without glyphosate were used in a field test during the summer of 1987. The study area is located at Sturgeon Lake, Chalk River, Ontario, Canada. The site was prepared for plantation establishment. Glyphosate was applied to a part of the field in August, 1986 at 9 l ha^{-1}. In May and June 1987, the whole field was treated with glyphosate at 1.5 l ha^{-1}. The basic experimental design was that of a completely randomised block, with each block measuring 4m^2. The two-month-old seedlings of *A. glutinosa* from the greenhouse were transferred in early July to the nursery for a four week hardening period. They were then outplanted on the glyphosate-treated and control plots on August 3, 1987. There were eight replicates for each treatment. The seedlings were harvested two months later and were subjected to the same evaluation as in the greenhouse test. The data from all the three experiments were subjected to analysis of variance (SAS, 1982). The individual means were compared by Duncan's Multiple Range test using SAS software.

RESULTS

In Vitro Growth

Glyphosate at 0.1, 1.0 and 10 ppm had no effect on the *in vitro* growth of either *Frankia* or *P. involutus* when compared to control (Fig. 1). The growth of both the organisms was significantly ($P < 0.001$) reduced by the 50 ppm and higher glyphosate

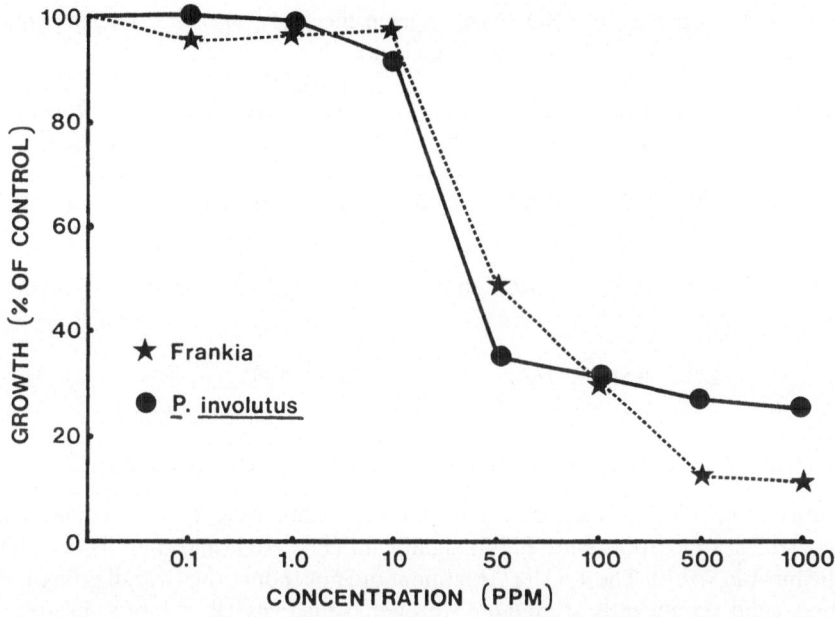

Fig. 1: The effect of different concentrations of glyphosate on the growth of pure cultures of *Frankia* and *Paxillus involutus* (statistically significant P ≤ 0.001) effects on both organisms at concentrations over 10 ppm glyphosate

treatments. *P. involutus* appeared to be slightly more susceptible to glyphosate only at the 10 and 50 ppm treatments. This effect was reversed at higher concentrations.

Greenhouse Test

The main effects of *P. involutus* inoculation, regardless of glyphosate treatment, were significant for five of the eight parameters evaluated (Table 1). Highly significant inoculation effects were obtained for shoot height (P = 0.003), root weight (P = 0.001), nodule weight (P = 0.037), and mycorrhizal infection (P = 0.001). Inoculation had no overall significant effect on root length, shoot weight, or nitrogenase activity. If we regard the main effects of glyphosate treatment without considering inoculation, the only significant effects were observed on shoot height (P = 0.007), shoot weight (P = 0.045) and, to a lesser extent, on total weight (P = 0.069). Significant interaction of the main effects occurred with respect to nitrogenase activity only.

In the absence of glyphosate treatment, *Alnus glutinosa* seedlings performed better when inoculated simultaneously with *Frankia* and *P. involutus* than with *Frankia* alone. Values obtained for all parameters evaluated were higher when both organisms

Table 1: Levels of significance (ANOVA) of the main effects of *P. involutus (Pi)* and glyphosate on *Frankia*-inoculated *Alnus glutinosa* seedlings in greenhouse studies

Source	Shoot Ht.	Root Lth.	Shoot Wt.	Root Wt.	Nodule Wt.	Total Wt.	Mycorrhizal short roots	Nitrogenase activity
Inoculation (Pi)	0.003	0.201	0.201	0.001	0.001	0.037	0.001	0.666
Glyphosate Levels	0.007	0.161	0.045	0.324	0.407	0.069	0.983	0.456
Inoc x Glyphosate Levels	0.229	0.516	0.684	0.912	0.227	0.652	0.983	0.006

were present (Table 2). The increase in root and nodule weights, nitrogenase activity of mycorrhizal short roots were highly significant ($P \leq 0.05$) and that of shoot height significant at $P \leq 0.10$. The 1.5 l ha^{-1} treatment did not reduce the overall growth of the seedlings, and significantly stimulated nitrogenase activity ($P \leq 0.05$) and total dry weight ($P \leq 0.10$) in the non-mycorrhizal seedlings. This 1.5 l ha^{-1} glyphosate treatment appeared to have a dampening effect on the tripartite association as compared to the no herbicide treatment. Nitrogenase activity in the dual inoculated plants was significantly reduced ($P \leq 0.05$), thus reversing the effects of the tripartite association. The 9 l ha^{-1} glyphosate treatment reduced shoot height and increased root weight over controls but the values were significant only when seedlings were inoculated with *Frankia* alone. Further reduction in the mycorrhizal effects was evident at the 9 l ha^{-1} treatment. Mycorrhizal infection was not affected by either of the glyphosate treatments.

Field Test

By the time of the harvest, 45 percent of the seedlings on the non-glyphosate treated plots had perished, whereas all survived in the herbicide-treated plots. The control plots were invaded by weeds, including wild berries, fireweed and other herbs. Growth of the seedlings was only a fraction of that observed in the greenhouse studies.

Of the overall effects, both *P. involutus* inoculation and glyphosate-treatments were significant for many of the parameters evaluated (Table 3). Highly significant effects of *P. involutus* inoculation were observed for root length (P = 0.000), shoot weight (P = 0.017), root weight (P = 0.015), nodule weight (P = 0.003), total weight (P = 0.003), and mycorrhizal short roots (P = 0.000). The main effects of glyphosate treatment were highly significant on root length (P = 0.043), root weight (P = 0.000), nodule weight (P = 0.001), and total weight (P = 0.000). The effect on shoot weight was less significant (P = 0.108).

Table 2: Effect of glyphosate on seedling growth, mycorrhizal incidence and nitrogenase activity of *Alnus glutinosa* under greenhouse conditions

Glyphosate Levels	Growth Parameters															
	Shoot Ht. (cm)		Root Lth. (cm)		Shoot Dry Wt. (g)		Root Dry Wt. (g)		Nodule Dry (mg)		Total Dry Wt. (g)		Mycorrhizal short roots (%)		Nitrogenase activity μM C_2H_4/h/Pl	
	Fr	Fr+Pi	Fr	Fr+Pi	Fr	Fr+Pi	Fr	Fr+Pi	Fr	Fr+Pi	Fr	Fr+Pi	Fr	Fr+Pi	Fr	Fr+Pi
0	82.3b*x	99.1ax	21.1ax	22.1ax	17.5ax	22.0ax	2.5bx	4.6ax	200bx	420ax	20.2ax	27.0ax	0bx	46.0ax	10.9by	29.2ax
1.5 l/ha	84.2ax	86.8ax	20.9ax	20.8ax	24.3ax	28.1ax	2.5bx	4.5ax	200bx	310ax	27.1b*x	33.0ax	0bx	45.1ax	31.6ax	14.8by
9 l/ha	66.3by	83.8ax	19.0b*x	20.9ax	21.0ax	21.2ax	3.3by	5.0ax	221bx	290ax	24.5ax	26.4ax	0bx	45.2ax	25.6axy	21.3axy

Fr = *Frankia*, Pi = *Paxillus involutus*

Values are the means of 8 seedlings. All values followed by the same letters (a,b within rows and x,y within columns) for a parameter are not significantly (P ≤ .05) different from each other. * indicates (P ≤ 0.10).

Table 3: Levels of significance (ANOVA) of the main effects of *P. involutus (Pi)* and glyphosate on *Frankia*-inoculated *Alnus glutinosa* seedlings in field studies

Source	Shoot Ht.	Root Lth.	Shoot Wt.	Root Wt.	Nodule Wt.	Total Wt.	Mycorrhizal short roots	Nitrogenase activity
Inoculation (Pi)	0.257	0.000	0.017	0.015	0.003	0.003	0.000	0.0439
Glyphosate Levels	0.477	0.043	0.108	0.000	0.001	0.0000.	810	0.545
Inoc. x Levels	0.689	0.936	0.121	0.584	0.024	0.513	0.810	0.484

As observed in the greenhouse study, *P. involutus* plus *Frankia* inoculation gave better growth values than *Frankia* alone for all parameters measured in the absence of glyphosate (Table 4). The effects of *P. involutus* were only significant ($P \leq 0.05$) for root length, shoot weight, total weight, and mycorrhizal short roots. The dual inoculation had similar effects at the 1.5 and 9 l ha^{-1} treatments for all parameters except nitrogenase activity. At the 1.5 l ha^{-1} glyphosate treatment, the significant ($P \leq 0.05$) effects of *P. involutus* were observed for root length, root weight, nodule weight and mycorrhizal short roots, while at the 9 l ha^{-1} treatment, such effects were observed for root length and mycorrhizal shoot roots. Herbicide treatments generally resulted in healthier seedlings with significant effects ($P \leq 0.05$) on short, root, and nodule weights. Though not significant, nitrogenase activity in the field was enhanced by herbicide treatment only when *Frankia* alone was present. Seedlings which did not receive *P. involutus* were inoculated with a naturally occurring species. However, the infection rate ranged from 7.7 percent to 9.2 percent, as compared to about 37 percent for the *P. involutus* inoculated seedlings.

DISCUSSION

The results obtained in this study show that growth, nodulation and nitrogenase activity were enhanced when the seedlings were inoculated with *Frankia* and *P. involutus* simultaneously, rather than with *Frankia* alone. The effectiveness of the dual inoculation was also evident in the presence of glyphosate. It is a well-known fact that mycorrhizae improve phosphorus uptake from the soil. Since phosphate supply limits nitrogen fixation, mycorrhizal inoculation is likely to stimulate nitrogen fixation (Barea and Azcon-Aguilar, 1983). An increased uptake and translocation of N$_2$ from the soil by mycorrhizae may accelerate plant growth and nutrition. Mejstrik and Benecke

Table 4: Effect of glyphosate on seedling growth, mycorrhizal development and nitrogenase activity of *Alnus glutinosa* under field conditions

Glyphosate Levels	Shoot Ht. (cm)		Root Lth. (cm)		Shoot Dry Wt. (g)		Root Dry Wt. (g)		Nodule Dry (mg)		Total Dry Wt. (g)		Mycorrhizal short roots (%)		Nitrogenase activity μM C_2H_4/h/Pl	
	Fr	Fr+Pi	Fr	Fr+Pi	Fr	Fr+Pi	Fr	Fr+Pi	Fr	Fr+Pi	Fr	Fr+Pi	Fr	Fr+Pi	Fr	Fr+Pi
0	18.5ax	20.6ax	30.1ax	41.1bx	1.3ay	2.5bx	0.96ay	1.33ax	4.37ax	4.58ax	2.30ay	3.87bx	7.7ax	36.5bx	0.10ax	0.15ax
1.5 l/ha	20.3ax	22.3ax	29.1axy	39.5bx	2.4ax	2.5ax	1.9b*x	2.6a*y	20.18ay	51.68by	4.43ax	5.29ax	9.2ax	37.8bx	0.21ax	0.14ax
9 l/ha	20.6ax	20.7ax	27.8ay*	38.6bx	2.1axy	2.4ax	1.9ax	2.2ay	8.35ax	19.60ax	4.02ax	4.70ax	8.6ax	36.4bx	0.17ax	0.14ax

Fr = *Frankia*, Pi = *Paxillus involutus*

Value are the means of 8 seedlings. All values followed by the same letters (a,b within rows and x,y within columns) for a parameter are not significantly (P $\leq$ 0.05) different from each other. * indicates (P $\leq$ 0.10.)

(1969) showed that an ectomycorrhizal association on the roots of *Alnus viridis* greatly enhanced the ability of the plant to absorb phosphate. The uptake of phosphate was found to be directly related to the degree of mycorrhizal frequency.

Herbicides, in general, are known to inhibit the *in vitro* growth of mycorrhizal fungi. Kelley and South (1980) reported that 16 herbicides had stimulatory, inhibitory, or no effect on the growth of mycorrhizal fungi *in vitro*. Chakravarty and Sidhu (1987) reported similar findings with other herbicides including glyphosate. It is clear from our results that low concentrations had either no effect or showed a slight initial inhibitory tendency or did not affect mycorrhizal development. In both greenhouse and field, no seedling mortality and mycorrhizal development were observed even at the highest rate (9 l ha^{-1}). In the field conditions, 45 percent of the seedlings failed to survive in non-herbicide treated plots because of weed competition. Seedlings were always vigorous in the glyphosate-treated plots in both greenhouse and field conditions. In addition to eliminating weed competition, glyphosate may act as a fertiliser because of its nitrogen and phosphorus content. In the field, the toxicity of glyphosate was reduced to a great extent even at the highest concentration (9 l ha^{-1}). The consensus is that pesticides, when applied at field rates, do not adversely affect the major biochemical cycles (Tu and Miler, 1976; Drandarevski et al., 1977; Wainwright, 1977). The toxicity of forestry herbicides and their biodegradability by micro-organisms are both important for evaluating their environmental impact and efficacy. Glyphosate is known to be decomposed by microbes in the soil (WSSA, 1983). The field data suggest the dissipation of glyphosate as a result of its absorption by target species, microbial degradation and leaching.

Root inhibition is a primary effect that herbicides have on plant growth (Probst et al., 1975; Zilkah et al., 1978). In our experiments, there was no inhibition of shoot or root growth by glyphosate at any level. It is interesting to note that, in the field, seedlings that did not receive *P. involutus* were infected by the naturally occurring mycorrhizal fungi within two months after planting, although the infection rate was very low. This suggests that glyphosate did not eliminate naturally occurring mycorrhizal inocula. It is very difficult to destroy all the microflora, especially one as abundant as ectomycorrhizal fungi in temperate forests. Some of the ectomycorrhizal fungi can survive with glyphosate up to 1,000 ppm, whereas recommended field application varies from 4-18 ppm (Chakravarty and Sidhu, 1987).

It is clear from our greenhouse test results that the nitrogenase activity was not affected by glyphosate at 9 l ha^{-1}, although seedlings inoculated with *Frankia* alone exhibited a higher activity in the presence of glyphosate at 1.5 l ha^{-1}. The nitrogen and phosphorus content of glyphosate (WSSA, 1983) could be responsible for this stimulation.

Nodule function (nitrogenase activity) was significantly higher under a tripartite situation in the absence of glyphosate, whereas in its presence the nitrogenase activity was lower. Mycorrhizal seedlings are better able to absorb herbicides from the soil than non-mycorrhizal ones (Chakravarty and Sidhu, personal communication). This inhibitory effect could, thus, arise from residues of glyphosate taken up by the mycorrhizal root system.

Under field conditions, glyphosate did not affect nitrogenase activity. This could be attributed to the fact that the time lapse between herbicide application and the transplanting of seedlings was long enough to allow for complete degradation of the herbicide in the soil. However, long-term field evaluations might be needed.

Herbicides have a great potential in forestry use. It is clear from our results that the use of glyphosate at the recommended dose for forest weed control does not appear to pose high risks to ectomycorrhization or nodulation of alders in the field. Herbicides with a negative effect on seedling growth and mycorrhizal formation can be identified and avoided. At the same time, herbicides with positive effects on mycorrhizae formation and on seedling growth may prove extremely useful beyond the goal of weed control.

ACKNOWLEDGEMENT

We are grateful to D. Lachance for technical assistance and to the Natural Sciences and Engineering Research Council of Canada, Ottawa, for the Visiting Fellowships (P.C. and P.S.).

REFERENCES

Barea, J.M. and Azcon Aguilar, C., 1983, Mycorrhizae and their significance in nodulating nitrogen fixing plants. *Adv. Agron.,* 36:1-54

Bond, G., 1983, Taxonomy and distribution of non-legume nitrogen fixing systems. In: *Biological Nitrogen Fixation in Forest Ecosystems: Foundations and Applications* (J.C. Gordon and C. Wheeler, eds.), pp. 55-87, Martinus Nijhoff, Dr. W. Junk, The Hague.

Chakravarty, P. and Sidhu, S.S., 1987, Effect of glyphosate, hexazinone and triclopyr on *in vitro* growth of five species of ectomycorrhizal fungi, *European J. Forest Pathol.,* 7: 204-210.

Chatarpaul, L. and Carlisle, A., 1983, Nitrogen fixation: A biotechnological opportunity for Canadian forestry, *For. Chron.,* Oct., 249-259.

Daft, M.J., Clelland, D.M. and Gardner, J.C., 1985, Symbiosis with endomycorrhizae and nitrogen fixing organisms, *Proc. Royal Soc.* (Edinburgh), 85B: 282-298.

Drandarevski, C.A., Eichler, D. and Domsch, K.H., 1977, Behaviour of triforine in soil and its influence on microbial soil processes. *Z. Pflanzenschutz.,* 84: 18-30.

Godbout, C. and Fortin, J.A., 1983, Morphological features of synthesized ectomycorrhiza of *A. crispa* and *A. rugosa, New Phytol.,* 96: 249-262.

Green, T.L., McNabb, H.S. and Mize, C.W., 1979, Symbiosis among *Alnus* spp: Actinorhizae and mycorrhizae. In: *Symbiotic Nitrogen Fixation in the Management of Temperate Forests* (J.C. Gordon, C.T. Wheeler and D.A. Perry, eds.), pp. 476-477, Oregon State University, Corvallis, Oregon.

Harley, J.L. and Smith, S.E., 1983, *Mycorrhizal Symbiosis,* Academic Press, New York, pp. 483,

Kelley, W.D. and South, D.B., 1980, Effectives of herbicides on the *in vitro* growth of mycorrhizae of pine (*Pinus* spp.), *J. Weed Sci.,* 28: 599-602.

Lalonde, M. and Calvert, H.E., 1979, Production of *Frankia* hyphae and spores as an infective inoculant for *Alnus* species. In: *Symbiotic Nitrogen Fixation in the Management of Temperate Forests* (J.C. Gordon, C.T. Wheeler and D.A. Perry, eds.), pp. 95-110, Oregon State University, Corvallis, Oregon.

Malik, N. and Vandenborn, W.H., 1986, Use of herbicides in forest management, Can. For. Serv. NoFC, Edmonton, Alberta, Infor. Rep. NOR-X-282.

Marks, G.C. and Kozlowski, T.T., 1973, *Ectomycorrhizae - Their Ecology and Physiology,* Academic Press, New York, pp. 444.

Marx, D.H., 1969, The influence of ectotrophic mycorrhizal fungi on the resistance of pine roots to pathogenic infections. In: Antagonism of mycorrhizal fungi to root pathogenic fungi and soil bacteria, *Phytopathology,* 59: 153-163.

Mejstrik, V. and Benecke, V., 1969, The ectotrophic mycorrhizae of *Alnus viridis* (Chaix) D.C. and their significance in respect to phosphorous uptake, *New Phytol.,* 68: 141-149.

Probst, G.W., Gealab, T. and Wright, W.L., 1975, Dinitroamilines. In: *Herbicides - Chemistry, Degradation and Mode of Action* (P.C. Kerney and D.D. Kaufman, eds.), pp. 453-500, Marcel Dekker Inc., New York.

SAS Institute Inc., 1982, SAS *User's Guide: Statistics,* Cary,: SAS Institute Inc. North Carolina, pp. 584.

Silvester, W.B., 1977, Dinitrogen fixation by plant associations excluding legumes. In: *A Treatise on Dinitrogen Fixation Section IV: Agronomy and Ecology* (R.W.F. Hardy and A.H. Gibson, eds.), pp. 141-190, John Wiley and Sons, New York.

Tarrant, R.F. and Trappe, J.M., 1971, The role of *Alnus* in improving the forest environment, *Plant Soil,* (Spec. Vol. 1971): 335-348.

Tu, C.M. and Miler, J.R.W., 1976, Interactions between insecticides and soil microbes, *Residue Rev.,* 64: 17-65.

Wainwright, M., 1977, Effects of fungicides on the microbiology and biochemistry of roots - A review, *J. Pflanzenernacher Bodenkd.,* 140: 587-603.

WSSA (Weed Science Society of America), 1983, *Herbicide Handbook* 5th Ed. WSSA, Illinois, U.S.A.

Zilkah, S., Bocion, P.K. and Gressel, J., 1978, Target tissue for napropamide inhibition: Effects on green and white callus cultures and seedlings, *Weed Sci.,* 26: 711-713.

V

GENETIC ENGINEERING OF FOREST SPECIES

26

GENETIC ENGINEERING OF TREE SPECIES

H.K. SRIVASTAVA

INTRODUCTION

Forests are not only responsible for climatic stability, soil and water, conservation but are also a source of goods and services (including firewood and fodder) to the rural and tribal populations and the timber, pulp and fiber industry. There appears to be a widening gap today between the supply and demand of these materials. With the use of Remote Sensing Technology, the forest cover in India was estimated to be 13.94 percent of the total land mass which has now been revised to 19.52 percent. Out of this only 10.99 percent (35.79 Mha) represents closed forest. However, under the Government of India forest policy of 1952, 33 percent of the land mass should be under forest cover, of which 60 percent has to be in the hills. In order to restore the forest cover to 33 percent (108 Mha), nearly 72.21 Mha of land needs to be brought under tree cover.

*H.K. Srivastava**, Department of Biotechnology, Block - II, C.G.O. Complex. Lodi Road, New Delhi 110 003

The forest cover is declining at alarming rates throughout the developing world. While the forest cover is progressively shrinking, the gap between the demand and supply of wood and wood-based industry is widening. This has resulted in a real crisis. In order to meet this challenge, there is an urgent need to mount major programs of restoring forest cover. Large-scale plantations will not only meet the growing demands of food, fodder and fuelwood, but will also ensure long range ecological security. In the former case, the underlying strategy should be to enhance biomass production

* by identifying location-specific elites and multiplying them on a mass scale by micro-propagation methods;
* Such a strategy, when taken in conjunction with the use of optimum of plant nutrients, nitrogen fixation and mycorrhizae can provide rapid and high yielding tree/shrub cover.

The average yield of biomass production from our forest is very low as compared to the yield level generally obtained in Europe, USA and South America. A good example to be cited here is from the Brazilian experience on *Eucalyptus*. From the plantation of 77,000 ha, one private group in Brazil was able to harvest 35m³/ha/yr. The same group later selected some elite material in *Eucalyptus* and as a result, the biomass production from the same plantation has been enhanced three fold to the tune of 100m³/ha/yr. This is one of the successful examples to illustrate the role of selection in enhancing biomass production. About 2,000 woody tree/shrub species have been exploited for a wide range of useful products in addition to the production of timber, fodder and fuelwood. The multipurpose trees and shrubs are deliberately grown and managed for preferably more than one intended use. This may be economically and/or ecologically motivated. It would be prudent to consider the utilisation aspect of forest products. About 10 percent of the biomass production by the forestry sector goes to meet the demand of the paper and pulp industry in the country. The remaining 90 percent of the biomass production is used as fodder, fuel, etc. The existing shortage of fuelwood and fodder is perhaps a reflection of the fact that whatever has been done to enhance the yields of multipurpose species, has not been enough. In agricultural production, particularly in major crops like wheat and rice, it has been established that only 15 percent of the total area under crop cultivation produces almost 75 percent of the agricultural output, whereas the rest 85 percent of the area produces the remaining 25 percent. The analogy to be made here is that a similar exercise could be undertaken in the forestry sector or agro-forestry, the selected areas could be taken up for intensive cultivation of multipurpose forestry species to give a significant boost to the biomass production to cater to the needs of rural and urban communities.

There is an urgent need to increase the productivity by planting multi-purpose species. Virtually all rural, agricultural or forestry development programs now include the expansion or introduction of multipurpose species, especially those that are both fast growing and highly adaptive to local environmental conditions. Unfortunately, the demand for such valuable plant material commonly exceeds supply. Presently, the supply of proper material in the form of seed or seedlings is difficult to obtain; also, there is inadequate information about the planting and management practices of various species under diverse forest and agro-climatic conditions.

FOREST GENETICS IN TREE BREEDING

The long life and large size of trees have always been the major barrier to progress in forestry, especially in the fields of forest genetics and tree breeding.

In forestry, some selected tree breeding has been practised. Initially, work was concentrated in Europe and North America but was lately taken up in Australia, Japan, Brazil, Africa and India. At the outset, Indian workers like Champion recognised the importance of genetic variation in the forest trees as early as 1925 and since then, a vast amount of information has been collected on the biology, cytogenetics and phenology of many indigenous species. However, the genetic input in India could not move beyond the identification of "plus" trees on phenotypic basis in a few important species. In recent years, a wide range of research related to tree breeding of a number of species has been started at many universities and research organisations. Thus, the possibility of genetic improvement in forest trees has now increased.

While the most advanced breeding work has been achieved in industrial species for saw-timber and pulp wood (eucalyptus, pines, poplars and spruces), some work is now in progress with fine hard woods (oak, teak, walnut) also. Nowadays, the most pressing demand for tree breeding lie in the rapid burgeoning of interest in multipurpose trees and in agro-forestry systems for social rural development. Recognising the definition of breeding strategy given earlier, the general objective could be the optimum management of resources to meet various needs; however, within this philosophical objective, there are several specific objectives that include:

Genetic Gain: This implies a gain in the population mean value with advancing generation resulting in the increased yield of the various products and services outlined above. These gains may be for each major environmental types separately or for a country or the region as a whole.

Production of Improved Planting Stock: Once the breeding population mean has reached to an acceptable level of improvement over the starting population value, bulk material is required for routine planting. This may be seeds directly obtained from seed orchards or vegetative propagules obtained through cuttings or layering.

Conservation of Genetic Variation: Inter-generational justice demands that current foresters bequeath to their successors material that permits them to obtain continued gain in all characters or to make change in the selective direction (if markets, processing methods, management systems or pests and disease changes). While a worldwide conservation lobby may succeed in conserving particular genetic resources of many species, tree breeders themselves must be aware of the need to conserve variability in breeding populations for future known and unknown requirement.

Security: This implies a physical security of the breeding material itself (requiring protection in duplication in two or more locations) and financial security brought out by being independent of other supply of planting stock. This is particularly important where exotic species are used, especially if seed supplies are unreliable or if hard currency is required to pay off for imports.

Demonstration: An essential part of the tree breeding strategy is the demonstration of the achieved gains to politicians, administrators, foresters and all others who are involved in the plantation program.

BIOTECHNOLOGY IN FORESTRY

Plant tissue culture and other emerging areas of biotechnology can contribute significantly in alleviating demands of food, fodder, fuelwood and health care placed upon India by its growing population, which will reach 1.5 billion by A.D. 2020. The need is to genetically modify the tree species which have direct relevance to the existence of mankind. Thus, trees will have to be modified for early and high yields of paper pulp, fiber or timber and for disease and pest resistance. We will have to engineer trees so as to reclaim about 100 Mha of wastelands in India. Coming to the fact that India is losing 1.3 Mha of the forest cover annually, rapid and mass propagation of elite tree species will have to be taken up for reforestation. Some of the reasons for the incredibly low average yields of our forests are:

* The use of wastelands for plantations which are the only land available for this purpose
* Plantations are generally based on wild stocks which are highly heterogeneous as far as yield is concerned
* Minimal integration of forest genetics and tree breeding
* Usage of outdated forestry management practices, and
* The practical method of propagation of many tree species is through seeds. This has two basic disadvantages:
(a) existing genetic variability in the seeds that produces extremely variable progenies, and
(b) it takes many years before seeds are produced.

As a result of these problems, the breeding program in tree species is slow and deficient. The other conventional methods of vegetative propagation, for example, rooting of cuttings, grafting, budding, layering, etc., produce genetically identical plants.

Biotechnology provides immense prospects for bringing about improvements in forestry species in terms of biomass production for fodder, fuel and timber. There are two systems of reproduction prevalent in forestry species. Sexual (propagation through seeds) and asexual (vegetative propagation and apomixis). Both systems together determine the genetic structure of the population. They, thus, form an integral part of the genetic system of the species. Whereas asexual/vegetative propagation gives rise to isogenic trees, the sexual system reproduces heterogenic trees. A successful system from the point of view of variation and evolution in a tree species would be to encourage a good balance of sexual and asexual methods of reproduction. The main factors limiting biomass production from tree and shrub species are water-logging, drought, soil salinity, alkalinity, diseases and pests. Conventionally, these problems are dealt through better water and land management practices and spraying of chemicals. Due to dwindling water resources and increasing cost of chemicals, etc., these methods are becoming less useful especially in the poor and developing countries. Plant tissue culture provides potential for shortening the time required in the production of improved cultivars through plant breeding. By the tissue culture technique, existing useful plants can be cloned rapidly, and by regenerating plants which are tolerant to various adverse conditions, it would be possible to reduce the cost of production and crop losses and increase the yields substantially. The detailed work plan in this regard includes:

* Selection of appropriate explant

* Selection of cryoprotectant for the preservation of the explant

* Regeneration of plantlets from the frozen explants through tissue culture

* Transfer of *in vitro* produced plants to the soil, their field evaluation, and

* Biochemical, cytological and morphological evaluation of plants for genetic stability.

Microspore and anther culture have been used as alternative methods for genetic improvement in plants. The significance of haploids in genetics and plant breeding has been realised for a long time. However, their exploitation remained restricted because of extremely low occurrence in nature. The technique of androgenesis has been successfully extended to numerous economically important plant species. The advantages of using haploids in crop improvement include rapid achievement of homozygosity, mutational and cytogenetical studies and creating recombinants as well as variants. Because of chromosomal instability during anther culture, haploids have the potential of rapid transfer of alien chromosome or genes into the genome of other cultivars in wide breeding programs. Chromosomal instability can be further increased by mutagenic treatment. Selection of mutants which are easily evaluated at the haploid level can be carried out under a given selection pressure. Another useful method of somatic embryogenesis offers tremendous potential for large-scale production of plants at low cost. Somatic embryos in large numbers can be produced in suspension cultures by using bioreactor technology which can further cut down labor, time and cost. Thus, mass plant propagation by tissue culture techniques has developed into an important industry with considerable potential for the future. A biotechnology company in Israel has recently overcome the problem of the high cost of producing plants from tissue culture by evolving an automated plant tissue culture system for mass propagation of plants. A new biological/mechanical vitromatic system integrates a bioreactor (fermenter) with a bioprocessor system. An automated transplanting machine transfers propagules to soil mix in greenhouse trays.

A specific work plan for developing intensive R & D to enhance biomass production may be confined to the following:

* Establishment of Pilot Plant Tissue Culture Units
* Selection of elite individuals and germplasm conservation
* *In vitro* multiplication of the elites
* Nitrogen fixing organisms and mycorrhizae
* Effect of macro- and micro-nutrients on plant growth
* Complete package of practices
* Long range R & D problems
  - Protoplast fusion/regeneration
  - Somatic embryogenesis
  - Synthetic/artificial seeds
  - Somaclonal variation
  - Haploids and double haploids
  - Triploids

In the country, some progress has already been made in terms of achieving high frequency regeneration of somatic embryos in eucalyptus and bamboo species. Efforts are being made to solve the problems of embryo synchronisation, proliferation, maturation, singulation and germination at the production level.

GENETIC ENGINEERING IN FOREST TREES

Many research groups have already commenced genetic engineering experiments using recombinant DNA technology to bring improvements in forest tree species in terms of biomass production and disease resistance. In 1986, a group from the University of California at Santiago introduced the 'fire fly' gene for the light generating enzyme luciferase into tobacco plants. In 1987, Dr. D.J. Durzan of the Department of Pomology at the University of California reported the insertion of this gene into the cells of conifers. While this will probably not result in Christmas trees that light up when they are cultured with the substrate luciferase, the construct can be used as the measure of gene expression. By applying the techniques of genetic engineering to forestry, Durzan's group has demonstrated that *Agrobacterium tumefaciens* can be used for gene transfer in Douglas-fir. Dr. R. Sedcroff recently reported gene transfer into loblolly pine, while other researchers have used *Agrobacterium* to insert a gene for glyphosate herbicide tolerance into a hybrid poplar. Forest biotechnology is still at the infancy stage and various techniques are being tried to improve the propagation and characteristics of forest trees. Dr. Durzan's group has used somatic polyembryogenesis to generate multiple embryos from a mass of cells for the cloning of conifers. The embryos which are produced can be encapsulated in a gel and stored until needed for the mass production of desirable trees. The cloned cells *in vitro* can also be used to implant genes carrying desired traits or to screen for specific characteristics. Another group at the University of Kentucky has recently succeeded in transferring *Bacillus thuringiensis* (BT) toxin gene into conifer and further used restriction fragment length polymorphism (RFLP) in *Populus deltoides*. Thus, in a matter of one or two years, the ability to genetically manipulate plants has been extended from tobacco through crop plants and cereals to the possibility of regenerating forest trees.

India's afforestation program is affected by the problems of alkaline, saline and eroded soils and low yields. Genetic engineering offer new ways to increase forest productivity, high timber quality and other desirable characteristics such as resistance to pests, pathogens, stress and herbicides. The Ti-plasmid of *A. tumefaciens* is an established vector for genetic engineering of a wide range of gymnosperms and dicotyledonous angiosperms. It is recently shown that foreign DNA can be introduced into a plant genome by direct gene transfer into protoplasts without the help of *A. tumefaciens*. The incorporation of osmoregulatory gene (proline AB gene) and insect resistance gene (BT toxin gene) into the plant genome will help in the development of drought, stress and pest resistant varieties.

CONCLUSIONS

Genetic engineering in trees is still in its infancy. Molecular biologists must identify silviculturally important genes from the unexploited gene pools. Fundamental questions remain unanswered about the feasibility of some of these techniques. Tree improvement involves three stages:

* creation of variability
* selection of desired traits, and
* testing of plants for yield performance.

Genetic engineering is expected to contribute in the first two stages only. More vectors need to be searched out to carry the foreign genes into the plant cells. The scientists must develop reliable methods for regenerating plants from single cell culture. Little is known about the response of the plants to the introduction of foreign or alien genes, if, for instance, yield or heterotic vigour may suffer. Progress in plant genetic engineering is hampered by limited knowledge of plant molecular biology. The successful application of genetic engineering to plants will require a fundamental breakthrough in the understanding of gene expression and regulation as well as increased knowledge of plant physiology and metabolism. Although it is too early to assess with accuracy either the potential or limitations of genetic engineering for tree improvement, other techniques based on the ability to culture and regenerate plant cells, are already producing a short-cut (compressing breeding cycle) in the selection of breeding programs. Careful choice for monogenic versus polygenic traits, etc. has to be made. Success could be assured in case and monogenic inheritance, but much work needs to be done on the molecular biology of polygenic traits. The incorporation of single gene trait(s) such as resistance to pests and diseases is now exhibiting promise. However, traits such as salt, moisture, temperature, stress resistance so far appear to be much difficult tasks to be accomplished.

REFERENCES

Champion, H.G., 1925, Contributions towards a knowledge of twisted fiber in trees, *Indian For. Rec. Silvic.,* 11: 11-80.

Champion, H.G., 1933, Importance of origin of seed used in forestry, *Indian For. Rec. Silvic.,* 17: 77.

Dogra, P.D., 1981a, Forest genetics - research and application in Indian forestry, I, *Indian For.,* 107: 191-262.

Dogra, P.D., 1981b, Forest genetics - research and application in Indian forestry, II, *Indian For.,* 107: 263-288.

Morel, G., 1960, Producing virus-free *Cymbidium, Amer. Orchid Soc. Bull.,* 29: 495-497.

Morel, G., 1965, Clonal propagation of orchids by meristem culture, *Cymbidium Soc. News,* 20: 3-11.

Morel, G. and Martin, C., 1952, Guerison de Dahlias atteints d'une maladie a virus, *C.R, Acad. Sci. (Ser.* D) 235: 1324-1325.

Redenbaugh, K., Viss, P., Slade, D. and Fujii, J.A., 1987, Scale-up artificial seeds. In: *Plant Tissue and Cell Culture* (C.E. Green, D. Somers, W.F. Hacket and D.D. Biesboer, eds.), pp. 473-493, Alan R. Liss, New York.

Skoog, F. and Miller, C.O., 1957, Chemical regulation of growth and organ formation in plant tissues cultured *in vitro, Symp. Soc. Exp. Biol.,* 11: 118-131.

Vasil, I.K., 1987, Developing cell and tissue culture systems for the improvement of cereal and grass crops, *J. Plant Physiol.,* 128: 193-218.

Vasil, I.K. and Vasil, C., 1986, Regeneration in cereal and other grass species. In: *Cell Culture and Somatic Cell Genetics of Plants Vol. 3, Plant Regeneration and Genetic Variability* (I.K. Vasil, ed.), pp. 121-150, Academic Press, Orlando.

27

GENETIC MANIPULATION OF WOODY SPECIES

Malathi Lakshmikumaran

ABSTRACT

Advances in biotechnology have made it possible to genetically transform several plant species. Genetic transformation of forest tree species for improvement would be useful as compared to the conventional methods of breeding. One of the pre-requisites for the introduction of foreign genes into tree species is the development of methods for gene transfer. Many methods have been developed for the introduction of foreign genes into crop plants. The most effective method is *Agrobacterium* mediated gene transfer. In plants, a number of specific genes have been transferred using *Agrobacterium*, such as herbicide resistance gene, kanamycin resistance gene, luciferase gene and gene coding for *Bacillus thuringensis* toxin. A few tree species have also been transformed using *Agrobacterium* strains. The notable ones are loblolly pine, sugar pine, Douglas-fir and poplar. *Agrobacterium* mediated transformation of different plant species is reviewed.

Malathi Lakshmikumaran * Tata Energy Research Institute, 90 Jor Bagh, New Delhi - 110 003, India.

INTRODUCTION

The transfer of foreign genes into plant cells has become a commonly used technique. The different methods are: direct gene transfer (Portrykus et al., 1985a,b), *Agrobacterium* mediated transfer (Horsch et al., 1985; Ooms et al., 1987) and protoplast fusion (Melchers, 1980; Muller-Gensert and Schieder, 1987). The most widely used method depends upon the natural gene transfer system from *Agrobacterium tumefaciens*.

Agrobacterium Mediated Gene Transfer

Agrobacterium tumefaciens is a gram-negative soil bacterium which is capable of causing crown gall disease in a wide range of dicotyledonous plants. *Agrobacterium* infects the wounded cells which then proliferate to form a tumor (called crown gall). The bacteria attach to the walls of the plant cells at the wound site. They are stimulated by phenolic compounds released by these cells to transfer a portion of the Ti-plasmid called T-region into the plant cells (Stachel et al., 1985). The T-region (Fig. 1) is integrated into the plant nuclear genome and stably maintained (Lemmers et al., 1980). It directs the synthesis of opines which are not found in normal plant cells. These opines are metabolised specifically by the *Agrobacterium*. These opines stimulate the undifferentiated growth of the infected tissue to form tumors or galls. *Agrobacterium* can thus be considered as a natural plant engineering system since it can transfer genes into plants.

The Ti-plasmid of *A. tumefaciens* is approximately 200 Kilo bases (Kb) long. The Ti-plasmid contains two regions that are essential for plant transformation; the virulence (*vir*) region and the T-DNA border sequences (Fig. 1). The *vir* region codes a group of bacterial genes whose products are required for processing the T-DNA from the Ti-plasmid and for its transfer and integration into the plant genome (Ooms et al., 1980; Horsch et al., 1986).

The T-DNA border sequences are imperfect direct repeats of 25 base pairs (bp) that flank the T-DNA and are required for the transfer of the DNA into plant cells (Nester et al., 1984). The right border sequences seem to be essential for the transfer of T-DNA to the plant genome (Wang et al., 1984). Since the T-DNA is transferred to plant tissues on infection, any foreign DNA cloned into T-DNA is also integrated (Herrera-Estrella et al., 1983). The disadvantage of Ti-plasmid is that it causes tumor formation in plants. So the Ti-plasmid vectors that have deletions of the *onc* genes have been developed. Two types of Ti-plasmid have been constructed for gene transfer.

The co-integrate type Ti-plasmid has been constructed by Zambryski et al. (1983), in which the border sequences flank a pBR322 copy. The acceptor vector pGV3850 is derived from nopaline-type Ti-plasmid pTiC58. All the T-DNA genes responsible for tumorous growth are deleted and substituted by pBR322. It contains

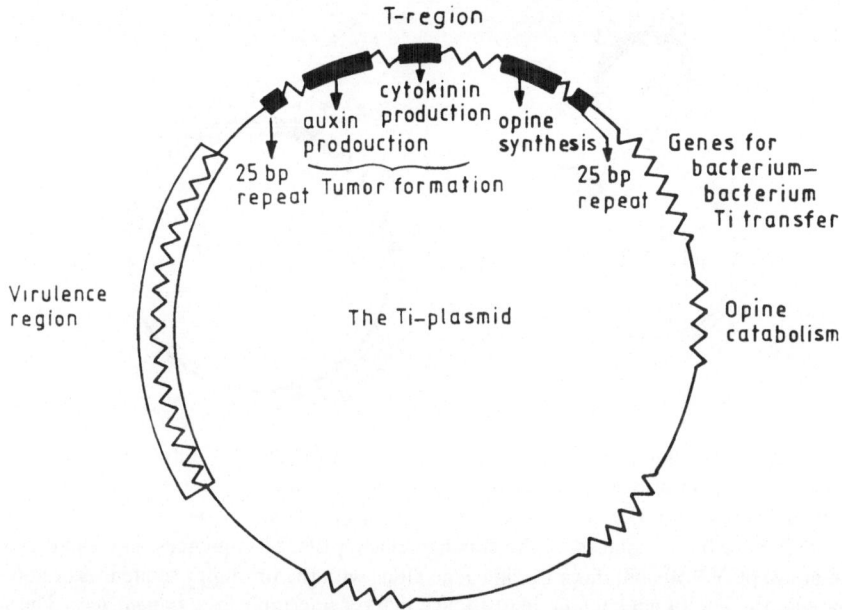

Fig. 1: *Ti-plasmid:* The virulence region and T-region required for tumor formation are indicated. The virulence region is not transferred but enables the plasmid to induce tumor. The imperfect 25 base pair direct repeats are on either side of the T-DNA. Regions outside the T-DNA govern the catabolism of opines.

the T-DNA border regions and all contiguous Ti-plasmid sequences outside the T-DNA region. The T-DNA sequences near the right border encoding the nopaline synthase enzyme are present as a marker to identify transformed cells. It is one of the most versatile acceptor plasmids. Any foreign gene cloned in pBR322 can be easily inserted between the T-DNA borders by homologous recombination to form a co-integrate plasmid. This plasmid is used to transform plants (Fig. 2). *Agrobacterium* strains containing pGV3850 were virulent on a variety of plant hosts. The vector, being non-oncogenic, allowed the full differentiation of transformed cells into transgenic plants (Zambryski et al., 1983).

A binary vector has two plasmids. One is the versatile vector which is able to replicate both in *E. coli* and *A. tumefaciens*. It also contains the T-DNA border sequences flanking the plant selectable marker and also carries unique cloning sites. The second one is a helper Ti-plasmid that contains the *vir* genes which are capable of acting in trans on the T-DNA. This plasmid allows the transfer of the DNA between the border sequences to the plant nuclear genome (Hoekema et al., 1983; An et al., 1985).

A number of plant species have been transformed using *A. tumefaciens*. Many different methods have been developed for the introduction of Ti-plasmid from *A. tumefaciens* to host plant cells.

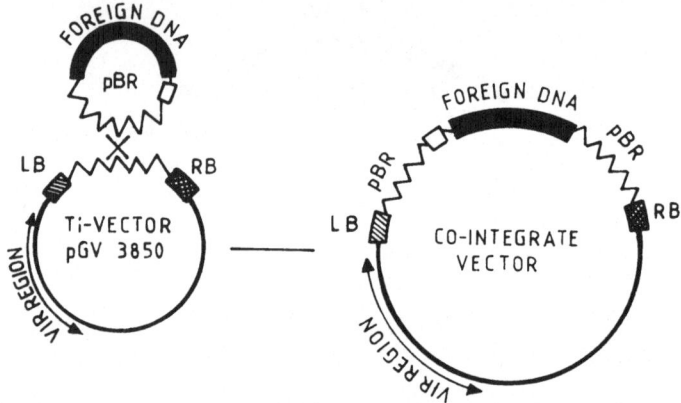

Fig. 2: Vector Ti-plasmid pGV3850 and introduction of foreign DNA into it by co-integration: The plasmids are not drawn to scale - pGV3850 contains all the Ti-plasmid sequences of the nopaline C58 except for the internal oncogenes of the T-DNA. The oncogenic genes between the T-DNA have been replaced by the cloning vehicle pBR322 sequences. The small plasmid shown above pGV3850 contains a foreign gene along with the promoter sequences of nopaline synthase and an additional kanamycin resistance marker selectable in *A. tumefaciens*. The small plasmid also contains sequences homologous to pBR322. This small plasmid can be inserted into pGV3850 by a single cross-over between the homologous pBR sequences. Recombination results in the relative reversal of the foreign gene in the cointegrate plasmid. *Agrobacterium* strains containing such plasmid have the ability to insert foreign DNA into the chromosomal DNA of the plant cells.

1. Co-cultivation: This is a very efficient transformation system. It involves the incubation of *A. tumefaciens* bacterial cells with plant tissues - preferably protoplasts and subsequent regeneration. This system has been used for the transformation of tobacco (Marton et al., 1979), poplar (Fillatti et al., 1987) and many other plants (Velten and Schell, 1985).

2. Leaf Disc Transformation: Discs are punched from leaves and incubated with a virulent *A. tumefaciens* strain harboring a modified Ti-plasmid containing a selectable marker. After treatment, the leaf discs are incubated on shoot regeneration medium and then transferred to the antibiotic medium for selection. The transformation of tobacco (Horsch et al., 1985) and tomato (McCormick et al., 1986) was obtained using this comparatively simpler method.

3. Fusion with *Agrobacterium* Spheroplasts: This involves the incubation of freshly isolated *Agrobacterium* spheroplasts with plant protoplasts. In this method, Ti-plasmid transfer takes place due to the fusion of protoplasts and spheroplasts. This system was used for the transformation of tobacco, using kanamycin resistance gene (Hain et al., 1984).

Gene Transfer in Forest Tree Species

Conventional breeding methods of forest trees are difficult because of long sexual generation time. Genetic engineering of forest tree species will be useful, as genetic changes are possible in a short period of time. A number of tree species have been transformed with foreign genes.

Pine is one of the commercially important species for fiber production. A number of wild type strains of *Agrobacterium* were used for infecting pines. Two strains (M2/73 and U3) were able to induce gall formation in loblolly pine. Opines were identified in the transformed pine tissues (Sederoff et al., 1986). This clearly indicates that loblolly pine can be transformed using *A. tumefaciens*. A number of pine species, Douglas-fir and incense-cedar were transformed using three strains of *Agrobacterium*. Gall formation and opine synthesis were used to assay gene transformation (Stomp et al., 1988).

Sugar pine shoots were transformed with binary *Agrobacterium* strains containing the chimeric gene neomycin phosphotransferase under the control of nopaline synthase promoter (Klee et al., 1985). NPT II activity was detected in the transformed tissues.

Douglas-fir (*Pseudotsuga menziesii*) was transformed using the *Agrobacterium* mediated gene transfer. Two strains of *Agrobacterium* (K12x562E and K12x167) that contain a chimeric bacterial kanamycin resistance gene were used for the transformation of Douglas-fir. Southern blot analysis of DNA from the transformed plants confirmed the presence of foreign genes in the transformed cells (Dandekar et al., 1987).

Genetic transformation of poplar was demonstrated by Parson et al. (1986). Recently, a transformation and regeneration system was developed for *Populus* hybrids (Fillatti et al., 1987). *Populus* NC-5339 (*P. alba* x *P. grandidentata*) was transformed using binary oncogenic strain of *A. tumefaciens*. The vector contained two neomycin phosphotransferase II (NPT II) and one bacterial 5-enol pyruvyl shikimate 3-phosphate (EPSP) synthase *(aro A)* genes. The transformed cells were assayed for NPT II enzyme activity. The presence of EPSP synthase was confirmed by Western blotting. The *aro A* gene confers tolerance to the herbicide glyphosate. To determine whether the transformed *Populus* plants were resistant to glyphosate, the control (untransformed) plants and transformed plants were sprayed with glyphosate. The transformed plants showed higher levels of tolerance to glyphosate as compared to control plants. This indicated that the introduction of genes of agronomic importance is possible in woody species.

REFERENCES

An, G., Watson, B.D., Stachel, S., Gordon, M.P. and Nester, E.W., 1985, New cloning vehicles for the transformation of higher plants, *EMBO J.,* 4: 277-284.

Dandekar, A.M., Gupta, P.K., Durzan, D.J. and Knauf, V., 1987, Transformation and foreign gene expression in micropropagated Douglas-fir (*Pseudotsuga menziesii*), *Biotechnology*, 5: 587-590.

Fillatti, J.J., Sellmer, J., McCown, B., Haissig, B. and Comai, L., 1987, *Agrobacterium* mediated transformation and regeneration of *Populus*, *Mol. Gen. Genet.*, 206: 192-199.

Hain, R., Steinbiss, H.H. and Schell, J., 1984, Fusion of *Agrobacterium* and *E. coli* spheroplasts with *Nicotiana tabacum* protoplasts: Direct gene transfer from microorganisms to higher plants, *Plant Cell Rep.*, 3: 60-64.

Herrera-Estrella, L., De Block, M., Messens, E., Hernalsteens, J.P., Van Montagu, M. and Schell, J., 1983, Chimeric genes as dominant selectable markers in plant cells, *EMBO J.*, 2: 987- 995.

Hoekema, A., Hirsch, P.R., Hooykaas, P. and Schilperoot, R.A., 1983, A binary plant vector strategy based on separation of *vir* and T-region of the *Agrobacterium tumefaciens* Ti-plasmid, *Nature (Lond.)*, 303: 179-180.

Horsch, R.B., Fry, J.E., Hoffmann, N.L., Eichholtz, D., Rogers, S.G. and Fraley, R. T., 1985, A simple and general method for transferring genes into plants, *Science*, 227: 1229-1281.

Horsch, R.B., Klee, H.J., Stachel, S., Winans, S.C., Nester, E.W., Rogers, S.G. and Fraley, R.T., 1986, Analysis of *Agrobacterium tumefaciens* virulence mutants in leaf discs, *Proc. Nat. Acad. Sci.*, USA, 83: 2571-2575.

Klee, H.J., Yanofsky, M.F. and Nester, E.W., 1985, Vectors for transformation of higher plants, *Biotechnology*, 3: 637-642.

Lemmers, M., de Benekeleer, M., Holsters, M., Zambryski, P., Depicker, A., Hernalsteens, J.P., Von Montagu, M. and Schell, J., 1980, Genetic identifications of functions of T-DNA transcripts in octopine crown gall, *EMBO J.*, 1: 147-152.

Marton, L., Wullems, G.J., Molendijk, L. and Schilperoort, R.A., 1979, *In vitro* transformation of cultured cells from *Nicotiana tabacum* by *Agrobacterium tumefaciens*, *Nature (Lond.)*, 277: 129-131.

McCormick, S., Niedermeyer, J., Fry, J., Barnason, A., Horsch, R. and Fraley, R., 1986, Leaf disc transformation of cultivated tomato (*Lycopersicon esculentum*) using *Agrobacterium tumefaciens*, *Plant Cell Rep.*, 5: 81-84.

Melchers, G., 1980, Protoplast fusion, mechanisms and consequences for potato breeding and production of potatoes + tomatoes. In: *Advances in Protoplast Research*, (L.Ferency and G.L. Farkas, eds.), pp. 283-286, Pergamon Press, Oxford.

Muller-Gensert, E. and Schieder, O., 1987, Interspecific T-DNA transfer through plant protoplast fusion, *Mol. Gen. Genet.*, 208: 235-241.

Nester, E.W., Gordon, M.P., Amasino, R.M. and Yanofsky, M.F., 1984, Crown gall: A molecular and physiological analysis, *Ann. Rev. Plant Physiol.*, 35: 387-413.

Ooms, G., Burrell, M.M., Karp, A. and Hille, J., 1987, Genetic transformation in two potato cultivars with T-DNA from disarmed *Agrobacterium*, *Theor. Appl. Genet.*, 73: 744-750.

Ooms, G., Klapwijk, P.M., Poulis, J.A. and Schilperoort, R., 1980, Characterization of Tn 904 insertions in octopine Ti-plasmid mutants of *Agrobacterium tumefaciens*, *J. Bacteriol.*, 144: 82-91.

Parson, T.J., Sinkar, V.P., Stettler, R.F., Nester, E.W. and Gordon, M.P., 1986, Transformation of poplar by *Agrobacterium tumefaciens*, *Biotechnology*, 4: 533-536.

Portrykus, I, Shillito, R.D., Saul, M.W., and Paszkowskii, 1985a, Direct gene transfer-state-of-the-art and future potential, *Plant Mol. Biol. Rep.*, 3: 117-128.

Portrykus, I., Saul, M.W., Petruska, J., Paszkowskii, J. and Shillito, R.D., 1985b, Direct gene transfer to cells of monocots, *Mol. Gen. Genet.*, 199: 183-188.

Sederoff, R., Stomp, A., Chilton, W.S. and Moore, L.W., 1986, Gene transfer into loblolly pine by *Agrobacterium tumefaciens*, *Biotechnology*, 4: 647-649.

Stachel, S.E., Messens, E., Van Montagu, M. and Zambryski, P., 1985, Identification of the signal molecules produced by wounded plant cells that activate T-DNA transfer in *Agrobacterium tumefaciens*, *Nature (Lond.)*, 318: 624-629.

Stomp, A.M., Loopstra, C., Sederoff, R., Chilton, S., Fillatti, J., Dupper, G., Tedeschi, P. and Kinclaw, C., 1988, Development of a DNA transfer system for pines. In: *Genetic Manipulation of Woody Plants* (J.W. Hanover and D.E. Keathley, eds.), pp. 231-241, Plenum Press, New York.

Velten, J. and Schell, J., 1985, Selection-expression plasmid vectors for use in genetic tranformation of higher plants, *Nucleic Acid Res.*, 13: 6981-6998.

Wang, K., Herrera-Estrella, L., Van Montagu, M. and Zambryski, P., 1984, Right 25 bp terminus sequence of the nopaline T-DNA is essential for and determines direction of DNA transfer from *Agrobacterium* to the plant genome, *Cell*, 38: 455-462.

Zambryski, P., Joos, H., Genetello, C., Leemans. J., Van Montagu, M. and Schell, J., 1983, Ti plasmid vector for the introduction of DNA into plant cells without alteration of their normal regeneration capacity, *EMBO J.*, 2: 2143-2150.

28

GENETIC TRANSFORMATION SYSTEM IN DOUGLAS-FIR *Pseudotsuga menziesii*

Pramod K. Gupta[*],
Abhaya M. Dandekar[**] **and**
D.J. Durzan[**]

ABSTRACT

Tumors have been induced on *in vitro*-grown shoots/seedlings of *Pseudotsuga menziesii* using two strains of *Agrobacterium tumefaciens*. These strains contain a derivative of the wild type Ti plasmid PtiA6 that contains a chimeric foreign gene encoding resistance to the antibiotic kanamycin. Tumors that were grown on phytohormone-free medium synthesized octopine. Transformed cells demonstrated the presence of the foreign DNA sequence and expressed kanamycin phosphotransferase activity

Promod K. Gupta - To whom all correspondence may be addressed [*]Division of Biochemical Sciences, National Chemical Laboratory, Pune - 411 008, India.

Abhaya M. Dandekar and *D.J. Durzan*[**] Department of Pomology, University of California, Daviz, U.S.A. 95616.

NCL Communication No. 4404

[APH(3')II]. The luciferase gene from the 'firefly'(*Photinus pyralis*) was directly transferred to embryonal Douglas-fir protoplasts through electroporation. Transient luciferase gene expression was observed after 36 hours of electroporation. These results show the potential of genetic engineering in Douglas-fir, one of the most important conifer species in North America.

INTRODUCTION

Pseudotsuga menziesii (Mirb) Franco (Douglas-fir), one of the most important timber species in the world (Preston, 1976), is a member of the family Pinaceae. It is one of the biggest trees, occasionally attaining a height of 75 m on the West Coast and up to 40 m on the Rocky Mountains. Douglas-fir dominates the most productive forest lands of western North America (Preston, 1976). The productivity of Douglas-fir has increased to about 70 percent over that in the natural forest on the same site-class (Farnum et al, 1983). This is due to better silvicultural practices and breeding programs. However, even these intensively managed plantations are achieving less than 50 percent of their potential productivity (Farnum et al, 1983). It is likely that progress can continue with existing techniques. However, many of the gains in forest tree productivity in the next few decades will come from a more effective utilization of land resources because of economic, climatic (environmental) or soil limitations. The emphasis in the future will be to obtain and sustain optimal productivity under all conditions, especially in "tree-poor" countries of the Third World (Powledge, 1984). This means that goals will be more complex and difficult and progress in producing new varieties will require the cooperative effort of scientists from various disciplines, e.g., molecular biology, tissue/cell culture, plant breeding, etc. Modern biotechnology approaches that utilise recombinant DNA techniques along with tissue/cell culture, offer a new set of tools that can be used to improve the existing germplasm of Douglas-fir and other important forest crops.

During the past two years, great strides have been made in the introduction and expression of foreign genes in plants. A significant milestone was crossed when several laboratories reported the introduction, expression and genetic stability of foreign genes at the whole plant level (Fraley et al, 1985). The strategy for this important advancement involved the harnessing of the *Agrobacterium* Ti-plasmid system as a vector, the use of foreign genes as selectable genetic markers and the regeneration from cell culture of whole plants harboring the foreign gene. This approach opens the door to the application of this technology to forest trees. *Agrobacterium tumefaciens* has a broad host range and has been shown to infect a wide range of gymnosperms and dicotyledonous angiosperms including Douglas-fir. The Ti-plasmid of *A. tumefaciens* has long been proposed as a vector for the genetic engineering of plants and only in recent years has its potential begun to be realised. Normally, the transfer of a defined segment of Ti-plasmid into the genome of a plant results in the formation of a crown-gall tumor (DeCleene and Lay, 1976; Chilton et al., 1977). However, morphologically normal plants can be produced if the tumor-inducing genes are removed (disarmed

Ti-plasmid) from the transferred DNA (T-DNA; Fraley et al., 1985). Chimeric genes constructed by fusing a bacterial antibiotic-resistance gene to the control signal of the T-DNA encoded nopaline synthetase gene have been transferred and expressed at the whole plant level using disarmed Ti-plasmids. All plants carrying the chimeric genes developed normally, set seeds and the inheritance of the antibiotic-resistance trait was shown to be Mendelian (DeBlock et al., 1984).

The direct gene transfer technique has also been developed for plant transformation in response to a limited host range of *A. tumefaciens* -mediated DNA transfer. One of the most useful techniques of direct gene transfer is electroporation. Electroporation refers to the process of applying a high intensity of electric field to reversibly permeablise the cell membrane. Electroporation of cell membranes has proved useful for cell fusion as well as for gene transfer into plant and animal cells. Stable transformation and gene expression has been achieved through electroporation of plant cells with a foreign gene (Fromm et al., 1986). Transformants have been shown to regenerate and stably pass on the foreign gene to their progeny (Shillito et al., 1985). In this paper, we are focusing our efforts on such a multi-disciplinary biotechnological approach, with the aim of developing a gene transfer system in Douglas-fir and subsequently using this system to transfer genes for resistance to environmental stress, insects, pests, etc.

MATERIALS AND METHODS

Shoot and seedlings were grown *in vitro* by methods described earlier (Gupta and Durzan, 1985). 40-50 mm long shoots/seedlings were used for *Agrobacterium* infection. Shoots and seedlings were infected, first by wounding the epidermal tissue, either by piercing with a sterile needle several times at one spot, or by cutting the surface with a scalpel (Table 1). Inoculations were performed by applying the bacterial suspension culture to the wound surface. The excess suspension was removed by soaking on sterile filter paper. The infected shoots/seedlings were then cultured on DCR basal medium (Gupta and Durzan, 1985) with two percent sucrose and 0.6 percent Difco Bacto agar. Cultures were incubated in continuous light at $28 \pm 2°C$. Primary tumors were grown on DCR basal medium with carbenicillin (50 mg l^{-1}), casein hydrolysate (500 µg l^{-1}), glutamine (50 µg l^{-1}), sucrose (three percent) and agar (0.7 percent). Normal callus was initiated by inoculating untransformed Douglas-fir stem segments from micropropagated shoots or seedlings on DCR medium with 2,4-D (3.0 µg l^{-1}) and BAP (0.2 µg l^{-1}). Cultures were incubated in continuous white light (2.8, 2.0 and 0.5 uW cm^{-2} nm^{-1} blue, red and far-red spectrum, respectively).

Bacterial Growth

AELB (1 percent tryptone, 0.5 percent yeast extract) and AB (Chilton et al., 1974) were used for liquid culture of *A. tumefaciens* strains. Before infection, strains were grown on a reciprocating shaker at 28°C in liquid culture for 24 h.

Opine Assay

Opines were extracted from the callus (μgl^{-1}) in acidified methanol (10 drops conc. HCl and 100 ml of methanol). 20 μl was spotted on Whatman No. 3 paper for the assay. 5 ul each of standard arginine (1 mM) and octopine (1 mM) were used as standard and an aqueous solution of methylene green was used as a marker which migrated just behind arginine. Paper electrophoresis was performed according to the methods described by Otten and Schilperoort (1978). UV light gives a yellow-green fluorescence from octopine and arginine spots.

Kanamycin Phosphotranferase [APH (3') II] Assay

Crude cell extraction was carried out according to the method described by Reiss et al. (1984). Non-denaturing polyacrylamide gel electrophoresis (without SDS) was performed according to Schrier et al. (1985).

DNA Isolation and Southern Blot Analysis

Total DNA was isolated from one gm of tissue by a method described earlier (Dandekar et al., 1987). DNA was cleaved with Pst 1, fractioned on 0.8 percent agarose gel and hybridised with a nick translated 923 bp Pst 1 fragment of Tn5 that contains the structural sequence of the [APH (3') II] gene (Southern, 1975).

Electroporation

Protoplasts were obtained from an embryonal cell suspension of Douglas-fir, as described earlier (Gupta and Durzan, 1988b). The luciferase gene from the 'fire fly' *Photinus pyralis* was used as a reporter of gene expression by light production. Gene constructs of the fire fly luciferase structural gene containing the 35s cauliflower mosaic virus promoter (Ow et al., 1986), present on a plasmid pDO432, were used for electroporation. Protoplasts (10^6 ml^{-1}) were suspended in 800 ul of buffer [(10 mM Hepes, 140 mM KCl, 5mM CaCl$_2$, six percent myo-inositol (w/v)] with or without three percent polyethylene glycol (PEG 600) (w/v) mixed with 600 ug salmon sperm DNA as carrier and with or without 60 ug supercoiled plasmid pDO432. Electroporation was carried out according to published procedures (Fromm et al., 1985). An electric pulse was delivered from a 500 μF capacitor charged to 200 v. This corresponds to an electric field intensity of 500 v/cm and RC times constant of 35. Luciferase assay was carried out as described earlier (Ow et al., 1986). Data are expressed in light units relative to that produced by a fire fly tail extract.

RESULTS

Different methods and different strains were used for infection, as described in Table 1 and 2. The two strains of *Agrobacterium tumefaciens* K12x562E and K12x167 are derivatives of the wild type Ti-plasmid pTiA6 that contain a chimeric bacterial antibiotic-resistance gene. These two strains induced tumors, both on micropropagated shoots and *in vitro* seedlings. The *A. tumefaciens* K12x562E contains pTiA6 : CGN562E that has a chimeric kanamycin resistance gene with a promoter for octopine synthetase gene and was constructed by Dandekar et al. (1987). The strain K12x167 contains PTiA6:CGN167 that also contains a chimeric kanamycin-resistance gene with the promoter for 35s transcript of cauliflower-resistance mosaic virus and was constructed by Dandekar et al. (1987). Twenty percent of micropropagated shoots and 15 percent of *in vitro* seedlings developed tumors with strain K12x562E whereas with strain K12x167, five percent of the micropropagated shoots and 40 percent of the *in vitro* seedlings developed tumors. Tumor formation was obtained within 5-6 weeks near the base of micropropagated shoots and within 9-10 weeks near the shoot and root junction (hypocotyl) on seedlings with both the strains of *A. tumefaciens* (K12x167 and K12x562E). Growth of tumors was significantly higher (mean weight 115 $\pm$ 6.25 mg) on seedlings than micropropagated shoots (mean weight 85 $\pm$ 3.56 mg; see Table 1.

Tumors were excised and cultured on DCR medium (see protocol) containing 500 ug ml^{-1} carbenicillin without phytohormones. Within 4-5 weeks, callus growth of tumors obtained from seedlings (405 $\pm$ 23.67 mg mean fresh wt.), was significantly

Table 1 : The effect of infection methods in gall formation by two strains of *Agrobacterium tume-faciens*

	Micropropagated shoots			In vitro grown seedlings		
	Gall formation percent		Fresh Wt. of gall mg	Gall formation percent		Fresh Wt. of gall mg
	Strain K12x562E	Strain K12x167		Strain K12x562E	Strain K12x167	
1. Piercing						
Apical Region	-	-	-	-	-	-
Basal Region	-	-	-	-	-	-
2. Splitting						
Aplical Region	-	-	-	-	-	
Basal Region	20	5	85$\pm$3.56	15	40	115$\pm$6.25

(-) : nil ; Apical region : just below the apex ; Basal region : 20-30 mm below the apex of the shoot, for seedling near the junction of the shoot and root (hypocotyl).

Table 2 : Genotype of *Agrobacterium tumefaciens* strains

Strain Designation	LB	Selectable genetic marker Promoter--Str.seq.--3'end	Oncogenes	Opine	RB
pMON 200 II	+	nos 5'APH(3')II nos 3'	+ +	Nop	+
K12xpCGn562E	+	ocs 5'APH(3')II ocs3'RBLB)$_2$	+ +	Oct	+
K12xpCGn167	+	CaMV35s5'APH(3')II ocs3'	+ +	Oct	+
K12 (WT)	+		+ +	Oct	+

Nos   -   nopaline synthetase
Nop   -   nopaline
Ocs   -   octopine synthetase
Oct   -   octopine
LB   -   Left border sequence
RB   -   Right border sequence

CaMV35s - cauliflower mosaic virus 35s transcript.
5' - promoter sequence
3' - polyadenylation sequence
Oncogenes - three T-DNA genes
  involved in the synthesis of
  auxin (indole acetic acid) and
  cytokinin (ribosylzeatin).

higher when compared to that of micropropagated shoots (171 $\pm$ 8.31 mg mean fresh wt.). However, the growth of the callus, after a second sub-culture from micropropagated shoots and after a third sub-culture from seedlings was inhibited and the callus displayed a tendency to turn brown on phytohormone-free medium.

Opine Detection

Opine analysis was carried out by extraction in acidified methanol (see Materials and Methods) and opines were separated by high voltage paper electrophoresis (Otten and Schilperoort, 1978). Two spots were detected one corresponding to arginine and the other to octopine, when compared with the mobility of standard arginine and octopine. Octopine was detected in all tumor calli obtained after infection with strains K12x562E and K12x167. Octopine was not detected in the normal callus (untransformed callus) or uninfected tissue.

Kanamycin Phosphotransferase Activity [APH(3') II]

Tumors displayed variability in their growth response in the DCR medium containing kanamycin. Some tumors displayed stable growth on concentrations as high as 200 ug ml^{-1}. The normal (untransformed) callus was unable to grow at this concentration of the antibiotic. The enzyme kanamycin phosphotransferase [APH(3')II] encoded by transposon Tn5 confirms resistance on its host cells to both the antibiotics kanamycin and neomycin (Davis and Smith, 1976). The [APH(3')II] enzyme catalyses the phosphorylation of kanamycin, using ATP as a substrate (Haas and Dowding, 1975). This reaction has been performed directly after separation of proteins by nondenaturing polyacrylamide gel electrophoresis (Laemmli, 1970) and layering this gel with a secondary agarose gel containing the substrate kanamycin and (Y-^{32}P) ATP.

The [APH(3')II] enzyme converts kanamycin into its ^{32}P labelled derivatives, which is immobilised on Whatman P81 paper (Schreier et al., 1985). The P81 paper was subsequently treated with proteinase K to suppress other phosphorylated proteins but not phosphorylated kanamycin. Radioactive bands (^{32}P labelled kanamycin) were visualised by autoradiography. The results demonstrated the presence of a radioactive band (phosphorylated kanamycin) from the tumor callus obtained after infection with both *A. tumefaciens* strains. The electrophoretic mobility was similar to that of the standard [APH(3')II] enzyme isolated from *E. coli* carrying Tn5. No radioactive band was detectable in the normal callus (untransformed).

Genomic DNA Analysis

To verify the integration of the kanamycin resistance [APH(3')II] gene into the chromosome of Douglas-fir, we carried out a Southern blot analysis (Southern, 1975). The total DNA from the tumor callus and the control (untransformed) callus was isolated and digested with restriction enzyme Pst I and transferred to nitrocellulose paper. The 923 bp Pst I fragment was used as a hybridisation probe, containing most of the coding region of the [APH(3')II] enzyme (bp 1732 to 2655 of Tn5). Genomic fragments isolated from the tumor callus hybridised to the probe DNA. The DNA fragment isolated from the control (untransformed) callus did not hybridise.

Ninety percent viable protoplasts were obtained from three-day-old embryonal suspension cultures of Douglas-fir. After electroporation, the viability of protoplasts had reduced to 45-55 percent. The luciferase activity was measured after 36 hours of plating of electroporated protoplasts. No activity was detected without plasmid pDO432 DNA, either in the presence or absence of polyethylene glycol (PEG). The luciferase activity was detected in electroporated protoplasts with pDO432 DNA. The activity of luciferase was higher in the presence of PEG.

DISCUSSION

These results suggest that the foreign gene has been stably transferred and expressed in Douglas-fir cells. Southern blot hybridisation reveals the integration of kanamycin-resistant [APH(3')II] gene in the plant DNA.

These results show that the *A. tumefaciens* strain carrying the recombinant derivative of the octopine producing Ti-plasmid pTiA6, successfully infects Douglas-fir. Excised tumors have successfully grown on medium without phytohormones and synthesised opine (octopine), demonstrating that the cells were transformed with *Agrobacterium* T-DNA. These results confirm that biotype I *Agrobacterium* infects Douglas-fir, a point that is not apparent in the studies reported by De Cleene and De Ley (1976) in their review.

The growth of the gall callus on medium containing kanamycin, as well as the detection of [APH(3')II] gene product demonstrate the transcription of chimeric kanamycin resistance gene and its translation into an active protein [APH(3')II]. The level of activity of the [APH(3')II] enzyme could be a function of the promoter activity.

We have observed that although a culture initiated on hormone-free media have rapid growth, further growth is inhibited on sub-culturing (second and third sub-culture) on the same medium. This may be due to changes in endogenous phytohormone levels which may not be sufficient for autotrophic growth of tumor cells on sub-culture. Sederoff et al. (1986) were unable to initiate or maintain hormone autotrophic growth of tumor callus obtained after infection of loblolly pine with *A. tumefaciens*. Regeneration of plantlets from transformed or unstransformed callus has not yet been obtained. However, the callus stage can be bypassed by infecting with disarmed vector using expalnts of cotyledons or leaf segments, as was found to be successful for regenerating some dicotyledonary plants (Horsch et al., 1985). We have established a rapid method for plantlet regeneration from cotyledons of Douglas-fir seedlings (Gupta and Durzan, 1988a). We have infected explants of cotyledons using a disarmed vector, in order to regenerate. transformed plantlets which have been screened by growth in kanamycin.

Several genes such as herbicide-resistance, apical dominance gene, osmoregulatory gene, lignin degradation gene, insect and pest resistance gene have been identified, which are very important in the genetic improvement of trees. Recently, genetic transformation and foreign gene expression has been achieved in loblolly pine (Sederoff et al., 1986) and in poplar (Parson et al., 1986). Field testing of genetically engineered glyphosphate-resistant *Populus* plants are in progress, at Calgene Inc. Davis, California, (Fillati et al., 1987).

In Douglas-fir, insect attack is a major problem at the time of cone formation. The cloning of the insect resistance gene with the promoter, which expresses itself at the time of cone formation, has a great impact on obtaining insect resistance cones with healthy seeds. These results demonstrate the possibility and potential of genetic engineering of conifer cells, specially of Douglas-fir.

REFERENCES

Chilton, M.D., Currier, T.C., Frrand, S.K., Bendich, A.J., Gordon, M.P. and Nester, E.W., 1974, Stable incorporation of stable DNA into higher plant cells : The molecular basis of crown gall tumorigenesis, *Cell, II* : 263-271.

Chilton, M.D., Drummond, M.H., Merlo, D.J., Scaiky, D., Montoya, A.L., Gordon, M.P. and Nester, E.W., 1977, *Agrobacterium tumefaciens* and P58 bacteriophage DNA not detected in crown gall tumors, *Proc. Natl. Acad. Sci., USA.* 71 : 3672-3676.

Dandekar, A.M., Gupta, P.K., Durzan, D.J. and Knauf, V., 1987, Transformation of foreign gene expression in micropropagated Douglas-fir (*Pseudotsuga menziesii*), *Biotechnology*, 5 : 587-590.

Davis, J. and Smith, D.I., 1976, Plasmid determined resistance to anti-microbial agents, *Ann. Rev. Microbiol.*, 32 : 469-518.

DeBlock, M., Herrera-Estrella, L., Van Montagu, M., Shell, J. and Zanbryski, P., 1984, Expression of foreign genes in regenerated plants and their progeny, *EMBO J.*, 3 : 1681-1689.

DeCleene, M. and De Ley, J., 1976, The host range of crown gall, *Bot. Rev.*, 42 : 389-446.

Farnum, P., Timmis, R. and Kulp, J.L., 1983, Biotechnology of forest trees, *Science*, 219 : 696-702.

Fillatti, J.J., Sellmer, J.C. and McCown, B.H., 1986, Regeneration and transformation of *Populus* tissue, *Hortscience*, 21 : 773-774.

Fraley, R.T., Rogers, S.G. and Horoch, R.B., 1985, Genetic transformation in higher plants.

Fromm, M., Taylor, L.P. and Walbot, V., 1986, Stable transformation of maize after gene transfer by electroporation, *Nature, (Lond.)* 319 : 791-793.

Gupta, P.K. and Durzan, D.J., 1985, Shoot multiplication from mature trees of Douglas-fir and sugarpine, *Plant Cell Rep.*, 9 : 177-179.

Gupta, P.K. and Durzan, D.J., 1988a, Establishment and multiplication of juvenile and mature trees of Douglas-fir and sugarpine, *Acta Hortic.* (In Press).

Gupta, P.K. and Durzan, D.J., 1988b, Somatic proembryo formation and transgenic expression of luciferase gene from protoplasts of Douglas-fir, *Plant Sci.* (In Press).

Haas, M.J. and Dowding, J.E., 1975, Aminoglycoside modifying enzymes. In : *Methods in Enzymology XLIII* (S.P. Colowick and N.O. Kaplen, eds.), pp. 611-627, Academic Press, New York.

Horsch, R.B., Fry, J.E., Hoffmann, N.L., Eichholtz, D., Rogers, S.G. and Fraley, R.T., 1985, Simple and general method for transferring genes into plants, *Science*, 227 : 1229-1231.

Laemmli, W.K., 1970, Cleavage of structural proteins during the assembly of the head of bacteriophage T4, *Nature, (Lond.)* 227 : 680-685.

Otten, L.A.B.M. and Schilperoort, R.A., 1987, A rapid micro-scale method for the detection of lysopine and napoline dehydrogenase activities, *Biochem. Biophyl. Acta.*, 527 : 497-500.

Ow, D.W., Wood, K.V., Deluca, M., Jeffrey, C., Dewet, D., Helinski, D.R. and Howell, S.H., 1986, Transient and stable expression of the firefly Luciferase gene in plant cells and transgenic plants, *Science*, 234 : 856-860.

Parson, J.J., Sinkar, V., Reinhard, F.S., Nester, E.W. and Gorden, M.P., 1986, Transformation of poplar by *Agrobacterium tumefaciens, Biotechnology*, 4 : 533-536.

Powledge, T.M., 1984, Biotechnology touches the forest, *Biotechnology*, 2 : 763-772.

Preston, R.J., 1976, *North American Trees* 3rd ed., the MIT Press, Cambridge, Mass.

Reiss, B., Sprengel, R., Will, H. and Schaller, H., 1984, A new sensitive method for qualitative and quantitative assay of neomycin phosphotransferase in crude cell extracts, *Gene*, 30 : 211-218.

Scherier, P.H., Seftor, E.A., Schell, J. and Bohnert, H.J., 1985, The use of nuclear encoded sequences to direct the light regulated synthesis and transport of a foreign protein into plant chloroplasts, *EMBO J.*, 4 : 25-32.

Sederoff, R., Stomp, A.M., Chilton, W.C. and Moore, L.W., 1986, Gene transfer into loblolly pine by *Agrobacterium tumefaciens, Biotechnology*, 4 : 467-469.

Shillito, R.D., Saul, M.W., Paszkoski, J., Muller, M. and Potrykus, I., 1985, High efficiency direct gene transfer to plants, *Biotechnology*, 3 : 1099-1105.

Southern, E., 1975, Detection of specific sequences among DNA fragments separated by gel electrophoresis, *J. Mol. Biol.*, 98 : 503-517.

29

BIOLOGICAL AND ECONOMIC FEASIBILITY OF GENETICALLY ENGINEERED TREES FOR LIGNIN PROPERTIES AND CARBON ALLOCATION

**Roger Timmis and
Patrick C. Trotter**

ABSTRACT

Soon, the only obstacle to applying recombinant DNA technology to trees will be our ignorance of the molecular basis of important traits. We propose criteria for selecting such traits and consider two examples with respect to these criteria.

In the example of lignin amount and composition, the molecular basis is well understood. Work is proceeding in at least two laboratories to isolate the genes that

Roger Timmis and *Patrick C. Trotter* * Weyerhaeuser Company, Tacoma, P.O. Box 111852, Washington 98911, U.S.A.

would allow the more easily-pulped angiosperm-type lignin to be produced in conifers, and total lignin production to be reduced. In terms of savings in chemical costs alone, it has been estimated that these genes would be worth US $ 6 billion (10^9) annually to the U.S. kraft pulp industry.

In contrast, the allocation of above-ground biomass to cambium rather than shoot apex meristems, so as to produce shorter, stouter and higher-yielding trees, is a much greater challenge. The potential payoff is very large and evidence for major single gene effects exists. But ignorance of the molecular basis and greater risk of side effects make this a much more exploratory and long-term, though very worthwhile, undertaking.

INTRODUCTION

Given the recent technical advances in recombinant DNA technology and subsequent plant regeneration (Sederoff et al., 1986; Dandekar et al., 1987; Gupta and Durzan, 1987), we accept here as only a matter of time, the availability of routine procedures for genetically transforming trees. We discuss instead the feasibility of overcoming the greater challenge of obtaining useful genes. In particular, we consider genes controlling traits that would directly increase yield or product quality, rather than protect against disease and stress, i.e., to the investment versus insurance traits distinguished by Stomp and van Buijtenen (1987).

It is appropriate to question why one should bother to genetically engineer trees at all. The technology is, after all, expensive and risky; it competes for resources directly with conventional breeding and vegetative propagation technologies which have the same objective of tree improvement, and it seeks to create new genetic variation in crops where this already abounds, underutilized world-wide.

The answer is that recombinant DNA technology is a tool which can add new capabilities to a tree improvement program. As such, it should be considered in an added-value sense wherever a conventional breeding program is in operation. The technology can permit the placement of specific genes rarely found in the natural population, or genes from other taxa, or the enhanced expression of specific native genes, in individuals of established genetic value. This can be done without the introduction of all the unknown or unwanted genes which come from cross fertilisation with another parent, and then must be eliminated by generations of backcrossing. In this paper, we use two examples of investment traits to show that such added-value is an achievable goal in a time-frame, matching that of a typical tree breeding cycle in temperate regions.

Choice of Traits

The ideal trait for engineering into a tree is one controlled by a single gene whose action has a significant effect on the economic value of the phenotype. Thus, traits ought to be chosen with four criteria in mind.

First, there must be some evidence that the trait is indeed controlled by a single gene or at least a manageably small number of genes and that these genes affect only the trait in question. Segregation ratios for the trait in sexual crosses, the occurrence of mutations uniquely affecting the trait, or a detailed understanding of the underlying process, for which biochemistry can provide necessary evidence. It would also be desirable for this gene to be dominantly expressed. Then it could be used in varied genetic backgrounds and spread through sexual reproduction.

Second, there should be some clues to the molecular basis of the trait. In order to make a start on identifying the gene(s) and later isolating it(them), the more information there is, the less this most costly phase of the research will be needed. Implied here is also a need to understand as far as possible how the gene's action will translate upwards to the tissue and whole-plant level, when integrated with a host of other interacting tree processes.

The third criterion is that this end-effect should be acceptably large in relation to the degree of change in the gene's action at the molecular level. In other words, if our understanding of tree processes were to be expressed in a complex mathematical model, then the parameter controlled by this gene would be one to which model output would be very sensitive under expected plantation conditions. Efficiency of light energy conversion per unit area of leaf is an example of the ability to maintain growth at lower temperatures in temperature regions (Graham et al., 1985).

Finally, a target trait should be one in which there is much scope for improvement. Therefore, it would be prudent to seek traits that are not of high survival value for the species in natural stands. For, the less a trait has been optimised by generations of natural selection, the more successful we are likely to be in developing a better version of it for domestic or industrial use. On this basis, better light conversion efficiency e.g., by improving the CO_2 (vs O_2) affinity of the primary carboxylating enzyme, RuBP carboxylase, would seem to be a poor choice. A changed wood chemistry, however, might be relatively easy to achieve.

The two traits selected for discussion below differ considerably in how well they match these four criteria, and hence the cost of isolating genes. Lignin chemistry is relatively well understood, but the molecular basis of carbon allocation is unknown.

LIGNIN CONTENT AND COMPOSITION

Size of the Opportunity

The process of lignin formation in the xylem cell wall appears to meet the criteria we have set forth for a candidate trait for genetic engineering. First of all, higher plants vary widely in lignin content and chemical composition. Sarkanen and Ludwig (1971) give a range of 15 to 36 percent of the weight of the wood for the lignin content of trees, the lower levels being reported for angiosperms and the higher levels being reported for gymnosperms.

The chemistry of the lignin polymer also differs among the major classes of higher plants (Higuchi, 1985). The lignins of grasses contain three kinds of structural units: guaiacyl, syringyl, and p-hydroxyphenyl. Angiosperm lignins contain mixtures of guaiacyl and syringyl units in various proportions (quaking aspenwood also contains some p-hydroxyphenyl units). Gymnosperm lignins contain virtually all guaiacyl units (although p-hydroxyphenyl units have been reported in some gymnosperm compressionwood samples).

While it is true that environmental factors can produce variability in the lignin content within species (Wardrop, 1971), it was demonstrated over 20 years ago that lignification in higher plants is under genetic control (Kug and Nelson, 1964). Furthermore, each developing cell possesses the genetic mechanism and the ability to control its level of lignification as its condition and environment warrant (Wardrop, 1971).

What would be the benefit of manipulating the process of lignin formation in commercial tree species? Consider just chemical pulping. In the kraft process, which is the predominate chemical pulping process used in the world today, the ease of lignin removal is directly proportional to the ratio of syringyl to guaiacyl units in the lignin (Chang and Sarkanen, 1973). This is because syringyl precursors produce a less cross-linked lignin polymer than guaiacyl precursors, and it is easier for the pulping chemicals to break down and dissolve such a polymer. Thus angiosperms, whose lignins contain both syringyl and guaiacyl units, can be pulped not only by the kraft process, but by virtually all the other established or emerging processes, with fewer chemicals and with less energy and time than gymnosperms, whose lignins consist solely of guaiacyl units (Rydholm, 1965; Chiang and Sarkanen, 1983; Chiang et al., 1987). However, pulps from gymnosperms are preferred for many applications because the fibers are generally longer and stronger than those from angiosperms (Rydholm, 1965). Thus, manipulating lignin composition in gymnosperms to be more like that of angiosperms would produce substantial economic gain by virtue of the savings in time, chemicals and energy required for pulping. The saving in chemicals alone has been estimated at close to US $6 billion annually for the total U.S. kraft pulping industry. Controlling the amount of lignin formed in the wood would yield economic gains of similar magnitude.

Information Available at the Molecular Level

In the years since the pioneering experiments of Kug and Nelson (1964), the biochemical pathway of lignin formation in higher plants has been worked out in considerable detail (for a review and detailed description, see Higuchi, 1985 and the references cited therein. What follows is drawn largely from those sources). This pathway, and a listing of the enzymes that catalyse the several steps, is shown in Fig. 1.

The first key biochemical event is the deamination of phenylalanine (which itself arises via the well-known shikimic acid pathway), a reaction that is catalysed by the enzyme phenylalanine ammonia lyase, designated PAL in Fig. 1. Tyrosine, also arising from shikimic acid pathway, can also be deaminated enzymatically by the action of the enzyme TAL. The TAL route is important in grasses, but not in trees (Fig. 1).

Next comes a series of reactions in which the aromatic nucleus is hydroxylated, first in the 4-position then in the 3-position, followed by methylation of the hydroxyls that were inserted at position 3. This produces *p*-coumaric acid and ferulic acid (which account for the *p*-hydroxyphenyl units in the lignins of grasses and aspenwood, and for the guaiacyl units found in the lignins of all higher plants. We will henceforth disregard the grasses to simplify the remaining discussion).

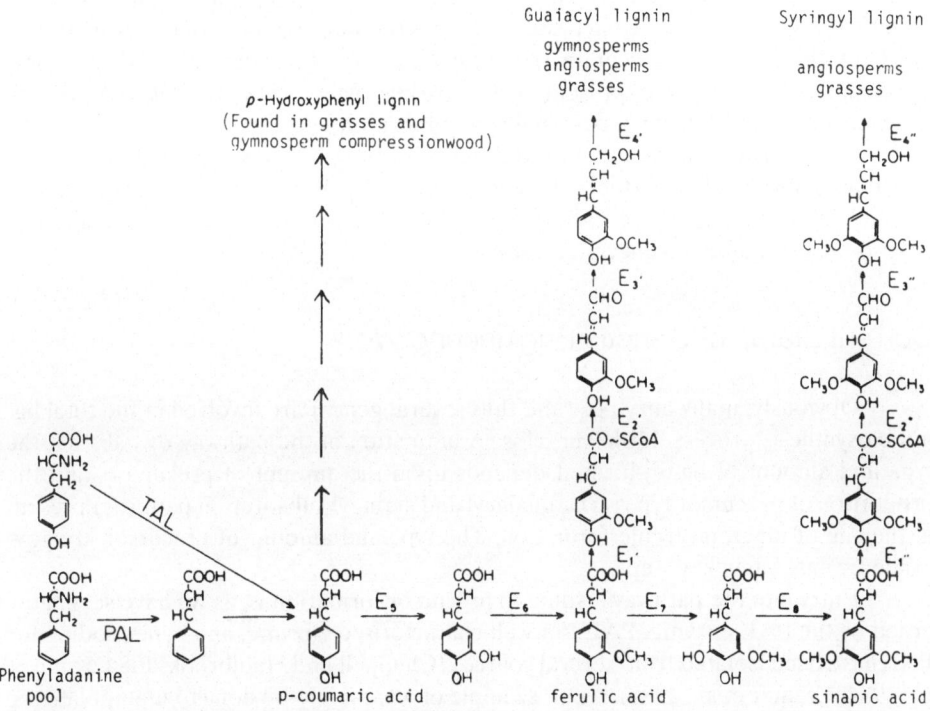

Fig. 1: Lignin Biosynthesis Pathway - Enzymes:(E1), (E1'), and (E1"), hydroxycinnamate:CoA ligase; (E2), (E2') and (E2"), hydroxycinnamoyl-CoA reductase; (E3), (E3') and (E3"), cinnamyl alcohol dehydrogenase; (E4), (E4'), peroxidase; (E5), p-coumarate 3-hydroxylase; (E6) and (E8), hydroxycinnamate O-methyltransferase; (E7), ferulate 5-hydroxylase; PAL, phenylalanine ammonia-lyase; TAL, tyrosine ammonia-lyase. (From Higuchi, 1985).

At this point comes an important juncture. Angiosperms, but not gymnosperms, possess a pair of enzymes (designated E7 and E8 in Fig. 1) that can hydroxylate the 5-position of the aromatic ring (E7), then convert that hydroxyl to a methoxyl group (E8). This leads to the formation of synapic acid, which accounts for the presence of syringyl units in angiosperm lignins. Feeding experiments have demonstrated that gymnosperms can process syringyl precursors into syringyl lignin (Nakamura et al., 1974); but they cannot make these precursors themselves.

So now we have the properly substituted aromatic acids present: ferulic acid in the gymnosperm system and synapic acid (along with ferulic acid) in the angiosperm system. These compounds are next activated to the corresponding CoA thioesters, which are, in turn, converted to alcohols via two consecutive reductive steps. The second of these steps, the conversion of coniferyl aldehyde and synapyl aldehyde to the corresponding alcohols, is catalysed by what Higuchi (1985) calls alcohol dehydrogenase enzymes (designated E3' and E3" in Fig. 1). This step is significant because it seems to be specific to lignifying tissues, whereas the steps up through the formation of the CoA thioesters can occur in the biosynthesis of flavonoids and other phenylpropanoids, and are thus present in the tissues where they are formed, as well (Higuchi, 1985; H.M. Chang, personal communication).

The final enzyme catalyzed step in lignin formation is dehydrogenation of coniferyl and synapyl alcohols to their phenoxy radicals by peroxidase enzymes (designated E4' and E4" in Fig. 1). These radicals then couple to form lignin macromolecules in a reaction that requires no enzyme catalyst.

Current Status of Genetic Engineering Work

Obviously, many enzymes (and thus several genes) are involved in the total lignin biosynthesis process. However, close examination of the pathway reveals that the type and amount of lignin formed depends upon the amount of precursors and the proportion of precursor types (i.e., guaiacyl and syringyl substitution patterns) present at the site of macromolecule formation. The type and amount of precursor are governed by only a few key steps.

Entry into the pathway leading to precursor formation is, as we have seen, governed by the PAL enzyme. PAL is a well characterised enzyme, and genes coding for PAL have been isolated from several sources (Chappell and Hahlbrook, 1984; Kuhn et al., 1984; Cramer et al., 1985). As an example of what might be done to manipulate the lignin content in conifers by genetic engineering, one might adopt an antisense strategy, perhaps at the m-RNA level, to reduce PAL activity and thus the amount of lignin precursors formed in the lignifying tissue (H.M. Chang and R.R. Sederoff, personal communication).

At the other end of the pathway, cinnamyl alcohol dehydrogenase enzymes (designated E3' and E3" in Fig. 1) catalyse the formation of coniferyl alcohol and synapyl alcohol. Although other alcohol dehydrogenase enzymes are present in higher plants, cinnamyl alcohol dehydrogenase (CAD) activity appears to be specific to lignifying tissue (Higuchi, 1985; H.M. Chang, personal communication). Isolating a gene for CAD, and particularly its promoter sequence, would give us a signal specific for the time and place of lignin formation that would be obviously useful for directing other genetic manipulations. Using the gene for another alcohol dehydrogenase previously isolated from pine as a probe, Professor R.R. Sederoff, now at North Carolina State

University, hopes to isolate the CAD gene from lignifying tissue (R.R. Sederoff, personal communication).

As for the type of lignin formed, this seems to be governed by the presence or absence of enzymes E7 and E8 in the lignifying tissue (Higuchi, 1985). E7 is ferulate 5-hydroxylase and E8 is a difunctional ortho-methyl transferase, so named because it can methylate the 3-hydroxyl group of 3,4-dihydroxycinnamic acid to form the guaiacyl unit, and the. 5-hydroxyl group of 5-hydroxyferulic acid to form the syringyl unit. Gymnosperms lack E7, and the gymnosperm ortho-methyl transferase will not methylate 5-hydroxyferulic acid. If the genes coding for these two angiosperm enzymes could be incorporated into the gymnosperm genome, then the gymnosperm capable of producing angiosperm-type lignin ought to result. To this end, a research group at Michigan Technological University, Houghton, Michigan, is working on isolating the genes for these enzymes from active lignifying tissue of aspenwood (V. Chiang, personal communication).

Risks and Time-Frame for Implementation

One could argue that since angiosperms have evolved a lignin-synthesising pathway which is somewhat different from gymnosperms, genetic engineering of gymnosperms to form angiosperm-type lignin might somehow prove detrimental to the engineered gymnosperms. This could be tested with suitably designed long-term feeding experiments. In 1974, Professor T. Higuchi's group at Kyoto University demonstrated that gymnosperms can indeed form angiosperm-type lignin by feeding syringyl precursors beyond the E7, E8 genetic block to freshly cut xylem tissues of *Gingko biloba* seedlings (Nakamura et al., 1974). More recently, Professor N. Terashima's group at Nagoya University repeated these experiments with essentially the same results on cut shoots of Japanese black pine (*Pinus thunbergii*; Terashima et al., 1987). However, both these sets of experiments were of short duration - the longest was three days - and so did not give information on the quality of the wood produced, or on long-term effects at the whole-plant level.

What one would ideally need to do is continually feed syringyl precursors to actively growing seedlings (or even larger trees) for at least two full growing seasons. If this could somehow be done, enough wood could be obtained for micro-pulping evaluation and for at least a partial assessment of wood quality (R. Megraw, Weyerhaeuser Company, Tacoma, Washington, personal communication). One would also obtain a view of the impact on the growth and development of the whole plant. To these ends, Dr. Vincent Chiang of Michigan Technological University, has set up some long-term feeding experiments using loblolly pine (*Pinus taeda*) seedlings. The seedlings were raised in agar media containing synapyl aldehyde, and then transferred into soil containing synapyl aldehyde solution.

Dr. Chiang's experiments rely on the seedling transport system to deliver the syringyl precursor to the lignifying tissue. Another way to achieve this is by feeding the precursors directly to active cambial tissue by carefully removing a patch of outer and inner bark, then quickly replacing it with a glycerol-water paste containing the precursor. The paste would be replenished throughout the growing season, and this would be continued until the experiment was terminated. Procedures for these "patch" applications of materials have been described in Nix and Brown (1987) and elsewhere.

So, at least some thought is being given to ways to assess, independently, the risk to the plants and to the resulting wood quality of genetic manipulation of lignin in gymnosperms. The next question is, when can we expect to implement such technology?

Given recent advances in our ability to genetically transform gymnosperm tissue and subsequent plant regeneration, we can anticipate that within the next two to three years, we will see the first successful recovery of a genetically transformed conifer plant. This will most likely be a demonstration-of-concept type of experiment where the foreign gene inserted in the conifer genome is some kind of convenient marker gene. However, this time-frame is in accordance with those set forth by both Professors Sederoff and Chiang to have lignin controlling genes in hand (personal communication). If we add a couple of additional years to prepare the appropriate constructs for insertion into the genome, and then to carry out the insertion and recovery of a plant, we could anticipate seeing the first transformant for altered lignin biosynthesis in about five years.

To this must be added an additional five to six years for field evaluation of the performance of the genetically transformed trees, and perhaps an additional year or two of wood quality and pulping evaluations. This totals to 12 or 13 years before planting stock could be available for commercial forestry operations.

CARBON ALLOCATION

The carbon fixed during photosynthesis is allocated to different parts of a plant, where it is used for the synthesis of new structure and maintenance of existing live tissue. The proportions thus allocated have a major effect on the total growth and form of the plant (Cannell, 1985). This allocation pattern is the result of a much more complex set of processes than in lignin biosynthesis. Nevertheless, there is evidence that single genes play a major role and there is scope for changing them to literally reshape trees for domestic use.

The sensitivity of whole-plant growth to carbon allocation is illustrated by the simple allometric equation: $W = aW_1^k$, where W is the plant dry weight, k is the fraction of assimilated carbon allocated to the plant part L of weight W_1, and a is a constant. Clearly, W will be quite sensitive to changes in the exponent k. Isolated trees maximise their growth and speed of site occupancy by reinvesting as much carbon as possible into the formation of new leaves (i.e., when W_1 is leaf weight in the above equation and k is large, relative to its value for other organs). In fact, Ledig (1969) showed that this reinvested fraction, combined with seasonally-adjusted net photosynthesis, accounted for 94 percent of the variation in seedling growth among eight loblolly pine families.

Trees in a crop, however, are not isolated for long. For most of their lives, they make up a closed-canopy forest in which annual per hectare light interception and mineral and water use remain stable for many years (Broms and Axelsson, 1985, and literature cited therein). From the forester's point of view there is, in this situation, no need for more leaves or roots except to replace aging or dead ones. Maximum harvestable crop growth will occur if carbon is allocated primarily to stems.

Carbon allocation to stems does take on a higher value following canopy closure, presumably as a result of reduced activity of lateral apices in the crown due to mutual shading of crowns. Cannell's compilation of forest production data from around the world shows that the stems take 40-60 percent of the current annual aboveground dry matter increment. He considered these to be underestimates due to frequent lack of accounting for woody litter-fall and tree mortality. In a 40-year-old Douglas-fir stand, about 30 percent of the above-ground dry matter increment went to the crown (Keyes and Grier, 1981). However, a genetically high value for this trait would not have been favoured by natural selection acting on individuals for the ability to produce and disperse seed (Ford, 1985 and Libby, 1987). Instead, tree crowns are programd to expand upward and outward to overtop their neighbours. The inevitable result is that competition-induced mortality occurs and the number of trees per hectare declines as the crop grows.

Crop behavior resulting from the evolved carbon allocation priority has been described by the 3/2 power law of self-thinning. This shows a rather universal limitation to the number of stems of a given size that can be grown on a given area of ground (Drew and Flewelling, 1977). Several investigators have recommended selecting for crop ideotypes rather than competitive ones in tree breeding programs (Ford, 1985) in order to avoid this limitation. More specifically, Libby (1987) and Van Buijtenen (1985) have advocated breeding for alternative tree shapes.

Magnitude of the Opportunity

What is the likely size of the benefit of changing carbon allocation in favour of cambial rather than shoot apex meristems? Keyes and Grier (1981) estimated that 28 and 30 percent of above-ground biomass went to the crown annually for high- and low-productivity sites of 40-year-old Douglas-fir, respectively. The remaining (approximately 70 percent) went to stemwood and bark.

Let us assume that allocation to the canopy was halved by genetic engineering, so that only 15 instead of 30 percent of above-ground production went there. The balance, 85 percent, would go to stem growth. Therefore, 15/70 or 21 percent more stem growth ought to occur each year - at least as long as the forest was in a stable state with regard to annual light interception and soil moisture/nutrient extraction rates. If this were the case, the stand growth would show a cumulative linear increase over the base case trend, as shown schematically and by line B in Fig. 2.

However, this oversimplified view does not allow for the fact that the larger diameter stems so produced will increase the amount of living tissue in the stand at a faster-than-usual rate. This is because added canopy biomass is matched by leaf fall

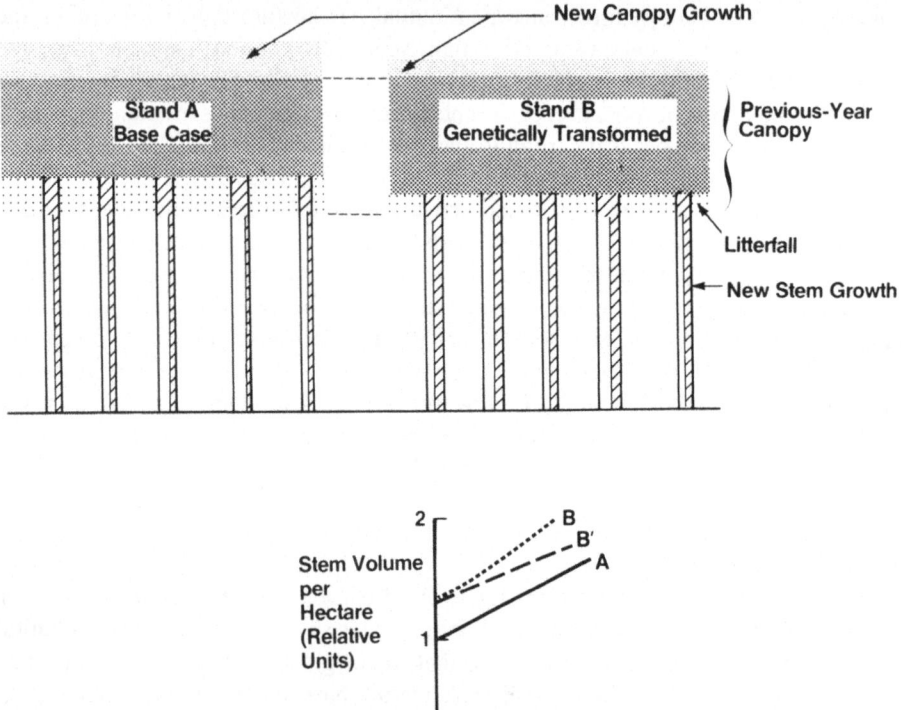

Fig. 2: Schematic representation of carbon allocation in stands of normal and genetically transformed trees. For comparison purposes the stands are assumed to begin with the same total biomass and average height. In the transformed case only half as much carbon is allocated to shoot apices; the balance goes to cambium. The result is that radial increment (shaded and exaggerated) is increased at the expense of height. Crown thickness and productivity, however, are not changed. The graph below shows the effect on timber production of the transformed stand (B) compared with the base case (A). Line B' shows how increasing respiratory burden lessens the growth advantage over time.

and branch shedding each year, whereas added stem biomass simply accumulates, as shown in Fig. 2. The added maintenance respiration of this stem tissue must be subtracted from the carbon reallocated from the canopy. Therefore, the 21 percent will be progressively reduced in subsequent years.

We can calculate the magnitude of this reduction by assuming that the increased stem biomass results mainly from diameter growth (versus height), and that the increased maintenance requirement M, is proportional to the stem surface area

S (Butler and Landsberg, 1981). Then,

$$\text{since} \quad \text{Weight,W} \propto \text{Volume,V}$$
$$\text{and} \quad S \quad \propto \quad \sqrt{V} \text{ (for a cylinder of fixed height),}$$
$$\text{and} \quad M \propto S$$
$$\text{therefore,} \quad M \propto \sqrt{W}$$

In other words, if, as in the example above, stem biomass increased initially from 1.0 to 1.21 units, maintenance respiration would increase from 1.0 to $\sqrt{1.21} = 1.10$ units, a 10 percent increase. The net effect of this increasing respiratory burden on the growth trend is described by the three equations given in Appendix.

Numerical simulations using these equations showed the general response pattern of line B' in Fig. 2. The initial magnitude of improved growth depends upon the fraction of carbon allocated to the canopy in the base case and the proportion of this that is reallocated in the genetically transformed case. The rate of decline of this response, on the other hand, depends upon the amount of living stem tissue assumed to be present to begin with and the specific maintenance rate thereof. The bigger these two numbers, the more rapid the decline in growth improvement. Two contrasting examples are shown in Table 1: a 33 percent annual increase declining linearly to 5 percent over 14 years and 24 percent declining exponentially to 5 percent over 4 years.

Table 1: Magnitude and duration of yield gain (without thinning) from carbon reallocation under two contrasting sets of assumptions

Scenario	Increase in stem biomass increment over base case (percent)									
Year	1	2	3	4	5	6	7	8	9	10
Optimistic	33	31	29	26	24	21	18	16	13	11
Pessimistic	24	15	9	5						

Assumptions:	Optimistic[1]	Pessimistic[1]
- Beginning stem biomass (t ha^{-1})	210	424
- Live fraction	0.02	0.05
- Specific maintenance rate (t ha^{-1} yr^{-1})	0.5	1.0
- Above-ground net primary production (t ha^{-1} yr^{-1})	25.0	13.7
- Fraction of above-ground NPP allocated to canopy in base case	0.4	0.3
- Fraction allocated to canopy in transformed case	0.2	0.15

1 The stem biomass, above-ground productivity and basecase canopy allocation values are taken from the Keyes and Grier (1981) fertile site for the pessimistic case. Stem biomass and productivity are estimated from yield tables for 35-year-old Douglas-fir stands for the optimistic case (Curtis et al., 1982).

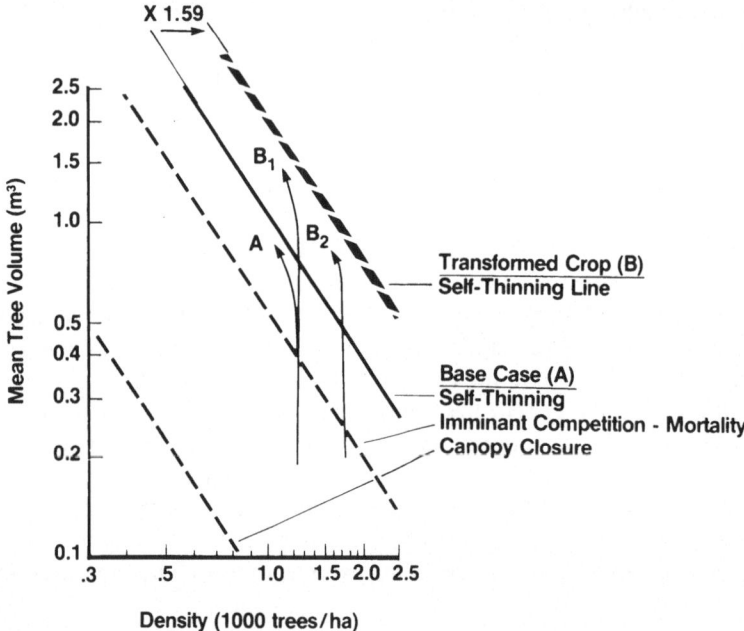

Fig. 3: Size-density relationships in genetically transformed and base case stands. Line A shows the expected reduction in density that must occur if mean tree volume is to increase allocation of carbon to stem wood, self thinning and consequent growth loss, will not occur until stems are considerably larger (B1) or at the same size in a stand with 1.59 times as many trees (B2).

This situation can be appreciated in relation to the self-thinning line in Fig. 3. Here the average stem biomass is plotted against the density of trees per hectare on logarithmic axes and no account is taken of time or rate of growth (Drew and Flewelling, 1979). Points on this line represent stands in which net assimilation and maintenance respiration are exactly in balance so that no further growth can occur without some trees dying. The two examples in Table 1 would be approaching the self-thinning line at different rates. The faster decline in the pessimistic scenario is due in part to its hypothetical starting point at a higher density relative to the maximum line, and in part due to the assumed live fraction and specific maintenance rates, both being higher at this point.

The typical unthinned Douglas-fir stand of Fig. 2 will track along the line A in Fig. 3. In a genetically transformed stand grown from trees in which biomass allocation to crown has been halved without changing the shape of the crown, the average area occupied by the crown of trees of a given average size will be reduced to 0.63 of the base case value. (This follows since $A \propto L^2$ [crown area is proportional to linear dimension squared] and Weight, $W \propto L^3$; then $A \propto W^{3/2}$, so that if W is halved, A is changed to $[0.5]^{3/2}$ or 0.63). Consequently, the density will be $1/0.63 = 1.59$ times higher for tree stems of any given average size. Hence, the genetically transformed stand ought to have a new, higher self-thinning line, as shown.

Thinning of the genetically transformed stand would best be carried out with respect to the new self-thinning line. This means that thinnings could be done to higher stocking levels, or else to the base case levels less frequently (see Drew and Flewelling, 1979, for rationale of thinning in relation to this line). In any case, an appropriate thinning regimen would allow us to avoid most of the decline in allocation-related growth gains discussed above.

In summary, the foregoing analysis on the yield gains from a 50 percent less crown-oriented carbon allocation shows that we could expect these to be in the range of 10 to 35 percent depending on physiological parameters and thinning regimen. Furthermore, the extra volume would be put on stems that were only about 80 percent as tall (i.e., $[0.5]^{1/3}$, since length is proportional to cube root of weight), which means that the logs would be about 25 percent greater in diameter. According to the prices of British Columbia second-growth logs quoted by Libby (1987), these stems would have a 17 percent greater value per m^3. Thus, the total value increase would be in the range of 30-58 percent.

There are also physiological advantages of tree stems being short and stout, which would translate into increase in net carbon production and an increase in the amount of this going to above-ground increment. The main one is that resistance to water flow through a given volume of stem is reduced by a factor of four if the stem length is halved and the cross-sectional area doubled (by analogy with electrical resistors in parallel versus series). Paltridge (1973) considered this effect in a theoretical analysis showing that maximum photosynthesis decreased markedly with ultimate tree height on water-limited sites. Also, new woody tissues have a lower nutrient content than foliage, so increased partitioning to wood decreased the average nutrient content of new dry matter increment. This decreased nutrient demand should lessen the proportion of assimilates taken by root (Cannell, 1985).

Evidence of Single Gene Control

Four lines of evidence suggest that single genes can have major effects on carbon allocation. The first is from studies of the progeny of a narrow-crowned Scots pine (*Pinus sylvestris*) in Finland. This open-grown tree has unusually short, thin branches arising at 90° from a tall stem (Fig. 4).

Some test sites, half of its open-pollinated progeny are narrower-crowned and faster growing, indicating that the trait is due to a single dominant gene (Karki and Tigerstedt, 1985).

The narrow-crowned trait is not the same as the stem allocation trait being considered in this paper, because it concerns the allocation of carbon between the terminal versus lateral shoot apices, not between shoot apices and cambium. Thus, it increases carbon allocated to stemwood only as an indirect effect of stronger apical

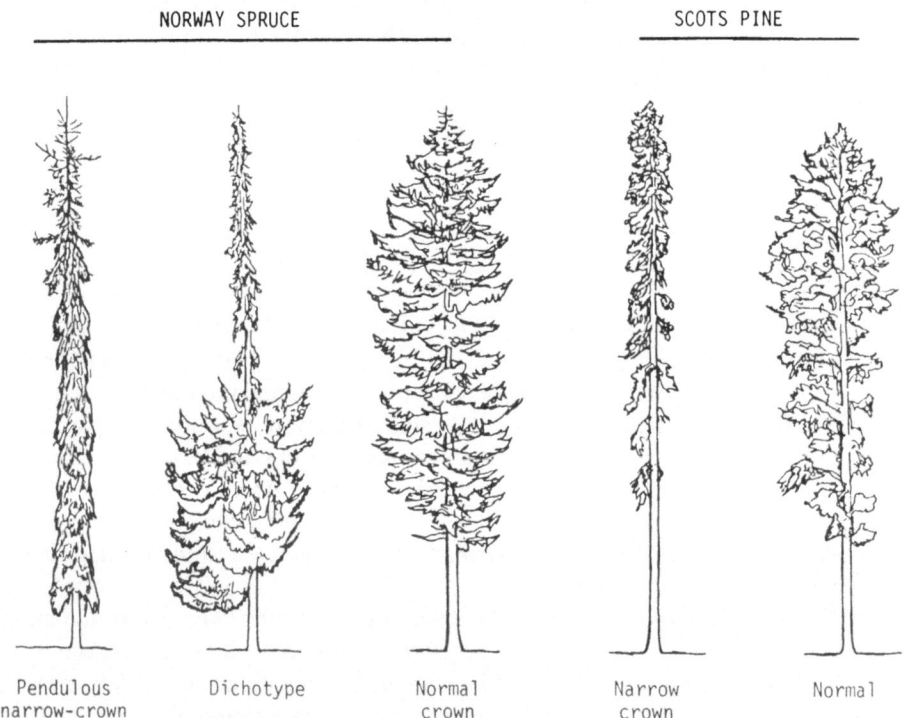

NORWAY SPRUCE   SCOTS PINE

Pendulous   Dichotype   Normal   Narrow   Normal
narrow-crown   crown   crown

Fig. 4: Single gene controlled crown forms in conifers (from Karki and Tigerstedt,1985).

control reducing lateral branch extension and, therefore, branch thickness. The advantages (and some disadvantages) of higher possible stocking levels and reduced thinning requirement (Karki, 1985) are obtained, but the same growth rate improvement would not be because the crowns are taller. Nevertheless, the point is that a single gene controls carbon allocation through (presumably) a change affecting phytohormone production or response at the molecular level.

The second piece of evidence is from dichotypes of Norway spruce (*Picea abies*), also found in Finland (Fig. 4). These are cases in which the crown form changes abruptly from the normal to a pendulous narrow type due to a mutation in the apical meristem. In this case, the narrow-crown form appears to be associated with drooping of branches and probably has a different physiological basis than in the pine. This pendulous narrow-crown form (not the dichotype) can be seen occasionally in many conifer species. What the dichotype demonstrates is the potential for single gene effects to cause major changes in carbon allocation, again in this case, through crown shape.

Single gene effects on crown and leaf form can also be seen among the progeny of Douglas-fir inbreeding. In this case, the trees are double recessives segregating in some crosses with a 3:1 ratio. While not being useful as parents in further breeding, some of these individuals with few and short branches provide experimental material for the study of the molecular basis of morphology and consequent carbon allocation.

The fourth type of evidence has been produced recently by the genetic transformation of plants with genes causing auxin over-production (Klee et al., 1987a). These workers introduced a gene encoding tryptophan monooxygenase from *Agrobacterium tumefaciens* attached to the powerful 19S promoter gene from cauliflower mosaic virus, into petunia plants. The transgenic plants exhibited an almost complete apical dominance, greater internode distance, leaf curling, a doubling of the amount of xylem and phloem production, and parthenocarpic fruit development (Klee et al., 1987b). It would be intriguing to observe the action of this gene construct in forest trees.

Obtaining Stem Allocation Genes

Given that small numbers of genes could indeed enhance value up to 55 percent through biomass reallocation, how might we find them? This would not be easy, and would require a research strategy combining several elements.

One such element would be to use recombinant DNA technology, as recommended by Klee's group, as a tool to investigate the effect of introducing hormone-related genes on tree phenotypes. In particular, the use of tissue- or organ-specific rather than constitutive promoters in such constructs would be important.

A second element would be the analysis of existing single gene crown form variants (including inbreds, dichotypes and dominant gene types) to pin down the molecular basis of the variant. This would include the classical approaches of tissue hormone analyses, exogenous application of hormones and inhibitors, etc. Information from these studies would be used to guide the genetic transformations and vice versa. One clue here is that in a similar columnar form of MacIntosh apple (containing the so-called *Co* gene), the shoots have been found to be highly tolerant to cytokinin *in vitro* (David James, personal communication). This, perhaps, suggests that the reduced branchiness arises from a block in the cytokinin response pathway.

A third element of our search for stem allocation genes would be comparative physiology studies of *Brassica oleracea*. These would have the objective of determining how this species had achieved such vastly different carbon allocation patterns in its five cultivated varietal forms: cabbage (which allocates to the terminal bud), kale (to leaves), kohlrabi (to stem), cauliflower (to inflorescences), and Brussel sprouts (to lateral buds). These forms point to the existence of an overriding sink strength control mechanism which can be turned on in the respective organs. Understanding this mechanism and what activates it, would be an important advance.

A fourth part of the research strategy would be to develop means of sabotaging the processes leading to reproductive structures. This would leave more biomass available for the leaves and stem (Cannell [1985] estimated about 5-10 percent more). It would have the added advantage of confining any introduced foreign genes to the commercially managed forest by preventing pollen and seed dispersal. In general, the degree of biochemical knowledge required to abort a process is much less than is needed to enhance it. The use of antisense RNA would be one of several approaches.

Finally, a search should be made, using subtractive hybridisation techniques, for a strong cambium-specific promoter in trees. This might enable other genes controlling carbon allocation to operate preferentially at the sites where wood is made. Constructs combining such a gene with one(s) controlling phytohormone relations could be quickly inserted and evaluated.

Risks and Time-Frame for Implementation

The main risk in undertaking a research program to genetically engineer carbon allocation is not that genes having a major effect will not be found or successfully inserted, but that they will have undesirable side-effects. These might lead, for example, to low wood quality or a changed seasonal growth pattern, both of which are closely linked to hormone production and transport in trees.

To guard against these possibilities, it would be essential to field-test transformants beyond the point of canopy closure (say for 6 to 10 years at close spacing) and then retest corrected transformants over a similar period. Thus, the time-frame for a commercial product would be perhaps 25 years, comprising 5 years for first transformants, added to 2 x 10 years (or 3 x 7 years) for field evaluation. While this is a long time, it is not much longer than that required to establish and bring to full production a new generation seed orchard in temperate regions. It would, of course, take considerably less time for species in tropical and subtropical regions. In compensation and in contrast to conventionally-bred tree traits, the gene construct would be saleable worldwide for use (with minor modification and testing) in other tree species.

CONCLUSIONS

The purpose of this paper was to assess whether it made sense to begin genetically engineering trees for traits conferring better yield and product value and, if so, where one might begin. The example of manipulating lignin chemistry for better pulping properties, we believe, demonstrates clearly that recombinant DNA technology can make a valuable contribution that is unlikely to be achieved using conventional methods and with minimal risk. No doubt, other aspects of wood chemistry will become the subject of equally promising endeavours.

The example of manipulating above-ground carbon allocation to favour stems over foliage and branches has even higher potential value. But this is not so clearly justified economically due to our ignorance of the underlying biochemistry and consequent uncertainty about side-effects. Yet, there are tantalising clues about the role of single genes and how one might approach the problem of isolating them. Therefore, work in this area should begin as an exploratory endeavor, leading in the long-term, to understanding a component of tree yield that appears to have much scope for improvement by changing single genes.

REFERENCES

Broms, E. and Axelsson, B., 1985, Variation in carbon allocation pattern as a base for selection in Scots pine. In: *Crop Physiology of Forest Trees*, (P.M.A. Tigerstedt, P Puttonen, and V. Koski, eds.), pp. 81-93, Univ. Helsinki Press, Helsinki.

Butler, S.R. and Landsberg, J.J., 1981, Respiration rates of apple trees, estimated by CO_2-efflux measurements, *Plant Cell Environ.* 4: 153-159.

Cannell, M.G.R., 1985, Dry matter partitioning in tree crops. In: *Attributes of Trees as Crop Plants*, (M.G.R. Cannell and J.E. Jackson, eds.), pp. 160-193, Inst. Terr. Ecol., Abbts Ipton.

Chang, H.M. and Sarkanen, K.V., 1973, Species variation in Lignin: Effect of species on the rate of kraft delignification, *TAPPI J.,* 56: 132-134.

Chappell, J. and Hahlbrook, K., 1984, Transcription of plant defense genes in response to UV Light or Fungal Elicitor, *Nature, (Lond)* 311: 76-79.

Chiang, V.L. and Sarkanen, K.V., 1983, Ammonium sulfide organosolv pulping, *Wood Sci. and Technology*, 17: 217-226.

Chiang, V.L., Cho, H.J., Puumala, R.J., Eckert, R.E. and Fuller, W.S., 1987, Alkali consumption during kraft pulping of Douglas-fir, Western Hemlock, *TAPPI J.*, 70: 101-104.

Cramer, C.L., Ryder, T.B., Bell, J.N. and Lamb, C.L., 1985, Rapid switching of plant gene expression induced by fungal elicitor, *Science*, 227: 1240-1243.

Curtis, R.O., Clendenon, G.W., Reukema, D.L. and DeMars, D.J., 1982, Yield tables for managed stands of coast Douglas-fir, *U.S. For. Serv. Gen. Tech. Rep.*, PNW-135, pp. 29.

Dandekar, A.M., Gupta, P.K., Durzan, D.J. and Knauf, V., 1987, Transformation and foreign gene expression in micropropagated Douglas-fir *(Pseudotsuga menziesii)*, *Biotechnology*, 5: 587-590.

Drew, T.J. and Flewelling, J.W., 1977, Some recent Japanese theories of yield-density relationships and their application to Monterey Pine Plantations, *For. Sci.*, 23: 517-534.

Drew, T.J. and Flewelling, J.W., 1979, Stand density management: An alternative approach and its application to Douglas-fir plantations, *For. Sci.*, 25: 518-532.

Ford, E.D., 1985, Increasing forest productivity and value by exploiting climatic variability. In: *Forest Potentials, Productivity and Value* (R. Ballard, P. Farnum, G.A. Ritchie, and J.K. Winjum, eds.), pp. 75-99, Weyerhaeuser Science Symposium 4.

Graham, R.L., Farnum, P., Timmis, R. and Ritchie, G.A., 1985, Using modeling as a tool to increase forest productivity and value. In: *Forest Potentials, Productivity and Value*, (R. Ballard, P. Farnum, G.A. Ritchie and J.K. Winjum, eds.), pp. 101-130, Weyerhaeuser Science Symposium 4.

Gupta, P.K. and Durzan, D.J., 1987, Somatic embryos from protoplasts of loblolly pine proembryonal cells, *Biotechnology*, 5: 710-712.

Higuchi, T., 1985, Biosynthesis of Lignin. In: *Biosynthesis and Biodegradation of Wood Components* (T. Higuchi, ed.), pp. 141-160, Academic Press, New York.

Karki, L., 1985, Genetically narrow-crowned trees combine high timber quality and high stem wood quality at low cost. In: *Crop Physiology of Forest Trees*, (P.M.A. Tigerstedt, P. Puttonen, and V. Koski, eds.), pp. 245-256, Univ. Helsinki Press, Helsinki.

Karki, L. and Tigerstedt, P.M.A., 1985, Definition and exploitation of forest tree ideotypes in Finland. In: *Attributes of Trees as Crop Plants* (M.G.R. Cannell and J.E. Jackson, eds.), pp. 102-109, Inst. Terr. Ecol., Abbots Ripton.

Keyes, M.R. and Grier, C.C., 1981, Above- and below-ground net production in 40-year-old Douglas-fir stands on low and high productivity sites, *Can. J. For. Res.*, 11: 599-605.

Klee, H., Horsch, R.B., Hinchee, M.A., Hein, M.B. and Hoffmann, N.L., 1987a, The effects of overproduction of two *Agrobacterium tumefaciens* T-DNA auxin biosynthetic gene products in transgenic petunia plants, *Genes & Development*, 1: 86-96.

Klee, H., Horsch, R. and Rogers, S., 1987b, Agrobacterium-mediated plant transformation and its further applications to plant biology, *Annu. Rev. Plant Physiol.*, 38: 467-486.

Kug, J. and Nelson O.E., 1964, The abnormal Lignins produced by the Brown-Midrib mutants of maize, *Arch. Biochem. Biophys.*, 10: 103-113.

Kuhn, D.N., Chappell, J., Boudet, A. and Hahlbrook, K., 1984, Induction of Phenylalanine Ammonia-Lyase and 4 Couseanate: CoA Ligase mRNAs in cultured plant cells by UV light or fungal elicitor, *Proc. Nat'l Acad. Sci.* 81: 1102-1106, USA.

Ledig, F.T., 1969, A growth model for tree seedlings based on the rate of photosynthesis and the distribution of phytosynthate, *Photosynthetica*, 3: 263-275.

Libby, W.J., 1987, Do we really want taller trees? Adaption and allocation as tree-improvement strategies, The H.R. MacMillan Lectureship in Forestry, University of British Columbia.

Nakamura, Y., Fushiki, H. and Higuchi, T., 1974, Metabolic differences between gymnosperms and angiosperms in the formation of syringyl lignin, *Phytochemistry*, 13: 1777-1784.

Nix, L.E. and Brown, C.L., 1987, Cellula kinetics of compression wood formation in slash pine, *Wood and Fiber Sci.*, 19: 126-134.

Partridge, G.W., 1973, On the shape of trees, *J. Theor. Biol.*, 38: 111-137.

Rydholm, S.A., 1965, Pulping Processes, pp. 1269, Wiley-Interscience, New York.

Sarkanin, K.V. and Ludwig, C.H., 1971, Lignins: Occurrence, Formation, Structure and Reactions, pp. 1, Wiley-Interscience, New York.

Sederoff, R., Stomp, A.M., Chilton, W.S. and Moore, L.W., 1986, Gene transfer into loblolly pine by *Agrobacterium tumefaciens*, *Biotechnology*, 4: 647-649.

Stomp, A.M. and van Buijtenen, J.P., 1987, Views on forest biotechnology: Report from a conference on tree improvement by genetic engineering, *Proc. of the 19th Southern Forest Tree Improvements Conference*, pp. 433-443, College Station, Texas.

Terashima, N., Fukishima, K. and Takabe, K., 1987, Heterogeneity in formation and structure of lignin visualized by microautoradiography, *Proc. of the 4th International Symposium on Wood and Pulping Chemistry*, Vol.1, pp. 267-272, Paris.

van Buijtenen, J.P., 1985, Increasing forest productivity and value by breeding for outstanding combinations of desirable characteristics. In: *Forest Potentials, Productivity and Value* (R. Ballard, P. Farnum, G.A. Ritchie and J.K., Winjum, eds.), pp. 233-251, Weyerhaeuser Science Symposium 4.

Wardrop, A.B., 1971, Occurrence and formation in plants. In: *Lignins: Occurrence, Formation, Structure and Reactions* (V.K. Sarkanen and C.H. Ludwig, eds.), pp. 19-41, Wiley-Interscience, New York.

APPENDIX

Equations to calculate diminishing returns from carbon reallocation

(1) $\Delta W's(t) = \Delta Ws - M(t)$

(2) $M(t) = M(t-1)\sqrt{1 + \dfrac{\Delta Ws - M(t-1) - \Delta Ws(to)}{\Delta Ws(to)}}$

(3) $M = Rm\,f\,Ws$

Where $\triangle$Ws' (t) is the current year's increase in stem biomass due to carbon reallocation after maintenance is taken into account

$$(t \ ha^{-1} \ yr^{-1})$$

$\triangle$Ws   is the reallocated amount (t ha^{-1} yr^{-1})

$\triangle$Ws(t$_0$) is the biomass allocated annually to stem growth in the base case (t ha^{-1} yr^{-1})

M(t)   is biomass-equivalent used in maintenance (as above) but in the previous years' tissue in the current year (maintenance of current year increment in current year is ignored for simplicity) (t ha^{-1} yr^{-1})

M(t-1)   is biomass-equivalent used in maintenance (as above) but in the previous year (t ha^{-1} yr^{-1})

Rm   is the specific maintenance rate (t ha^{-1} yr^{-1})

f   is the fraction of biomass that is living

Ws   is the total stem biomass in the stand (t ha^{-1})

VI

CONCLUDING REMARKS

30

CONCLUDING REMARKS

The alarming rate of deforestation due to population explosion of livestock and humankind has resulted in over-grazing, poverty-related destruction and over-cutting of the forests to meet the accelerated requirements of wood and non-wood needs from forests. This calls for :

- Strategies to ensure long-range ecological security and supply of goods and services to people and industry on a sustainable basis.
- A quantum jump in the production and productivity of wood for diverse uses both spatially and temporally.
- Using biotechnology as one possibility to boost productivity if backed by strong R & D base, involving intensive collaborative effort.

The members of the Workshop realised that biotechnology is an explosive science the impact of which needs to be assessed, in order that benefits accrue to mankind.

The following recommendations emerged as a result of the deliberations during the Workshop:

1. Creation of an R & D Base: For successful application of biotechnological approach to forestry, it is necessary to have an equally strong program on forest tree genetics and breeding, nutrition, tree physiology and silviculture. This would help in applying advanced forest breeding techniques to boost productivity and production of wood and non-wood products from trees and shrubs. It is recommended that such an R & D base be organised as expeditiously as possible.

2. Tie-up between Universities R & D Systems and Forestry Organisations: In order to obtain expeditious results, R & D on forestry need not be confined only to forestry institutions. It is recommended that efforts should be made to involve untapped talent in the Universities and the R & D system in a meaningful tie-up between them and the local forestry establishments, for field facilities for testing the products. This would reduce the time lag as well as the costs.

3. National List of Free Taxa: Based on proper protocols for the screening, testing and identification, species/provenances/ individual elites of trees and shrubs need to be identified for the application of biotechnology to circumvent specific problems in individual cases that are a barrier to higher yield.

It is recommended that a *National List* of such taxa be prepared, taking into account the different agro-ecological situations in the country, together with different land uses and end-uses and where superior stock material is in short supply and mass production is urgently called for. Such a list should be widely circulated among the funding agencies and researchers for intensifying work on their genetic system (breeding biology included), plant architecture (roots included), physiology, nutrition, etc.

4. Genetic Engineering in Forestry: In view of the general lack of mutants in forest trees, work on genetic engineering as relevant to forestry may be initiated on selected species/clones with a view to benefit forest industry by developing insect and pathogens resistance, better carbon allocations, superior quality and quantity of lignin, desirable pulp characteristics, improved wood texture and useful silvi-chemicals, as also resistance/tolerance against environmental stresses such as drought and acidification of atmosphere. Such transformed individuals may be segregated effectively both during their development and field-testing, so as not to contaminate through cross-pollination the natural forest stands and man-made plantations.

5. Tree Nutrition: Research needs to be intensified on nutritional aspects, together with silvi-biological nitrogen fixation and mycorrhizae to enhance tree productivity. Specific institutions should be identified to take up this work on priority basis.

6. Conservation of Tree Genetic Resources: It is recommended that conservation of forest tree/shrub genetic resources be undertaken through both *in situ* conservation of different forest types in selected ecosystems/biomes, and also by *ex situ* conservation through seed orchards, clonal orchards and seed, tissue and organ storage.

This needs to be done at selected places so as to cover, the entire forest tree wealth of the country. Efforts may be intensified to identify and conserve individual plus trees in natural forest stands/plantations as *National Trees*.

7. Long-term Support for Forestry R & D: Forestry, unlike agriculture, has a long gestation period and affects inter-generational equity. It is recommended that a long-range view be taken on forestry R & D and policies, and funding be ensured on a long-term basis. This would also bridge the gap between the techniques developed in the universities and research departments and the technology required for their large-scale application.

It is further recommended that pilot-scale facilities for upscaling the production of elite material and hardening facilities in glasshouses be established, preferably in the non-governmental sector, so as to enable expeditious implementation and release of information and planting material of elites.

8. Production and Supply of Quality Seed: Keeping in view the fact that people are increasingly involved in raising decentralised nurseries and planting trees under the agro-forestry programs, it has become necessary to ensure good productivity and assured returns from such plantings on fields or community and common lands. It is, therefore, recommended that the quality of the seed be ensured through the introduction of Seed Certification so that fair returns are not only ensured to the farmer but his stake in the program is also enhanced so as to create national wealth.

9. Involvement of People and Industry and Updated Policy on Forestry and Connected Areas: In view of the enormity of problems confronting the forestry sector and the extreme urgency to solve such problems for sustained ecological and economic security, it is recommended that the government may consider involving rural people at large to meet the local needs on the one hand, and forest-based industry on the other, for meeting the pressing needs of the industrial forestry of timber, pulp and fiber. Such involvement would enable a fresh look at the government policies on forestry, grazing, livestock, land use, planning and tenure, etc. and their updating, together with package incentives/disincentives.

10. Inter-Agency Task Force on Forestry: In view of the interest of most funding agencies in the area of biotechnology and in order to obtain tangible results in a short period, and avoid duplication, it is suggested that an inter-agency task force be appointed to identify and prioritise urgent national tasks to be accomplished, and allocate funds. Furthermore, such a task force would oversee implementation in order to obtain results expeditiously. The concerned agencies are Council of Scientific and Industrial Research, Department of Biotechnology (DBT), Department of Environment, Forests and Wildlife, Department of Science and Technology and University Grants Commission. A nodal department such as DBT be assigned the coordinating role and servicing of the task force.

11. Inter-country Task Force for Biennial Symposia on Modern Trends in Forestry R & D: The members, taking note of the tremendous cross-fertilisation of ideas between foresters, researchers, social workers and industry which resulted from the present symposium, recommend that efforts may be made to ensure such interactions, preferably every two years.

Furthermore, the members may note that the Indo-US-West German interaction has been mutually beneficial. It is recommended that the Indo-US-West German collaboration in this field be continued and a small inter-country task force be appointed, so that technology-transfer may be made in areas such as synthetic seed, research in juvenility, micropropagation, germplasm collection and storage, pilot-scale multiplication of elites using tissue culture and genetic manipulations by direct gene transfer or through *Agrobacterium.*

Contributors

M. R. Ahuja   Federal Research Centre for Forestry and Forest Products, Institute of Forest Genetics and Forest Tree Breeding, Sieker Landstrasse 2, D-2070 Grosshansdorf, Federal Republic of Germany

Nirmala Amatya   Tissue Culture Laboratory, Botanical Survey and Herbarium, Godawari, Kathmandu, Nepal

V. A. Bapat   Plant Biotechnology Section, Bio-Organic Division, Bhabha Atomic Research Centre, Trombay, Bombay 400 085, India

S. Bhaskaran   Hindustan Lever Research Centre, Andheri, Bombay 400 099, India

Sant. S. Bhojwani   Department of Botany , University of Delhi, Delhi 110 007, India

J. Burley   Oxford Forestry Institute, Oxford University, South Parks Road, Oxford OX1 3RB, England

P. Chakravarty   Canadian Forestry Service, Patawa National Forestry Institute, Chalk River, Ontario, Canada K0J 1J0

Uthai Charanasri   Bangkok Flowers Centre Co. Ltd., 34/19 Moo 7 Petchkasem Road, Nong Kang Plu., Nong Kam, Bangkok, Thailand

L. Chatarpaul   Canadian Forestry Service, Petawawa National Forestry Institute, Chalk River, Ontario, Canada K0J 1J0

A. N. Chaturvedi   Tata Energy Research Institute, 90 Jor Bagh, New Delhi 110 003, India

Abhaya M. Dandekar   Department of Pomology, University of California, Davis CA 95616, U.S.A.

Vibha Dhawan Tata Energy Research Institute, 90 Jor Bagh, New Delhi 110 003, India

P.D. Dogra Biomass Research Centre, National Botanical Research Institute (CSIR), Lucknow 226 001, India

Don J. Durzan Department of Environmental Horticulture, University of California, Davis CA 95616, U.S.A.

Promod K. Gupta Division of Biochemical Sciences, National Chemical Laboratory, Pune 411 008, India

B. M. Khan Biochemical Sciences Division, National Chemical Laboratory, Pune 411 008, India

T. N. Khoshoo Tata Energy Research Institute, 7 Jor Bagh, New Delhi 110 003, India

S. S. Khuspe Biochemical Sciences Division, National Chemical Laboratory, Pune 411 008, India

A. D. Krikorian Department of Biochemistry, Division of Biological Sciences, State University of New York, Stony Brook, New York 11794-5215, U.S.A.

P. Mohan Kumar Tata Tea Limited, Bishop Lefroy Road, Calcutta 700 020, India

Prakash P. Kumar Plant Physiology Research Group, Department of Biological Sciences, University of Calgary, Calgary, Alberta T2N 1N4, Canada

R. D. Lai Tata Tea Limited, Bishop Lefroy Road, Calcutta 700 020, India

Malathi Lakshmikumaran Tata Energy Research Institute, 90 Jor Bagh, New Delhi 110 003, India

R. E. Litz Tropical Research and Education Center, University of Florida, 18905 S.W. 280 St., Homestead, Florida 33031, U.S.A.

Robert D. Locy Native Plants Inc., 417 Wakara Way, Salt Lake City Utah 84108, U.S.A.

L. J. Maene Industriele Hogeschool van het Rijk C.T.L., Voskenslaan 270-9000, Gent, Belgium

A. F. Mascarenhas Biochemical Sciences Division, National Chemical Laboratory, Pune 411 008, India

E. M. Muralidharan Biochemical Sciences Division, National Chemical Laboratory, Pune 411 008, India

R. S. Nadgauda   Biochemical Sciences Division, National Chemical Laboratory, Pune 411 008, India

R. K. Pachauri   Tata Energy Research Institute, 7 Jor Bagh, New Delhi 110 003, India

V. R. Prabhudesai   Hindustan Lever Research Centre, Andheri, Bombay 400 099, India

Jitendra Prakash   A.V. Thomas Group of Companies, P.B. No. 520, Willingdon Island, Cochin 682 003, India

S. B. Rajbhandary   Department of Medicinal Plants, Ministry of Forests and Soil Conservation, Thapathali, Nepal

S. Ramachandran   Department of Biotechnology, C.G.O. Complex, Lodhi Road, New Delhi 110 003, India

P. S. Rao   Plant Biotechnology Section, Bio-Organic Division, Bhabha Atomic Research Centre, Trombay, Bombay 400 085, India

Keith Redenbaugh   Plant Genetics Inc., 1930 Fifth Street, Davis, CA 95616, U.S.A.

Steven E. Ruzin   Plant Genetics Inc., 1930 Fifth Street, Davis, CA 95616, U.S.A.

H.K. Srivastava   Department of Biotechnology, C.G.O. Complex, Lodhi Road, New Delhi - 110 003, India

P. Subramaniam   Canadian Forestry Service, Patawa National Forestry Institute, Chalk River, Ontario, Canada K0J 1J0

H. S. Thapar   Forest Research Institute and Colleges, P.O. New Forest, Dehradun 248 006, India

Trevor A. Thorpe   Plant Physiology Research Group, Department of Biological Sciences, University of Calgary, Calgary, Alberta T2N 1N4, Canada

Roger Timmis   Weyerhaeuser Company, Tacoma, P.O. Box 111852, Washington 98911, U.S.A.

Patrick C. Trotter   Weyerhaeuser Company, Tacoma, P.O. Box 111852, Washington 98911, U.S.A.

TAXONOMIC INDEX

Abies sp. 47
A. alba 215,217
A. balsamea 62
A. pindrow 5
Acacia 29,47,48,49,258,289
A. ampliceps 289
A. auriculiformis 289
A. bivenosa 289
A. catechu 5,29
A. holosericea 289
A. ligulata 289
A. koa 286-288
A. maconochieana 289
A. mangium 287
A. mearnsii 5
A. nilotica 5, 47,76,287
Subsp. *adstringens* 47
   cupressiformis 47
   hemispherica 47
   indica 47
   subalata 47
A. salicina 289
A. sclerosperma 289
A. stenophylla 287,289
A. tortilis 5
A. victoriae 289
Acer 298
Acer caesium 5
A. campbellii 5
A. negundo 300
A. saccharum 300

Adina 47
A. cordifolia 5
Aesculus indica 5
African oil palm (see also *Elaeis guineensis*) 119,120,121
Agathis 303
Agrobacterium sp.17,328,331,332,334,335, 340, 345
A. rhizogenes 123
A. tumefaciens 7,328,329,332-335,339-346,363
Ailanthus excelsa 5
Albizzia sp. 5,47
A. lebbeck 5,62,75,286,287
A. procera 5
Alder (see also *Alnus glutinosa*) 19,301,311,319
Alfalfa 57,58-60
Almond 32
Alnus sp. 5,258,309,311
A. glutinosa 309-317
A. nepalensis 6
A. viridis 310,318
Alstroemeria 269
Amanita muscaria 302
Amaryllis 246
Amoora 47
Anacardium occidentale 88,95
Anogeissus latifolia 5
A. pendula 5
Anthurium 246,269

A. andreanum 269
Apple (see also *Malus* sp.) 6,15,32,64,227
Arabian coffee (see also *Coffea arabica*) 65
Arachnis sp. 245,246
Araucaria sp. 303
A. cunninghamii 300
Arbutus menziesii 302
Arctostaphylos uva-ursi 302
Areca catechu 5
Arecanut (see also *Areca catechu*) 32,42
Artocarpus chaplasha 5
Arundinaria 49
Asparagus 59
A. officinalis 276
Aspen (see also *Populus* sp.) 219,222,355
Averrhoa carambola 114
Azadirachta indica 76

Bacillus thuringiensis 329,331
Bactris 135
Balsam fir (see also *Abies balsamea*) 62
Bamboo 6, 23,74,82,328
Bambusa sp. 49
B. arundinacea 6,74
B. vulgaris 6,74
Banana (see also *Musa paradisiaca*) 99
Beech (see also *Fagus sylvatica*) 218, 220,226,298
Betula utilis 48
Biota orientalis 76
Birch 302
Black cherry (see also *Prunus serotina*) 300
Black locust (see also *Robinia pseudo-acacia*) 301
Black walnut (see also *Juglans nigra*) 300
Boletus edulis 302
B. elegans 302
Bombax ceiba 5
Borassus 125
B. flabellifer 134
Box alder (see also *Acer negundo*) 300
Brassica 58

B. oleracea 250,251,363
Broccoli 60,277
Brussel sprouts (see also *Brassica oleracea*) 363

Cabbage (see also *Brassica oleracea*) 363
Cacao (see also *Theobroma cacao*) 65,92,110
Cajanus 151
Caladium 246
Calathea ornata 236,238,239
Callicarpa tomentosa 48
Camellia sinensis 88,89,90,260,261
Candida tropicalis 18
Cardamom (see also *Elettaria cardamomum*) 94
Caribbean pine 303
Carica cauliflora 101
C. papaya 90,101,110,276
Carica papaya x *C. cauliflora* 101
Carrageenan 59
Carrot (see also *Daucus carota*) 58
Cashew (see also *Anacardium occidentale*) 95
Cassia fistula 48
C. siamea 5
Casuarina 19,301,309
Casuarina equisetifolia 5
Cattleya 245,246
Cauliflower (see also *Brassica oleracea*) 269
Cedrus 47
C. deodara 5,75
Celery 57,58
Celtis australis 5
Cenococcum geophilum 302
C. graniforme 301
Chamaedorea costaricana 65
Cherry 15
Chir pine (see also *Pinus roxburghii*) 299, 303
Christmas palm (see also *Veitchia merrilli*) 65
Chrysanthemum 246

Chukrasia velutina 5
Citron (see also *Citrus medica*) 64
Citrus sp.15,32,63,90,100,101,110,111,112, 114,115,116
C. aurantifolia 64
C. aurantium 64
C. limon 64
C. maxima 101
C. medica 64
C. paradisi 64
C. reticulata 64
C. sinensis 64,100,116
Clingstone peach 193,195
Clove 32
Cocoa (see also *Theobroma cacao*) 14
Coconut (see also *Cocos nucifera*) 11,32,33,64,110,124,128,130,131,246
Cocos 124,130
C. nucifera 64,88,89,97,99,110,121
Coffea arabica 65,88,91,92,110,113
C. canephora 65,89,92
C. excelsa 92
C. liberica 92
Coffee (see also *Coffea arabica* var. Catimor) 14, 25,42,63,65,88,91,110,263
Colletotrichum sp. 115
C. gloeosporiodes 115
Corn 59
Cortinarius hemitrichus 302
Cotton 58
Cryptomeria japonica 5,32
Curcuma longa 88,89,94
Cynara scolymus 235

Dalbergia sp. 47,49
D. lanceolaria 75,286,287
D.latifolia 5, 6,75,286,287
D. sissoo 5,75,277,286,287
Date palm (see also *Phoenix dactylifera*) 12,17,64,98,119,120,124,128, 134,145
Daucus carota 111
Dendrobrium 245,246
Dendrocalamus 49
D. strictus 6,74,77,81

Dianthus 235
Dioscorea alata 254
D. bulbifera 254
D. rotunda 254
Diospyros melanoxylon 5
Dipterocarpus 47
Douglas-fir (see also *Pseudotsuga menziesii*) 62,298,328,331,335,339, 341, 342,346,357,360,363

Elms 19
Elaeis 128
E. guineensis 65,89,121,129,135
E. melanococca 121,129
Elettaria cardamomum 88,89,94
English walnut (see also *Juglans regia*) 62
E. coli 333
Eriobotrya japonica 64,112
Eucalyptus 11,23,32,33,34,47,48,53,59 74,77,79,80,81,82,257-260,262,299,302, 303,324,325,328
Eucalyptus sp. 5
E. camaldulensis 23,75
E. citriodora 23,62,76,77,80,259,277
E. globulus 5,76,259
E. grandis 75,79,259
E. tereticornis 47,75,77-79
E. torelliana 75,77-79
E. robusta 259
Eugenia sp., 112
E. malaccensis 64
Euphoria longan 113,114
European beech (see also *Fagus sylvatica*) 215,217
European larch (see also *Larix decidua*) 62

Fagus sylvatica 215,217,220
Feijoa sellowiana 279
Fennel 269
Ficus auriculata 157-161
F. beniamina 239,240
F. religiosa 77
F. lyrata 279

Firefly (see also *Photinus pyralis*) 340
Frankia 309-318
Fraxinus americana 62
F. pennsylvanica 300
Fusarium oxysporum 121

Gerbera 269
Gigaspora aurigloba 303
G. margarita 303
Ginger (see also *Zingiber officinale*) 95
Gingko biloba 355
Gladiolus 254,269
Glomus fasciculatus 300,301
Gmelina sp. 5
G. arborea 5
G. robusta 5
Grapefruit (see also *Citrus paradisi*) 64
Green ash (see also *Fraxinus pennsylvanica*) 298,300
Grevillea 258
Grewia optiwa 5
Griselinia littoralis 299
Gypsophila 246

Hardwickia binata 6
Helopeltis antonii 95
Heteropsylla cubana 289
Hevea 96
H. brasiliensis 65,88,89,96
Hibiscus integrifolia 5
Holoptelea sp., 5
Hoop pine (see also *Araucaria cunninghamii*) 298,300
Hopea 47
Howea forsteriana 65
Hybrid aspen (see also *P. tremula* x *P. tremuloides*) 215,217,219,222

Indian walnut (see also *Albizzia lebbeck*) 62

Jaboticaba (see also *Myrciaria cauliflora*) 64

Japanese black pine (see also *Pinus thunbergii*) 355
Jubaea, 135
Juglans 47,63
J. hindsii 62
J. nigra 300
J. regia 5,62

Kalanchoe 269
Kholrabi 269
Kydia calycina 5

Laccaria laccata 301,302
Larch (see also *Larix decidua*) 215,217, 226,302
Larix decidua 62,215,217
L. occidentalis 302
Lemon (see also *Citrus limon*) 64
Leucaena 6,279,289,290-292
L. diversifolia 62
L. diversifolia X *L. leucocephala* 290
L. (diversifolia X *pallida*) X *L. leucocephala* 290
L. leucocephala 23,51,53,77,251,252,254, 277,285-289,290,293,301
L. leucocephala X *L. diversifolia* 290
L. leucocephala X *L. esculenta* 290
L. leucocephala X *L. pallida* 290
L. pulverulenta X *L. leucocephala* 290
Lettuce, 58
Lilium 246,269
Lime (see also *Citrus aurantifolia*) 64
Liquidambar styraciflua 62,252
L. styraciflua 300
Liriodendron 63
L. tulipifera 65,300
Litchi chinensis 113
Loblolly pine (see also *Pinus taeda*) 59,300,328,331,346,355
Loquat (see also *Eriobotrya japonica*) 64

MacIntosh apple 363
Madhuca longifolia 5, 6

M. latifolia 6
Malay apple (see also *Eugenia malaccensis*) 64
Malus sp., 64
M. domestica 251,252
Mamillaria elongata 253
Mandarin (see also *Citrus reticulata*) 64
Mangifera indica 23,64,90,100,112
Mango (see also *Mangifera indica*) 32, 63,64,100,112,114,115
Maple 298
Melia azadirachta 5
Michelia champaca 5
Mimosa pudica 288
Morus 47
M. alba 23
M. indica 77,114
Musa 64,129,246
M. paradisiaca 90,99
Muscari armeniacum 254
Mycelium radicisatrovirens 300
Myrciaria cauliflora 64,112

Narcissus tazetta 254
Nerine 269
Nephrolepis 269
Nicotiana 208
N. tabacum 253
Norway spruce (see also *Picea abies*) 59,62,215,226,362
Nutmeg 32

Oaks 226,325
Oil palm (see also *Elaeis guineensis*) 59,65,97,98,130,134
Oriental arbor vitae (see also *Thuja orientalis*) 65
Oryza sativa 253
Ougeinia dalbergioides 293

Palm (see also *Chamaedorea costaricana*) 65,89,97
Papaya (see also *Carica papaya*) 110

Para rubber (see also *Hevea brasiliensis*) 65
Paradise Palm (see also *Howea forsteriana*) 65
Paulownia (see also *Paulownia tomentosa*) 65
Paulownia tomentosa 65
Paxillus involutus 309-318
Peach 32
Pear (see also *Pyrus* sp.) 32,64
Pepper (see also *Piper nigrum*) 94
Phalaenopsis 246
Phanerochaete chrysosporium 18
Phoebe attennata 5
Phoenix 125
P. dactylifera 23,64,89,120,128
P. sylvestris 128
Photinus pyralis 340,342
Picea 47
P. abies 61,62,215,217,362
P. contorta 302
P. glauca 62,205,206,209,210,213
P. monticola 302
P. pondorosa 302
P. sitchesis 302
P. smithiana 5,75
Pineapple 246
Pines (see also *Pinus* sp.) 15,59,298,299, 325,335
Pinus sp. 5,47,302
P. caribaea 301.
P. elliottii 301
P. gerardiana 75
P. lambertiana 62
P. patula 5
P. radiata 205,206,208,209,212,227,301,302
P. roxburghii 75,299
P. sylvestris 215,217,299,361
P. taeda 62,300,355
P. thunbergii 355
P. wallichiana 47,48,75
Piper nigrum 88,94
Pisolithus tinctorius 298,300-302,304

Pistacia vera 191
Pistacio (see also *Pistacia vera*) 188,194
Plaintain (see also *Musa*) 64,110
Plantanus occidentalis 300
Poncirus trifoliata 116
Pongamia pinnata 5
Poplar (see also *Populus*) 14,53,62,227, 325,331,334,346
Populus sp. 5,32,48,335,346
P. alba 277
P. alba x *P. grandidentata* 335
P. ciliata 62,75
P. deltoides 6,48,329
P. tremula 215-217,219-221,226,277
P. tremula x *P. tremuloides* 215,217
P. tremuloides 215-217,219-221
Potato (see also *Solanum* sp.) 246,269, 276,279
Prosopis sp. 5,6,29,48,49
P. alba 288
P. chilensis 288
P. cineraria 5,23,36,77,288
P. juliflora 288
P. tamarugo 288
Prunus sp. 182,190,192
P. cerasus 64
P. insitita 251,252
P. persica 192
P. serotina 300
Pseudotsuga menziesii 62,302,335,339,340
Pterocarpus 47
Pterocarya sp. 62
Pulberoboletus shoreae 303
Punica granatum 90,101
Putranjiva roxburghii 48,77
Pyrus sp. 64

Quercus sp. 5
Q. rubra 62

Radiata pine 205-209
Red maple 300
Red oak (see also *Quercus rubra*) 62
Redwood 15

Rhizobium 254,285,288,292,294,301,304
Rhizoctonia sylvestris 300
Rhizopogon 301,302
R. luteolus 301
Rhododendron 279
Rice (see also *Oryza sativa*) 253
Robinia pseudo-acacia 286,301
Robusta coffee (see also *Coffea canephora*) 65
Rose apple (see also *Syzygium jambos*) 64
Rubber 42,89,263
Rubus idaeus 235,252,279

Sago 32
Saintpaulia 269
Sal (see also *Shorea robusta*) 299
Salix 47
S. alba 5
S. babylonica 77
S. matsudana 277
S. tetrasperma 48,81
Salvadora persica 74,76,77,81
Sandal 23
Sandalwood (see also *Santalum album*) 65,145,153,154,155
S. album 6,23,65,76,145,146,150,151,155, 277
Sapindus trifoliatus 113
Sapium sebiferum 76
Scots pine (see also *Pinus sylvestris*) 215,217,226,361
Sesbania 6
S. grandiflora 77,286,288
S. sesban 286,288
Shorea 47
S. robusta 5,6,32,299
Silver fir (see also *Abies alba*) 215, 217,226,303
Sissoo (see also *Dalbergia sissoo*) 30
Sitka spruce 15
Solanum 253
S. laciniatum 251,252
S. quitoense 114

S. sisymbrifolium 253
Sour cherry (see also *Prunus cerasus*) 64
Sour orange (see also *Citrus aurantium*) 64
Soybean 58,59
Spinacea oleracea 208
Spruce 325
Stercularia sp. 5
Strawberry 6
Stylosanthes sp. 116
Sugar maple (see also *Acer saccharum*) 300
Sugar pine (see also *Pinus lambertiana*) 62,331,335
Suillus 302
Sweet gum (see also *Liquidambar styraciflua*) 62,298,300
Sweet orange (see also *Citrus sinensis*) 64,100,116 64
Sycamore (see also *Plantanus occidentalis*) 300
Syzygium cumini 48
S. jambos 64

Tamarindus indica 6,74,76,288
Tea (see also *Camellia sinensis*) 42,90,91,258-261,263
Teak (see also *Tectona grandis*) 23,30,325
Tectona 47
T. grandis 5,6,23,32,47,76,246
Terminalia sp., 5,47,49
T. belerica 48
T. chebula 48
T. ciliata 5
T. indica 5,77,81
Thelophora terrestris 300

Theobroma cacao 65,88,92,93,110
Thuja 258
T. orientalis 65
Tobacco (see also *Nicotiana*) 253, 328,329,334
Tomato 165,167,169,171-173,175,176, 334
Toona ciliata 48
Tsuga dumosa 5
T. heterophylla 302
Turmeric (see also *Curcuma longa*) 88,94

Vaccinium corymbosum 279
Vanda 245,246
Veitchia merrilli 65

Walnut (see also *Juglans hindsii*) 32,62,325
Watsonia 269
White ash (see also *Fraxinus americana*) 62
White cabbage 267
White spruce (see also *Picea glauca*) 62,205-207,211
Willow 14,227

Xerocomus bakshii sing.Singh 303

Yellow poplar (see also *Liquidambar tulipifera*) 65,298,300
Yucca 269

Zingiber officinale 88,89,95
Zizyphus jujuba 48
Z. nummularia 48
Z. rotundifolia 48